High-Power Electromagnetic Effects on Electronic Systems

For a complete listing of titles in the
Artech House Electromagnetics Series,
turn to the back of this book.

High-Power Electromagnetic Effects on Electronic Systems

D. V. Giri
Richard Hoad
Frank Sabath

ARTECH HOUSE
BOSTON | LONDON
artechhouse.com

Library of Congress Cataloging-in-Publication Data
A catalog record for this book is available from the U.S. Library of Congress.

British Library Cataloguing in Publication Data
A catalogue record for this book is available from the British Library.

Cover design by John Gomes

ISBN 13: 978-1-63081-588-2

Artech House
685 Canton Street
Norwood, MA 02062

10 9 8 7 6 5 4 3 2 1

Contents

Foreword

Modern civilization has become entirely dependent on the reliable functioning of electronics for its existence. This is true for electric power, communications, transportation, health, finance, and many other aspects of modern life. All of these capabilities are susceptible to the high-power electromagnetic effects of fields and their coupling to systems. Too few engineers are currently trained in protecting the capabilities on which we depend, leaving our civilization open to disruption from localized to continent-scale exposure to both natural and man-made electromagnetic effects.

In the United States and elsewhere, the most common way to deal with threats that occur infrequently is to ignore them until disaster strikes and then to try to overwhelm them with money and effort after they do occur, even though a modest amount of engineering design and preparation in advance of the disaster would have prevented it. This approach, while potentially disastrous on a local scale, will work for some threats, but for the continent-scale fields that can be produced by the exo-atmospheric detonation of a nuclear explosive, the damage to unprotected electronics would be so widespread that loss of life would be large and recovery would be very difficult.

Electromagnetic fields, coupling to conductors, and the effects of conducted voltages and currents on electronic systems are frequently taught separately, but addressing them in an integrated manner is essential to protecting modern systems. While protecting systems from high-power electromagnetic effects does not need to be difficult and expensive, it does require specialized knowledge and engineering discipline that is not usually gained in the course of an engineering education. Without that knowledge, the challenge of protecting complex systems can appear to be overwhelming. This book can fill a much-needed place by serving as the basis for an advanced course in understanding and protecting against high-power electromagnetic effects. It also serves as an excellent reference for the engineering practice of protecting and testing modern electronic systems and should be available to all engineers who design and protect electronic systems.

William R. Graham, PhD
Chairman, Commission to Assess the Threat to the United States from Electromagnetic Pulse EMP) Attack
March 2020
San Marino, California

Acknowledgments

D. V. Giri: I am thankful to my mentors, Professor R. W. P. King (1905 to 2005), Professor Tai Tsun Wu of Harvard University, and Dr. Carl E. Baum (1940 to 2010) of the Air Force Research Laboratory, who have laid the foundation for my education and for their constant guidance. At birth, we are all a blank slate. Many of my colleagues (too many to name here) from different parts of the world continue to write on this slate. I have extracted, structured, and organized the acquired knowledge from many experts. My contribution to this publication is not entirely my own. It is the result of my own and some collaborative work, especially with Dr. Fred M. Tesche, who has graciously permitted its use. The use of some of his work, especially in Chapters 5 and 7, performed for armasuisse of Switzerland is gratefully acknowledged. I am indebted to Dr. Tesche and armasuisse of Switzerland.

I wish to conclude by paying a tribute to my parents. My father was a self-made man. He used to say, "Education is the key to happiness in life." This is in resonance with a Sanskrit hymn that says, "Education leads to humility, humility leads to worthiness, worthiness begets wealth and the use of one's wealth for good purposes gives you happiness." He supported me and my siblings in our educational goals. I am who I am because of my parents. Their sacrifices and encouragement resulted in my arrival from India to Harvard University for graduate studies in 1969. With humility and reverence, I dedicate my portion of this book to my parents.

Richard Hoad: I acknowledge those technical experts who helped inspire me to dig deeper into the high-power electromagnetic topic. I could see early on that this particular area of technology would enable me to have a full and satisfying career, but Dr. Nigel Carter, Dr. Carl E. Baum, Dr. William Radasky, and my coauthor, Dr. D. V. Giri, all helped to stimulate or at times reignite my interest in the topic. My contribution to this book has strongly benefitted from their input and mentorship. I also thank my colleagues at QinetiQ for their contributions and encouragement for this work and in particular, Barney Petit, Gavin Barber, Paul Watkins, and David Herke.

Finally, but perhaps most importantly, I would like to thank my wife Rachel and my son Joshua for their strong support and encouragement.

Frank Sabath: I thank my mentor and good friend Professor Dr. Heyno Garbe, who supervised my first steps into the electromagnetic compatibility topic. Together we investigated the effects of high-power electromagnetic environments on systems and discussed appropriate models. Finally, he inspired me to pass the gained knowledge and experiences to students by giving lectures at the Leibniz University Hannover. I also thank my colleagues at the Bundeswehr Research Institute for Protective Technologies and CBRN Protection for their contributions and encouragement for this work and in particular, Dr. Daniel Nitsch, Dr. Martin Schaarschmidt, Dr. Stefan Potthast, Andre Bausen, and Jörg Maack.

Finally, I thank my wife Martina and my children Peter and Pauline for their patience, strong support and encouragement.

The authors thank the International Electrotechnical Commission (IEC) for permission to reproduce Information from its International Standards. All such extracts are copyright of IEC, Geneva, Switzerland. All rights reserved. Further information on the IEC is available from www.iec.ch. IEC has no responsibility for the placement and context in which the extracts and contents are reproduced by the author, nor is IEC in any way responsible for the other content or accuracy therein.

The authors thank the Institute of Electrical and Electronics Engineers (IEEE) for permission to reproduce selected extracts (identified in the text). All such extracts are copyright of IEEE. All rights reserved.

1

Introduction

1.1 Reliance on Electronics

In present-day society, we are increasing our reliance on widespread technological advancements that are largely underpinned by semiconductor-based electronic systems. The growth in modern technological systems that are enabled by electronics is unprecedented.

However, with technological sophistication, new risks to the functional reliability and even the safety of the electronics-based technologies are bound to emerge. One such risk factor described in detail in this book is that of high-power electromagnetic (HPEM) environments and the effects on electronics technologies that result from them.

1.2 HPEM Environment Overview

HPEM encompasses a wide variety of phenomena, both natural and man-made in origin, which together are called the HPEM environment. Natural lightning happens to be the only naturally occurring HPEM phenomena; the others, such as nuclear electromagnetic pulse (NEMP), high-power radio frequency (HPRF) phenomena including radio frequency (RF) broadcast, radar, HPRF directed energy (HPRF DE), and intentional electromagnetic interference (IEMI), are all made by humankind. The total electromagnetic spectrum is shown in Figure 1.1.

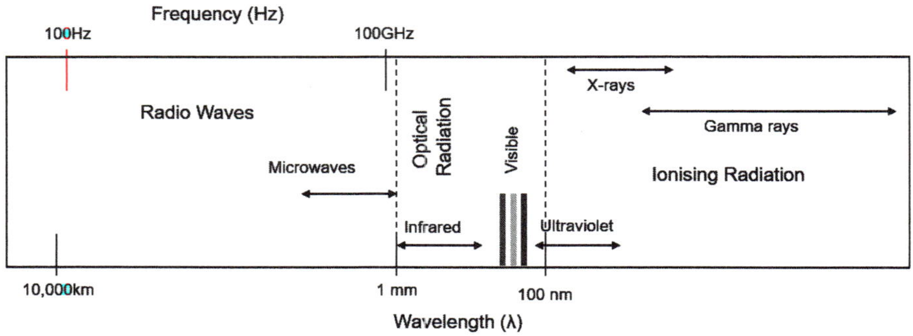

Figure 1.1 The total electromagnetic spectrum.

The focus of this book covers the HPEM phenomena associated with the lower end of the electromagnetic (EM) spectrum (i.e., the RF and microwave regions (a few hertz to a few tens of gigahertz (10^9 Hz))). This lower end of the spectrum has insufficient energy to cause ionisation and is therefore referred to as the nonionizing radiation region.

HPEM can be considered as an intense set of phenomena that have the potential to cause electromagnetic interference in electrical and electronic systems. The IEC definition of EM interference is [1]: "degradation of the performance of a device, transmission channel or system caused by an electromagnetic disturbance."

The technical understanding and characteristics of HPEM environments are now more fully understood, and a classification scheme for the different types of HPEM environment has recently been produced. A spectral representation of the HPEM environments was first presented by Giri and Kaelin in 1996 [2] and has been reworked several times. It is shown in Figure 1.2 and explained in more detail in subsequent chapters.

The physical dimensions of the system into which the electronic technology is configured have a bearing on the interaction of the HPEM environment with the system and therefore the resulting interference or effect. Put differently, the coupling and interaction of HPEM to electronic devices, circuits, equipment, and systems are very strong functions of frequency. Broadly speaking, large connected networks and infrastructures can be most affected by HPEM environments with frequency content below ~100 MHz, whereas isolated electronic equipment, which is not connected to infrastructure, is most affected by HPEM environments more than a few hundred megahertz. This is because the effect process is dominated by EM coupling, which is wavelength-dependent and described in more detail in subsequent sections.

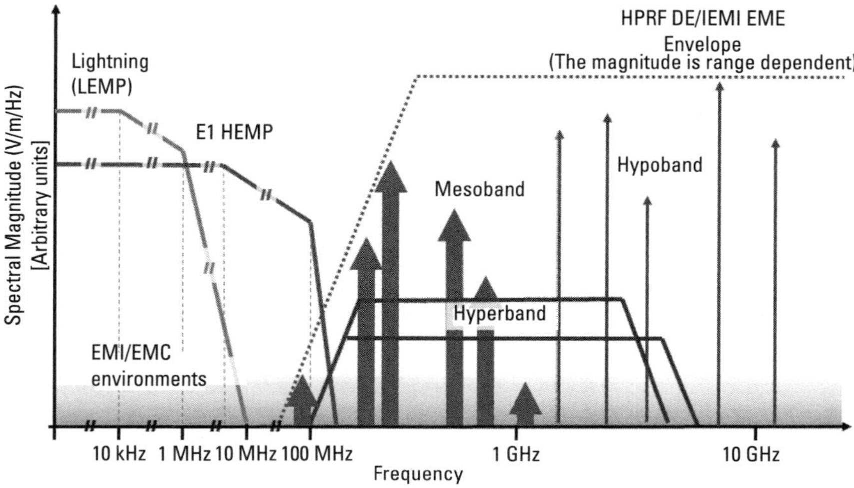

Figure 1.2 Comparison of the EMP, lightning, and HPRF DE/IEMI waveform spectra.

1.3 HPEM Effects Overview

The effect or impact of an incident HPEM environment on an electronics device, equipment, or system may be characterized as natural, accidental, or intentional. When such environments are created with a purpose of deliberately causing denial, disruption, or damage of electronic systems, they are most likely derived from HPRF DE or IEMI generators. Generally, the term HPRF DE is associated with military use of the generator technology whereas IEMI is associated with nonmilitary malicious actors. These phenomena can affect military assets, as well as civilian infrastructure.

The denial, disruptive, and damaging effects of such HPEM phenomena have been systematically studied by many researchers. The characterization and quantification of the HPEM-induced effects come from an understanding of the coupling and interaction between the HPEM environments and the system. The effect metric for a particular system can be a highly variable quantity because HPEM effect metrics are very highly sensitive to a very large number of variables. Good experimental practice and modeling and analysis are essential to minimize uncertainties in effect metrics. Still, a great deal of judgment or interpretation is required in the assessment of HPEM effects. Once the effects are understood then one of the goals can be to reduce the possibility of unwanted EM effects by desensitizing the system to the HPEM environment through the achievement of EMC or through implementing an EM hardening, protection or resilience solution.

1.4 A Brief History of EM Interference and Effects

Since the very first experiments in radio communication, conducted by Marconi in the late 1890s, interference phenomena have been known to exist. However, technical papers on radio interference only began to emerge in the 1920s [3], coincident with the greater proliferation of radio transmitters and receivers. This was largely a cochannel interference problem (i.e., one radio interfering with the reception of another radio) since radio regulation of bandwidth was not enforced at this time.

Other natural EM interference phenomena also contributed to the problem during this period. Examples of natural interference sources are lightning (direct and indirect effects, sometimes referred to as lightning electromagnetic pulse (LEMP)), static electricity (electrostatic discharge (ESD)), and precipitation static (P-static).

Later, in the 1930s, radio interference from electrical apparatus such as electric motors, electric railroads, and electric signs began to cause major problems to radio reception. These systems could be called emission sources or, more specifically, unintentional interference sources since the EM disturbance caused by them was not part of their main function. The disturbance was caused by some secondary effect such as harmonics produced by switching electrical arcs or self-oscillation in resonant circuits.

Primarily because of the need for world trade, international cooperation on managing EM interference began with the formation of two standards committees in 1933. These were the International Electrotechnical Commission (IEC) and the Special Committee on Radio Interference (CISPR).

However, the most significant increases in interference problems occurred with the invention of high-density electronic components such as the bipolar transistor (1950s), integrated circuits (1960s), and the microprocessor chip (1970s). During these periods military concerns and requirements provided the thrust for development of EMI protection and mitigation often associated with the term EMC [4].

The widespread introduction of semiconductors marked a new phase in the history of EMI and effects. Electronic circuits were capable of producing not only high levels of EMI but also electrical and electronic circuits, which were not receivers were also affected by the interference. The vast majority of interference cases were not caused by HPEM environments but relatively benign unintentional EMI from collocated systems. However, following the introduction of technologies which could produce HPEM environments, notably Radar, cases of HPEM effects began to emerge.

Two very important historical events that demonstrated HPEM effects on modern systems are the cases of the *USS Forrestal* [5] and the result of a high-altitude nuclear detonation Operation Starfish, conducted by the United States in 1962 [6]. A summary of these two cases is provided here; greater detail and further cases can be obtained in [7]:

> In 1967 off the coast of Vietnam, a Navy jet landing on the aircraft carrier U.S.S. Forrestal experienced the uncommanded release of munitions that struck a fully armed and fuelled fighter on deck. The results were explosions, the deaths of 134 sailors, and severe damage to the carrier and aircraft. This accident was caused by the landing aircraft being illuminated by carrier-based radar. The resulting EMI sent an unwanted signal to the weapons system. Investigations showed that degraded shield termination on the aircraft allowed the radar frequency to interfere with routine operations. As a result of this case, system level EMC requirements were revised to include special considerations for electro explosive devices.
>
> The Starfish nuclear device, with a yield of approximately 1 MT, was detonated about 400 km above Johnston Atoll during the night of 8th July 1962. The line of sight distance from the event detonation to the Hawaiian Island of Oahu was approximately 1400 km. On Oahu, problems were noted in the input circuits of radio receivers, surge arresters triggered unexpectedly on an aircraft with a trailing wire antenna, and 30 strings of streetlights reportedly failed simultaneously.

The Forestall disaster was a significant tragedy. It is perhaps important to point out that HPEM interaction with a device can leave no physical evidence whatsoever and this conspired to delay and cause doubt about the actual cause of the Forestall accident. Still, these cases, among others, prompted the military and civilian communities to become concerned about HPEM effects and their mitigation. Hence, standards and testing regimes were introduced to try to mitigate these problems.

1.5 A Systems-of-Systems Hierarchy

There is no doubt that electronics-based technologies have and will continue to pervade all levels of our modern society from whole complex infrastructure assets such as electronic banking to the personal handheld electronic devices such as cell/mobile telephones.

A system-of-systems hierarchy that will be used throughout this book is proposed in Figure 1.3. This hierarchy is important to the understanding of coupling and to effects.

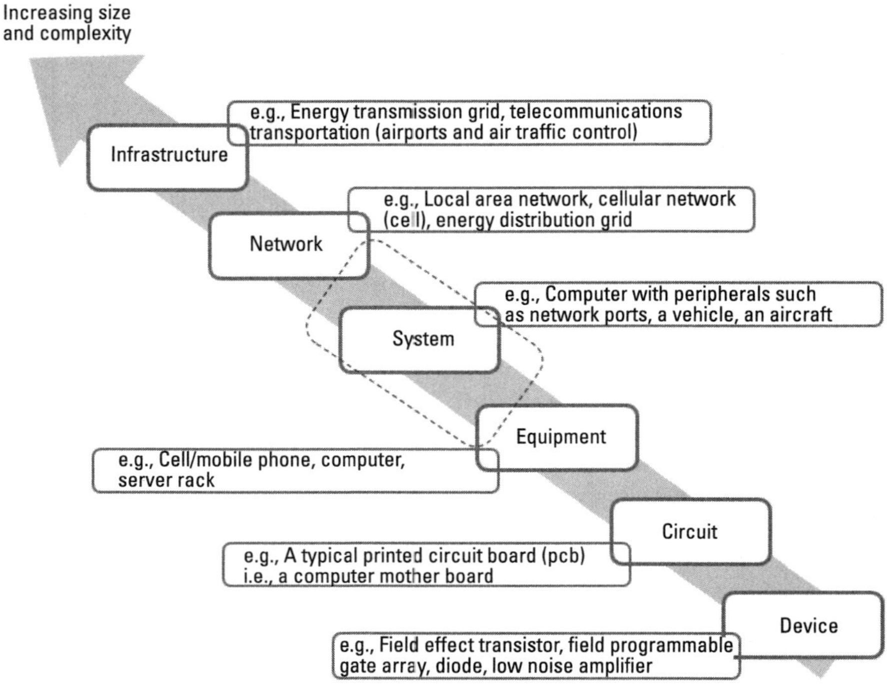

Figure 1.3 A system-of-systems hierarchy.

Descriptions of each element are provided next.

1.5.1 Device

A device is any active analogue or digital electronic component. In order for an electronic device to function it has to be implemented into a circuit, which at the very least delivers power to the device and perhaps provides an input and output. The device is often the point in the system where the effect actually occurs. HPEM effects testing at device level is difficult as it is necessary to replicate important elements of the circuit.

1.5.2 Circuit

A circuit is a collection of devices that when integrated together in a circuit carry out some desired function. Devices may be analog or digital devices and often there is a mixture of both types. Effects testing of circuits has been carried out but it is still relatively subjective to assess effects at the circuit level.

1.5.3 Equipment

Equipment can be considered to incorporate a circuit populated by devices provided within an enclosure with cable ports. Most effect experiments are based on samples of equipment. Typically, equipment must be connected to a system either wired or wirelessly in order to fully utilize its functionality. The assessment of effects at the equipment level is often the most practical.

1.5.4 System

A system is a collection of equipment that can be used together to deliver a function. A system has a defined boundary. System-level effect tests are usually the least abstract from the real world, but they can be practically difficult to carry out and may require large and expensive test facilities depending on the size of the system.

In Figure 1.3, the rectangular dashed outline area represents the boundary of an installation. An installation typically encloses a collection of equipment, systems, or a small network (e.g., an office building, data center, or other physical structure). The installation may be considered as a passive element in this hierarchy in that it cannot be directly affected by HPEM. However, it can modify the propagation of the HPEM environment through attenuation or coupling.

1.5.5 Network

A network is composed of interconnected systems, which could be wired or wireless. In practice, effects testing of whole networks is impractical and it is more typical to test systems and equipment that may compose a network individually and then simulate the network stimulus.

1.5.6 Infrastructure

An infrastructure is a large (many tens of kilometers), interconnected network usually fulfilling some important function. Infrastructure can employ many elements described above that may be present only some of the time. When infrastructure performs some critical or essential service such as electricity distribution over a metropolitan area, it is referred to as critical infrastructure. It is typically highly impractical to carry out effects testing at the infrastructure level.

At the device level, one can think of electronics technologies such as microprocessors and microcontrollers essentially as the electronics chip or brain of any electronics system. These devices and others support the functionality of the hierarchy right up to the infrastructure level. In recent times, there has been a trend to connect or network devices, equipment, and systems together using cable, optical fiber, or even wirelessly to form infrastructure-type functionality.

Figure 1.4 shows diagrammatically how the wavelength of each of the HPEM environment phenomena indicated in Figure 1.2 map on to the system-of-systems hierarchy shown in Figure 1.3.

The most important infrastructure requirements of our society are encompassed by the term critical national infrastructure (CNI). What is in scope of the CNI is broadly defined at a national or country level. One definition is [8]: "those facilities, systems, sites and networks necessary for the functioning of the country and the delivery of the essential services upon which daily life depends." The CNI can be categorized as comprising the following sectors:

- Communications;
- Emergency services;
- Energy;
- Financial services;
- Food;
- Government;
- Health;

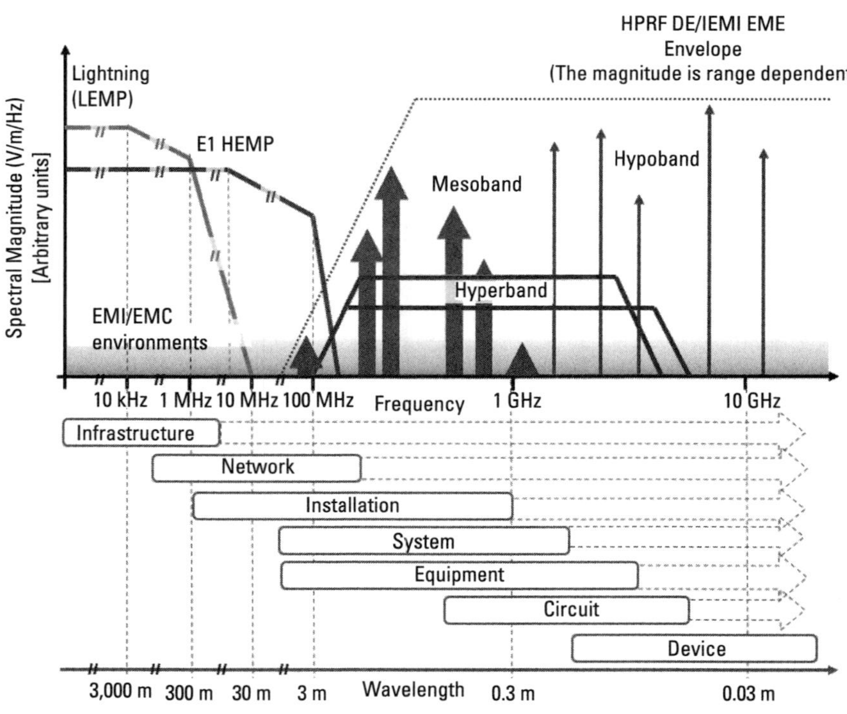

Figure 1.4 Mapping of the combined HPEMP EME onto the system of systems hierarchy.

- Transport;
- Water.

These sectors have undeniably become more dependent on the advancements in electronics-based technologies. Indeed, the societal reliance on infrastructure enabled via electronics based technologies is set to increase (e.g., through Smart Energy, Smart Cities, e-Healthcare, artificial intelligence, Internet of Things (IoT), autonomous vehicles), sometimes collectively termed as the 4th Industrial Revolution (4IR). In many nations, the delivery of the CNI is from commercial, private businesses and the functionality is provided on a commercial or quasi-commercial basis.

The effects from HPEM that can manifest as a degradation of function to the CNI could have very serious implications and perhaps lead to loss of life.

1.6 Summary

In this book, we describe some examples of the generation of HPEM environments. We then focus on HPEM coupling and effects citing examples where possible at many levels of the system-of-systems hierarchy presented earlier. We will draw a great deal on the large volume of published works and drawing conclusions where possible. We will show that the function of the device, circuit, and equipment that could be the victim of the HPEM disturbance has a profound impact on the severity of the effect. We will provide some insight on HPEM effect experimental techniques and the standards that can be used to control tests. We will discuss the validity of analytical techniques and computational modeling in an HPEM effects context.

The main objective of this book is to provide a reference text to the reader, which dispels myths, clarifies good experimental practice, and ultimately draws conclusions on the HPEM interaction with effects on electronics. It is hoped that this book will encourage the reader to consider the importance of HPEM phenomena as a threat to modern electronic-based technologies that underpin society and to therefore be preemptive in the consideration of HPEM resilience and protection.

References

[1] IEC 60050-161: International Electrotechnical Vocabulary (IEV) - Chapter 161: Electromagnetic Compatibility, Edition 1.0, August 27, 2007.

[2] Giri, D. V., and A. Kaelin, "Many Faces of HPEM – A Notional View of HPEM Environments," *AMEREM 1996*, Albuquerque, NM, 1996.

[3] Paul, C. R., *Introduction to Electromagnetic Compatibility*, New York: John Wiley & Sons, 1992.

[4] Prasad Kodali, V., *Engineering Electromagnetic Compatibility*, New York: IEEE Press, 2000.

[5] Leach, P. O., and M. B. Alexander, *Electronic Systems Failures and Anomalies Attributed to Electromagnetic Interference*, NASA Report 1374, National Aeronautics and Space Administration. Washington, D.C., July 1995.

[6] Tesche, F., "Discussion of EMP Paper by M. Rabinowitz," *IEEE Transactions on Power Delivery*, Vol. PWRD-2, 1987, p. 1213.

[7] IEC 61000-1-5, "High Power Transient Phenomena - High Power Electromagnetic (HPEM) Effects on Civilian Systems," 2004.

[8] CPNI, http://www.cpni.gov.uk/about/cni/. Accessed July 2018.

2

HPEM Environments

2.1 Introduction

In electromagnetics lexicon, an environment can be defined as "the totality of all electromagnetic (EM) phenomena across the electromagnetic spectrum."

The total EM environment is vast, stretching from a DC to the optical range and beyond; "DC to daylight." The EM spectrum is shown in Figure 2.1.

EM energy is radiated in packets and a single packet corresponding to a frequency f is given by hf where h is the Planck constant. It is interesting to note that the EM spectrum can approximately be classified into two groups of ionizing ($f > 10^{15}$ Hz or energy > 4 eV) and nonionizing ($f < 10^{15}$ Hz or energy <4 eV). The focus of this book is a subset of the total EM spectrum encompassing several HPEM phenomena, which are all well below 10^{15} Hz and hence nonionizing.

The HPEM environments considered in this book more typically cover a frequency range of a few kilohertz to perhaps 100 GHz [1]. Further, HPEM environments can be broadly classified into two groups, namely, anti-electronics and anti-personnel, as shown in Figure 2.2.

To date, anti-personnel use of HPEM technology has focused on millimeter-wave frequencies. However, the focus of this book is the anti-electronics HPEM environments where at least anecdotally 12 GHz is a practical limit. This is because this frequency range interacts more strongly (i.e., couples efficiently), with the systems-of-systems hierarchy presented in Chapter 1 by virtue of the wavelength dependence of coupling. In other words, coupling and interaction into a target system are very strong functions of frequency and this

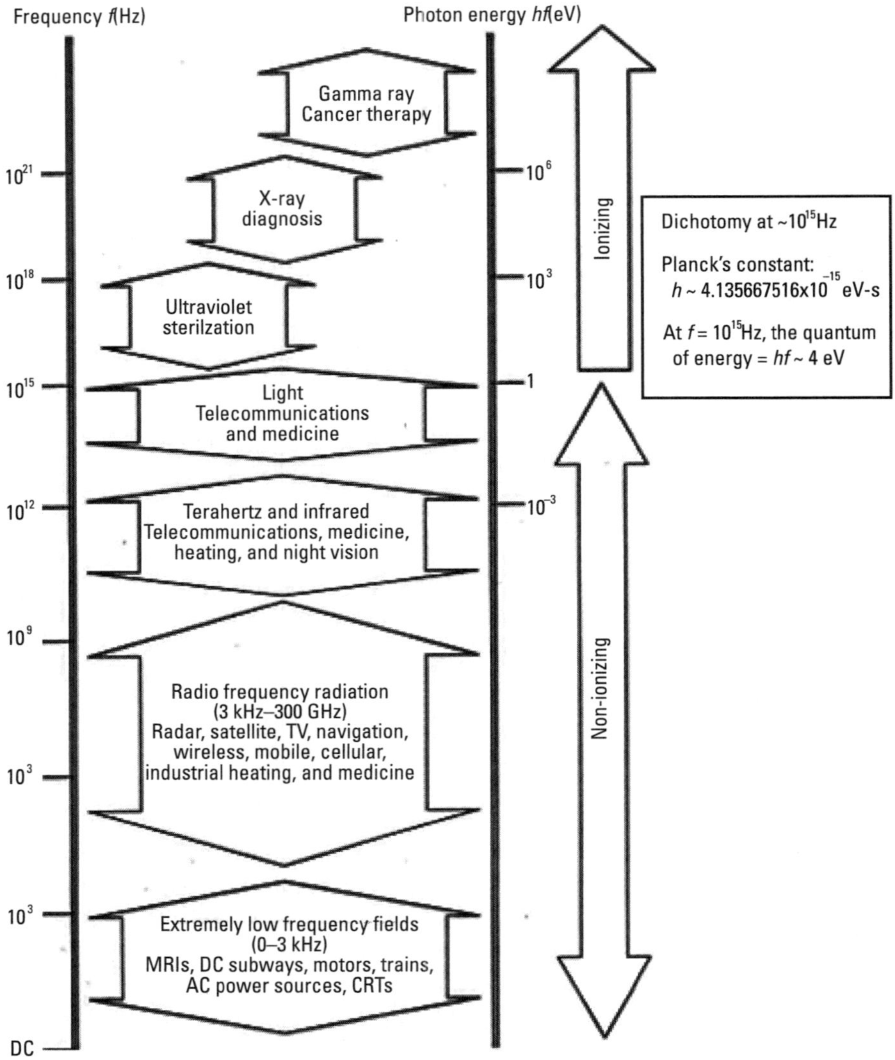

Figure 2.1 The EM spectrum.

will be discussed in detail in subsequent chapters. The anti-electronics HPEM environments can impact both civilian and military equipment [2].

The technical understanding of HPEM environments is fairly mature, at least mature enough to enable HPEM environment waveforms to be defined in a wide variety of standards. An environment definition is often the starting point for on effects assessment and EM protection or resilience design.

A diagram of some aspects of the HPEM environment relevant to the scope of this book is shown in Figure 2.3 where the electric field spectrum in its

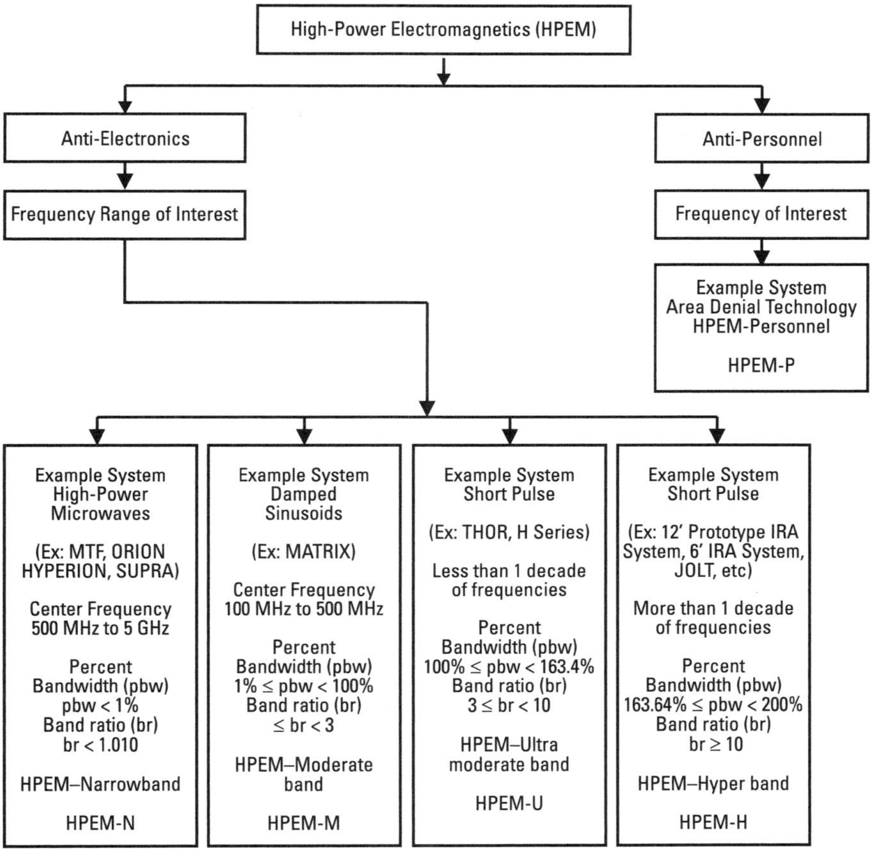

Figure 2.2 EM environments, effects, and protection (EEEP or E3P).

proper units of (V/m)/Hz. In this figure, the lightning environment is a calculation of a fairly small lightning discharge of 10 kA at two distances of 3 km and 300m from the strike [3].

Two curves for NEMP E1 are shown, the uppermost of the NEMP curves at 1×10^9 Hz is the analytical expression from IEC 61000-2-9 [4] and the other NEMP curve (lower curve at 1×10^9 Hz) is the quotient of exponentials form of the analytic expression from a now withdrawn German National standard as described in [5–7].

The black curve is the predicted output of a Mesoband HPRF DE system known as MATRIX. This system is discussed in [8].

The prototype impulse radiating antenna (IRA) and JOLT curves are from measurements made on those HPRF DE simulators at the distances specified (100m). These systems are both Hyperband generators and both of these systems are discussed in more detail in [9–13].

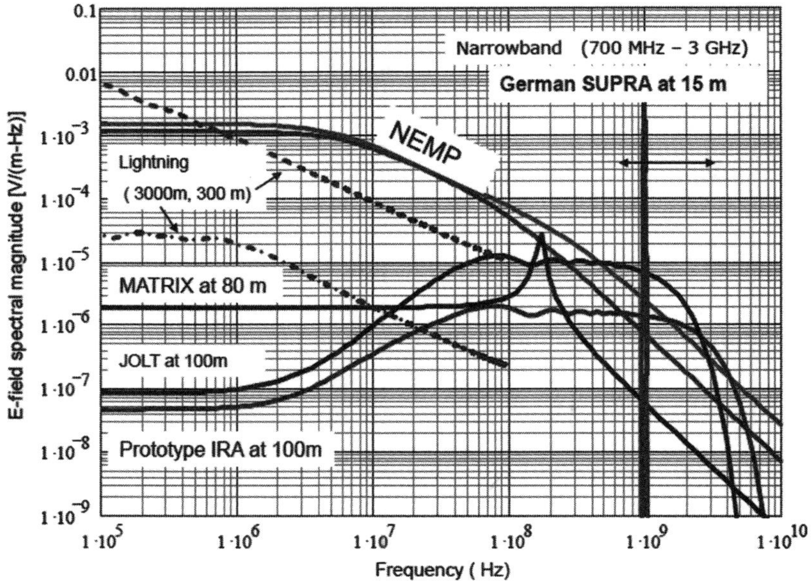

Figure 2.3 Aspects of the HPEM anti-electronics environments in actual units.

Finally, the curve 1×10^9 Hz is from the German narrowband or Hypoband HPRF DE simulator known as SUPRA. This system is discussed in greater detail in [14, 15].

In summary, an HPEM environment may be:

- A single pulse with many cycles of a single frequency (an intense narrowband signal that may have some frequency agility);
- A burst containing many pulses, with each pulse containing many cycles of a single frequency;
- A single pulse tending to an impulse with a very wide bandwidth (spectral content from hundreds of megahertz to several gigahertz);
- A burst of many impulse-like transient pulses.

The schematic in Figure 1.2 indicates that there are other HPRF DE/IEMI sources with different frequencies of operation for the Mesoband and Hypoband classes with potentially different spectral magnitudes. MATRIX, for example, can be designed to operate at four different dominant frequencies: 200 MHz, 300 MHz, 400 MHz, and 500 MHz [16].

This chapter provides a brief introduction to HPEM environments as this is necessary to design effects experiments that meaningfully represent the

HPEM environment that will be encountered by the device, circuit, equipment, and system in real-world scenarios.

2.2 Lightning

Lightning is the only HPEM signal made by nature. If the universe was created 13 billion years ago, it took another 8 to 9 billion years for our planet Earth and its atmosphere to form. Estimates put planet Earth at about 4 to 5 billion years old. It is our best estimate that natural lightning is as old as the Earth and its atmosphere itself. There are many types of lightning, though the two main types of concern are cloud to cloud and cloud to Earth (as seen in Figure 2.4). Lightning can obviously occur naturally or spontaneously or can be artificially triggered.

2.2.1 Overview

A very good overview of the lightning HPEM environment is given in [17, 18]. Figure 2.5 tries to capture the history of lightning in one chart.

In Figure 2.5, we have noted that the first case of lightning may be at the time of the Earth's atmosphere formation approximately 4 to 5 billion years ago. Electrical sparks were seen as early as 600 B.C., as seen in the chart by going around in a clockwise direction from the top. Early lightning rods were installed in the middle of the eighteenth century to protect structures such as churches and Benjamin Franklin's famous kite experiments followed. Throughout the nineteenth and twentieth centuries, a great deal of interest in lightning occurred and in particular work on measurements of lightning-induced currents and the HPEM-radiated environment produced by the lightning strike.

Figure 2.4 Two types of natural lightning (cloud to Earth and cloud to cloud).

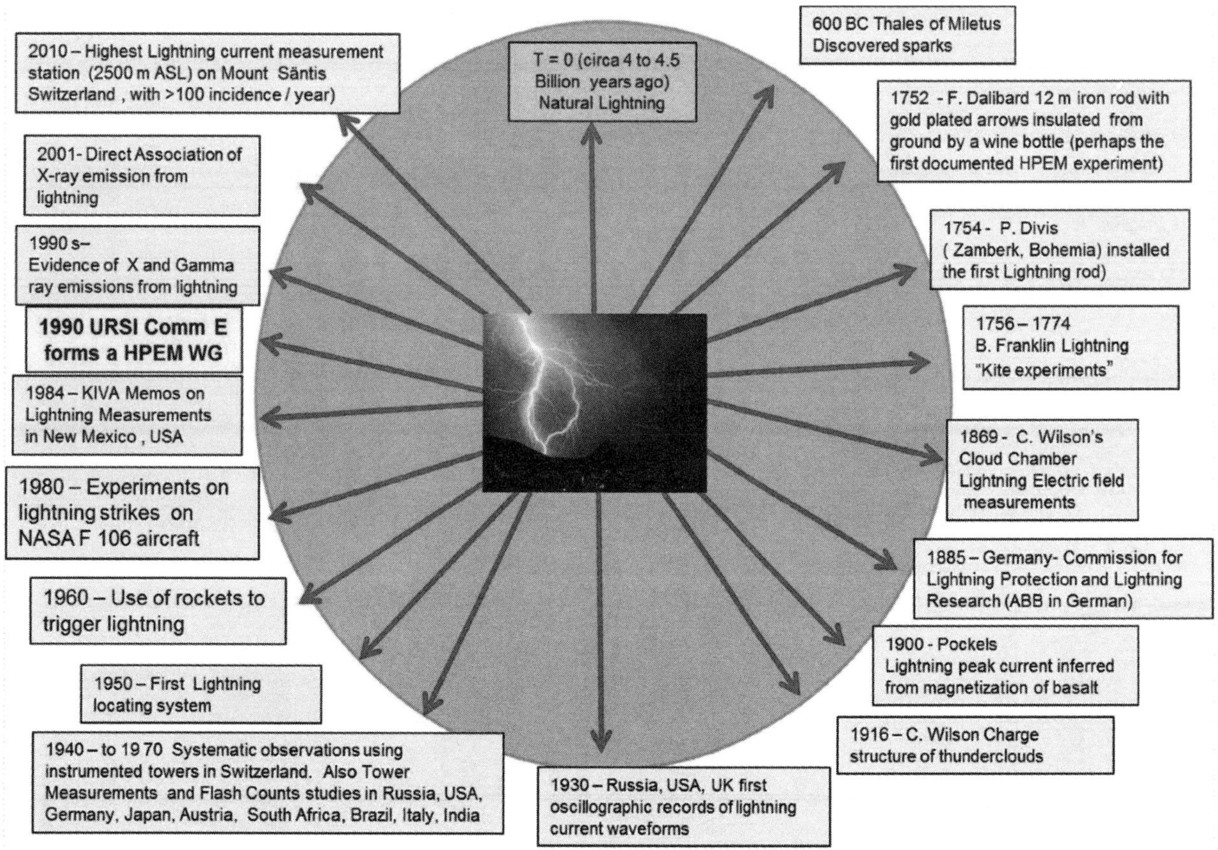

Figure 2.5 A chart of the history of lightning [19].

A good understanding of the transient current flowing in a natural lightning discharge is important for many reasons. Such an understanding leads to the clarification of:

- The physics of the discharge;
- Radiated EM fields;
- Interaction and coupling to electrical and electronic systems;
- Protection of systems from adverse effects of lightning.

Over many past decades, lightning parameter measurements have been made from mountain tops from different parts of the world such as the United States [20–24], Canada [25], China [26], and Santis Tower in Switzerland [27–31]. Generally speaking, channel current measurements are performed on structures or instrumented towers at top of mountains. From the measured currents, one may compute the radiated fields. Conversely, one may measure the radiated fields at a distance and deduce the parameters of the channel current.

In the United States, the Kiva 2 facility is an underground, electromagnetically shielded enclosure where triggered lightning measurements have been performed. It is located on South Baldy peak, which is the highest point in Magdalena Mountains in central New Mexico. Lightning-initiating rockets are launched from this facility. Basically, the facility is a steel tank that is 5.64m in diameter, 2.44m high with walls of 4.75-mm thickness, with a steel plate welded to the top of the cylinder. Instrumentation racks are housed on the interior. The parameters of interest are the rise time of the current, the rate of rise or dI/dt, the rise time of EM fields 50m away, and the optical emission of the lightning. An array of current viewing sensors were deployed and calibrated for measuring the current parameters. It was designed to be an unattended data recording of direct and nearby lightning.

Likewise, Santis Tower is a more recent instrumented tower in Switzerland that measures the lightning parameters. The tower is 124m tall and is shown in Figure 2.6.

For illustrative purposes, we show examples of the measured magnetic field waveform for the case of untriggered natural lightning in Figure 2.7.

We can see from Figure 2.7(b), that the peak current is 22 kA for this upward negative flash, and the peak dI/dt from Figure 2.7(c) is 56 kA/μs = 5.6×10^{10} Amps/sec. This results in a maximum rate of rise of (peak current/ peak derivative) [29] = 22 kA/{56 kA/μs)} = 400 ns. This is a case of employing the current and field sensors originally designed for HEMP measurements for lightning measurement.

An excellent summary of the modeling of lightning strikes to many instrumented towers in various parts of the world is available in [32].

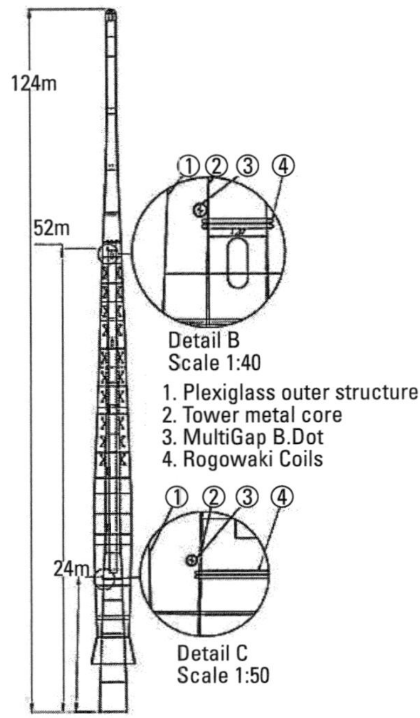

Figure 2.6 Instrumented Santis Tower in Switzerland [27]

In addition to lightning measurements from instrumented structures on mountain tops, lightning currents and fields due to nearby measurements have also been performed by airborne platforms. In the 1980s, both direct and indirect lightning attachments to aircraft were researched with an instrumented F-106 Vietnam-era aircraft [33], seen in Figure 2.8. This research effort spanned over 5 years and was sponsored by NASA Langley Research Center.

In Figure 2.8(a), the paint marks on the aircraft are the lightning attachment points. The B-dot sensor mounted on the nose boom of the aircraft, shown in Figure 2.8(b) was placed near a location where the scattered field from the aircraft has a null. In the case of nearby lightning, the sensor picks up the incident field exclusively since the scattered field is nulled out. This instrumented aircraft has flown in thunder clouds and collected data from direct attachment of lightning and measured the radiated fields from nearby lightning as well and reported in [34, 35].

2.2.2 Lightning-Radiated Environment

There are many cases of measured and calculated lightning waveforms from nearby lightning. For example, we have shown the radiated electric field from a 10-kA lightning discharge at two distances of 300m and 3,000m in Figure 2.2.

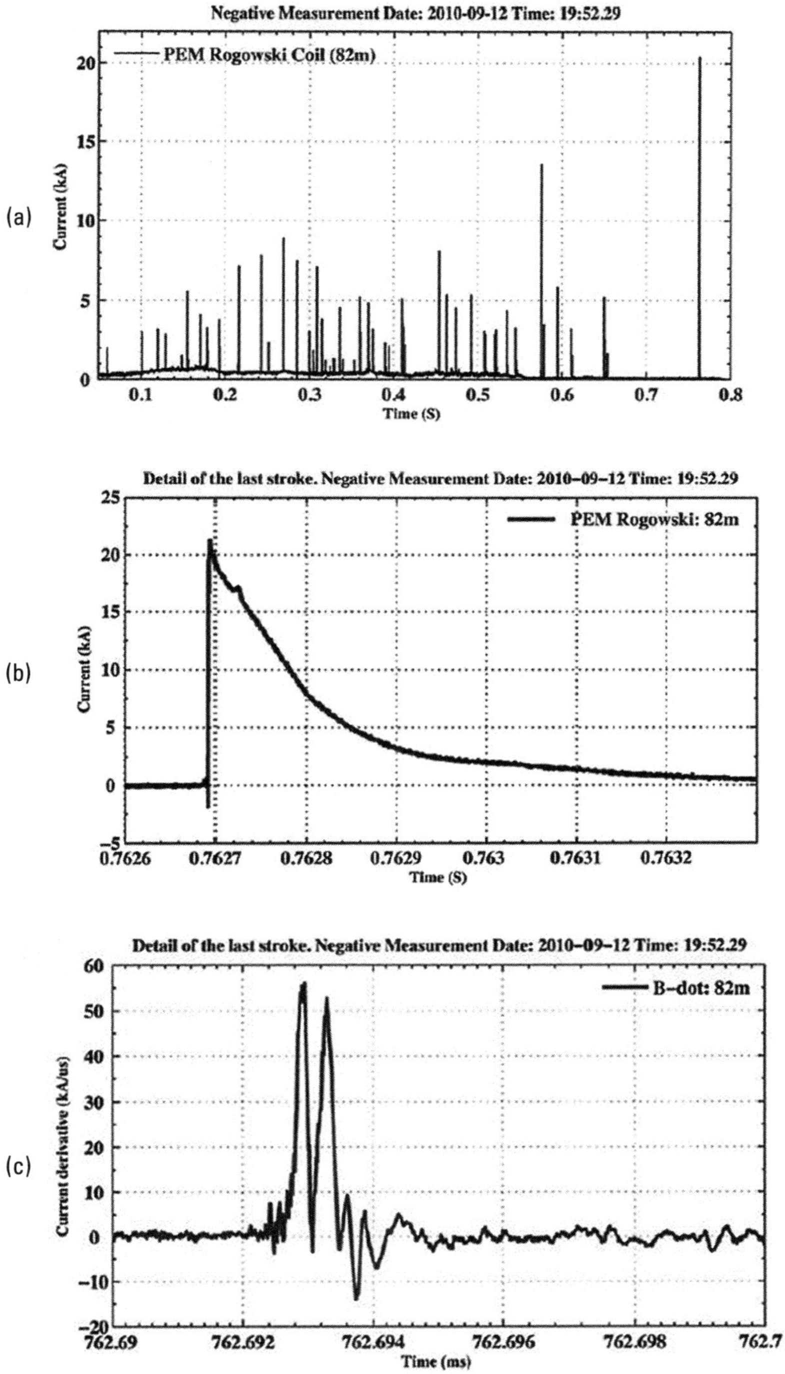

Figure 2.7 (a–c) Typical measured current waveforms due to natural lightning [27].

(a)

(b)

Figure 2.8 (a) F-106 (NASA 816) research aircraft during the Storm Hazard Program in 1982 [33]. (b) B-dot sensor on the nose boom of an F-106 aircraft [33].

One may conclude that the natural lightning can be a bigger threat at frequencies below 1 MHz and certainly HEMP takes over as a larger threat above 1 MHz. One other important difference between radiated lightning environment and HEMP is that lightning is a localized phenomenon impacting one aircraft or one office building, for example, while HEMP can affect a whole fleet of aircraft or large geographic region encompassing several cities in a single occurrence.

2.3 Nuclear EM Pulse

The term nuclear EMP (NEMP) is used to describe a variety of pulsed transient phenomena produced as a consequence of the explosion of a nuclear weapon. Of primary interest here is the EMP created by a high-altitude burst, HEMP.

HEMP is defined as occurring at a nuclear weapon burst height greater than 30 km above the Earth's surface [4]. For this altitude regime, the ionizing radiation, blast, and thermal effects produced by the nuclear burst do not reach the Earth's surface with sufficient magnitude to cause significant effects on electronics, but several types of intense EM signals referred to as HEMP E1, E2, and E3 do reach the Earth's surface and can cause significant effects.

The ionizing radiation produced by the nuclear detonation interacts with the atmosphere resulting in Compton scattering between electrons. These electrons spiral around the Earth's magnetic field, causing current flow in the atmosphere, setting up an effective antenna. The result of this interaction is a high-amplitude, fast rise time, transient EM field that has the following composition [36]:

- Early-time HEMP (E1): The initial part of the waveform defined by a high amplitude (tens of kilovolts per meter), fast nanosecond rise time, and a duration of tens of nanoseconds.
- Intermediate-time HEMP (E2): The middle part of the waveform defined by an amplitude of tens of volts per meter and a duration of milliseconds.
- Late-time HEMP or magneto hydrodynamic (E3 or MHD): The final part of the pulse defined as a low amplitude (mV/m or V/km) and a long pulse duration of the order of seconds.

Other subsets of NEMP include [37, 38]:

- Source region EMP (SREMP) or endo-atmospheric EMP: The EMP produced when the nuclear detonation is at lower altitudes (lower than 30 km), such as surface or air burst detonation. At small distances from the burst, other nuclear weapon effects such as blast, thermal, and ionizing radiation dominate systems interaction.
- System-generated EMP (SGEMP): SGEMP results from the direct interaction of nuclear weapon ionizing radiation output (gamma and X-rays) with a system that, in turn, generates electric currents in the structure, cables, and circuit boards of the system. Because nuclear weapon gamma and X-rays are attenuated by the atmosphere at low altitudes, SGEMP has special importance for systems outside the atmosphere, such as satellites. The term high-altitude nuclear explosion (HANE) is also used in connection with the broader nuclear weapon effects on space-based systems.

The HEMP E1 environment is prioritized for discussion in this book as it is the primary driver for HEMP effects at device level through to network level in the system of systems hierarchy.

As Glasstone and Dolan [39] pointed out, "Explosions of conventional high explosives can produce electromagnetic signals and so the generation of EMP from a nuclear detonation was expected. However, the extent and potentially serious nature of EMP effects was not realised for several years." The intensity of the HEMP environment was not experimentally observed until the early 1960s many years after the first nuclear detonations.

One well-documented event was the Starfish test performed by the United States in 1962 over the Johnston Atoll in the Pacific Ocean. This test was performed at an altitude of 400 km with a 1 MT nuclear weapon and generated HEMP E1 fields estimated at 5.6 kV/m, in Oahu, Hawaii, 1,400 km from ground zero. This HEMP caused burglar alarms and air raid sirens to be set off unintentionally, some streetlights were extinguished while others came on, a string of 30 lights failed in one area, and several fuses were blown [40]. At the time of the test, electronic systems were immature compared to today's standards and our subsequent reliance on them has increased to a level previously unimaginable.

In 1994, these types of effects were corroborated by Russian scientists [41] who had observed even more effects during testing in 1962 on more systems as their high-altitude nuclear tests were over land. Failures of several long line systems, including power and telecommunications, were reported. The failures of the protection devices on a 500-km-long telecom line are also well documented.

These tests confirmed the existence of the predicted HEMP phenomena and were a catalyst to the development of experimental techniques to characterize and quantify the HEMP environment and to develop test techniques to understand the implications of HEMP effects on systems.

2.3.1 HEMP-Radiated Environment

The actual radiated HEMP environment that is produced during a nuclear explosion is very often highly classified due to the dependence on design details of the nuclear weapon itself and on assumptions related to the weapon deployment. In the 1960s, published HEMP environments concentrated on the early-time HEMP E1 component of the HEMP composite environment.

In IEC 61000-2-9 [4], mathematical expressions for the double exponential waveforms used to describe each component of the HEMP waveform are provided. These expressions are shown in Figure 2.9. The combined HEMP timeline based on the analytical expression, including the intermediate time or E2 and late time or E3 components, is shown in Figure 2.10.

Early time HEMP (E1)

$$E_1(t)=\begin{cases}0 & \text{when } t\leq 0\\ E_{01}\cdot k_1\left(e^{-a_1 t}-e^{-b_1 t}\right) & \text{when } t>0\end{cases}$$

$E_{01}=50\,000$ V/m

$a_1=4\cdot 10^7\ s^{-1}$

$b_1=6\cdot 10^8\ s^{-1}$

$k_1=1{,}3$

Intermediate time HEMP (E2)

$$E_2(t)=\begin{cases}0 & \text{when } t\leq 0\\ E_{02}\cdot k_2\left(e^{-a_2 t}-e^{-b_2 t}\right) & \text{when } t>0\end{cases}$$

$E_{02}=100$ V/m

$a_2=1000\ s^{-1}$

$b_2=6\cdot 10^8\ s^{-1}$

$k_2=1$

Late time HEMP (E3)

$$E_i(t)=\begin{cases}0 & \text{when } \tau\leq 0\\ E_{0i}\cdot k_i\left(e^{-a_i\tau}-e^{-b_i\tau}\right) & \text{when } \tau>0\end{cases}$$

$\tau=t-1$

$E_{0i}=0{,}04$ V/m

$a_i=0{,}02\ s^{-1}$

$b_i=2\ s^{-1}$

$k_i=1{,}058$

$$E_j(t)=\begin{cases}0 & \text{when } \tau\leq 0\\ E_{0j}\cdot k_j\left(e^{-a_j\tau}-e^{-b_j\tau}\right) & \text{when } \tau>0\end{cases}$$

$\tau=t-1$

$E_{0j}=0{,}01326$ V/m

$a_j=0{,}015\ s^{-1}$

$b_j=0{,}02\ s^{-1}$

$k_j=9{,}481$

$$E3\,(t)=Ei\,(t)-Ej\,(t)$$

Figure 2.9 Analytical expressions of HEMP E1, E2, and E3 from IEC 61000-2-9.

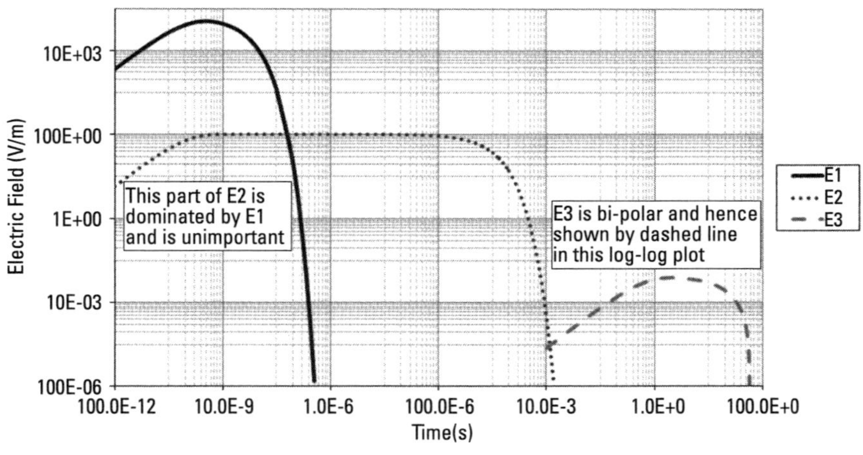

Figure 2.10 Combined HEMP timeline based on analytical expressions.

Arguably, the first openly published HEMP E1 waveform was the well-known Bell Labs EMP waveform [42], a description of the early-time E1 environment articulated as a 50-kV/m double-exponential waveform with a 4.6-ns rise time (t_r) and a pulse width of approximately 200 ns at full-width half-maximum (FWHM).

This description of the HEMP E1 environment has been updated a number of times by newer versions [43–45] as analytical and modeling techniques have improved. Table 2.1 shows some examples of the parameters of more recent unclassified HEMP standards.

Table 2.1

Comparison of HEMP E1 Environment Definition from Various Standards

Parameter	Bell Labs (1960s)	Baum (1992)	Leuthäuser* (1994)	VG 95371-10** (1995 and 2011)	IEC 61000-2-9 (1996)
tr (10%–90%)	4.6 ns	2.5 ns	1.9 ns	0.9 ns in 1995, changed in 2011 to 2.5 ns	2.5 ns
Peak E-field (E0)	50 kV/m	50 kV/m	60 kV/m	50 kV/m	50 kV/m
FWHM	184 ns	~23 ns	23.8 ns	23 ns	23 ns
k1	1.05	1.3	1.08	1.3	1.3
a1	4×10^6	4×10^7	2.2×10^9	4×10^7	4×10^7
b1	4.76×10^8	6×10^8	3.24×10^7	6×10^9	6×10^8
Energy density (J/m2)	0.891	0.114	—	0.144	0.114

*Note: The Leuthäuser [45] environment is a quotient exponential and not a double exponential as in the other cases.

**VG 95371-10 is a German military HEMP standard.

The evolution of all of the unclassified HEMP standards is summarized in [46]. These unclassified E1 environments are shown plotted in Figure 2.11.

The difference between the double exponential (DEXP) form and the quotient of exponentials (QEXP) forms are described next [47].

2.3.1.1 Double Exponential Representation of E1 HEMP

The double exponential description of the HEMP waveform has been used since the early days of HEMP research and was well described in [48]. The time-domain expression is given in (2.1):

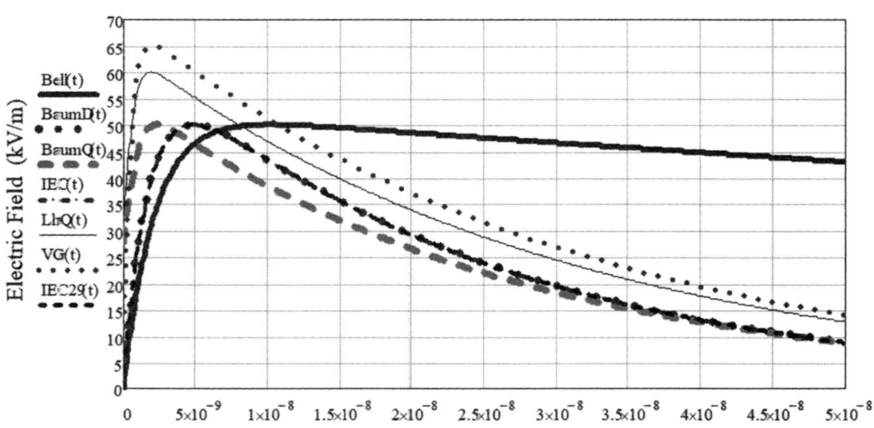

Figure 2.11 Time-domain waveforms of some published HEMP E1 standards.

$$E(t) = E_0\left[e^{-\alpha(t-t_0)} - e^{-\beta(t-t_0)}\right]u(t-t_0)\,(\mathrm{V/m}) \tag{2.1}$$

where E_0 = E-field intensity constant (V/m), α = decay constant (rad/s), β = decay constant (rad/s), $u(t - t_0)$ = unit step function, t_0 = time shift (s), and the corresponding frequency-domain expression is given by:

$$\tilde{E}(\omega) = E_0\left[\frac{1}{j\omega+\alpha} - \frac{1}{j\omega+\beta}\right] \tag{2.2}$$

where $\tilde{E}(\omega)$ = Fourier transform of $E(t)$ (V/m/Hz), $\omega = 2\pi f$ = radian frequency (radians/s), and $j = \sqrt{-1}$.

We can see from this expression then that:

- The first frequency break point is at $f = [\alpha/(2\pi)]$ Hz.
- The second frequency break point is at $f = [\beta/(2\pi)]$ Hz.
- The magnitude of the flat, low-frequency part of the spectrum is then given by:

$$E_1 = E_0\,{}^{(\beta-\alpha)}\!\big/_{(\alpha-\beta)} \tag{2.3}$$

E_1 is in (V/m/Hz); this is the low-frequency spectral magnitude.

The double exponential may be characterized in a number of ways. The equations are defined by the three variables, E_o, α, and β, which precisely define the double exponential waveform. However, the time-domain representation is typically characterized by quantities more easily related to the measured waveform. That is,

- Peak electric field, E_p (note: $E_p \neq E_o$);
- 10% to 90% rise time, t_r;
- FWHM;
- The e-fold decay time, when the amplitude reaches 1/e of E_p ($\approx$ 37%).

These parameters are chosen because they can be easily inferred from actual measurement data.

2.3.1.2 Quotient of Exponentials Representation of E1 HEMP

In order to correct for the discontinuity of the DEXP form, another analytic form was derived that is the reciprocal of sum of two exponentials, sometimes referred to as quotient double exponential, as shown in (2.4) and (2.5).

$$E(t) = \frac{E_o}{[e^{-\alpha(t-t_o)} + e^{-\beta(t-t_o)}]} u(t-t_o) \text{ V/m} \tag{2.4}$$

and the frequency-domain form is:

$$\tilde{E}(\omega) = \frac{E_0 \pi}{(\alpha+\beta)} csc\left[\frac{\pi}{(\alpha+\beta)}(j\omega+\alpha)\right] e^{-j\omega t_0} \text{ V/m/Hz} \tag{2.5}$$

This waveform has the advantage that it has continuous time derivatives of all orders for all times. The disadvantage of this expression is that it extends to $t = -\infty$. The parameter t_0 is used to adjust the amplitude of the signal to arbitrarily small values for $t \leq 0$.

Figure 2.12 shows a comparison of the DEXP and QEXP forms with the parameters noted for the Bell waveform and assuming that $t_0 = 5$ ns.

2.3.1.3 HEMP-Radiated Environment Summary

The overall trend in the evolution of the HEMP E1 definition appears to indicate from Table 2.1 that the newer HEMP E1 environment standards tend to have a faster rise time and a shorter fall time and therefore much less energy density than the original Bell Labs version.

The HEMP E1 environment defined in IEC 61000-2-9 has become the default generic unclassified description of the HEMP E1 environment. This definition is also used and referenced in the U.S. military standard MIL-STD-464A [49].

2.3.1.4 HEMP E1 Coverage

Effectively, the antenna that creates the HEMP environment is the Earth's magnetosphere. The HEMP E1 can be thought of as a plane wave propagating towards the Earth's surface at an angle relative to the magnetosphere's magnetic field pattern at the particular latitude and longitude relative to the nuclear detonation. This is known as the dip angle. The area illuminated by the HEMP E1 environment on and near the Earth's surface covers a very wide geographical area, as shown in Figure 2.13 [50].

Figure 2.13 is a sample E1 HEMP "smile" diagram that shows contours of peak incident electric field levels for a burst height of 75 km. Here contour levels are shown as fractions of the biggest peak level (nominally 50 kV/m),

(a)

(b)

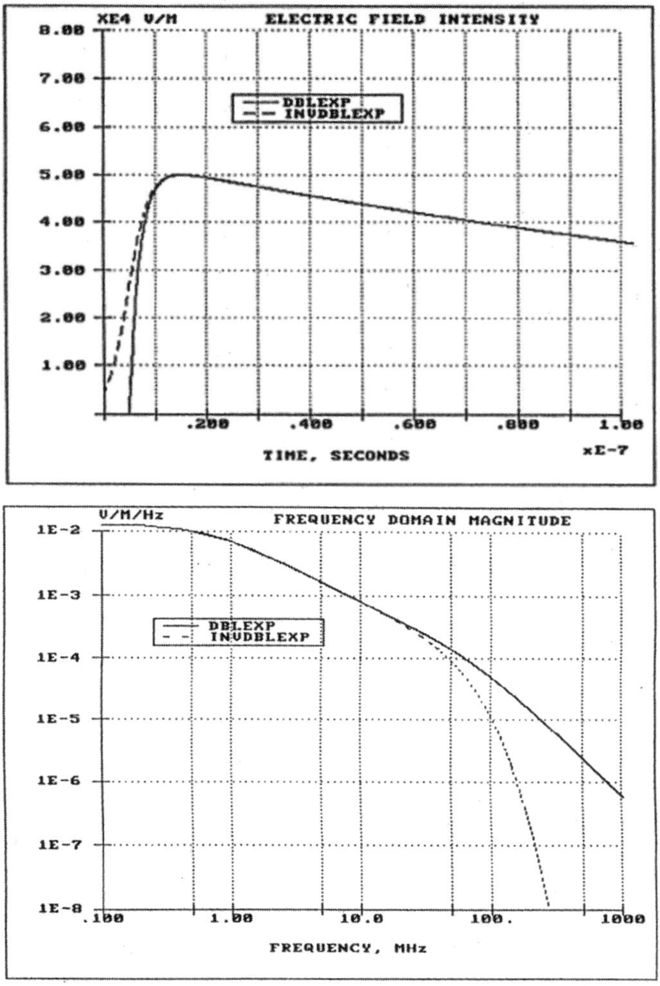

Figure 2.12 (a) Time-domain and (b) frequency-domain waveforms for HEMP comparing the DEXP and the QEXP models.

which is to the South of the burst point. The burst point is depicted as a cross in the figure just North of the "smile" contour.

It should be noted that E2 will immediately follow E1 and E3 follows E2. E2 and E3 are not shown. Practically speaking, the rising portion of E2 is much slower than E1 and therefore a large portion of the energy content of E1 is greater than E2 at low frequencies. This means that E1 dominates E2 in importance when considering effects testing. Systems, installation, and infrastructure are also likely to be tested and protected against lightning, which has a characteristic waveform similar to HEMP E2. However, the notion that protection

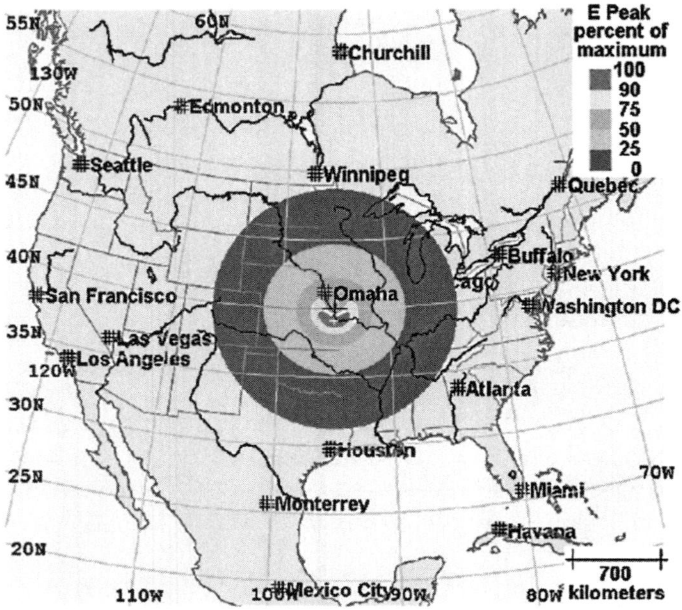

Figure 2.13 Example of HEMP E1 coverage. (Courtesy of Radaksy et al. [50].)

from lightning automatically provides protection for HEMP E2 can be flawed because the HEMP E1 that immediately precedes HEMP E2 may have compromised some of that lightning protection before the HEMP E2 occurs.

The HEMP E3 environment has been compared and contrasted with the natural geomagnetic storm environment [51, 52]. The standardized HEMP E3 environment (linked with the standardized HEMP E1 magnitude of 50 kV/m) produces an E3 magnitude in terms of rate of change of magnetic field (dB/dt) of ~1,500 nT/min (which has been equated to an E-field of ~40 V/km). At the time of this writing, there are ongoing efforts to refine the understanding of both geomagnetically induced current environment and the HEMP E3 environment.

2.3.2 HEMP-Conducted Environment

The radiated HEMP environment induces currents and voltages in metallic structures such as power lines, metallic data/telecommunications lines, pipes, and ducts. These currents and voltages propagating in conductors represent the conducted environment. This means that the conducted environment is a secondary phenomenon, derived from the HEMP-radiated environment [53].

The early-time HEMP (E1) environment is a high-amplitude electric field that couples efficiently to antennas and to any exposed lines such as power and data/telecommunications lines.

The intermediate-time HEMP (E2) environment only couples efficiently to long conductors in excess of 1 km.

The late-time HEMP environment (E3) is only important for coupling to very long (more than a few kilometers) transmission lines such as power and data/telecommunications lines.

For long conductors external to a facility, arguably the biggest influence on the HEMP conducted environment depends on whether the conductor is elevated above the ground or buried in the Earth. Other factors to consider include conductor length, the surface ground conductivity (for depths between 0m and 5m, for HEMP E1), conductor load impedance, the angle of elevation, and polarization. The HEMP-conducted environment is therefore often defined in a statistical manner based on the probability of producing particular levels of current for different conditions.

2.3.3 Open-Source Accounts of HEMP Disturbances

Some of the effects observed from actual atmospheric high-altitude nuclear tests conducted by the Russians and Americans in the 1960s were briefly discussed in Chapter 1.

The Starfish test, performed by the United States in 1962 over Johnston Atoll in the Pacific Ocean, produced HEMP effects in Oahu, Hawaii, 1,400 km from ground zero. In 1994, Russian scientists reported that a test in 1962 showed that HEMP effects were similarly observed.

At the time of these tests, electronic systems were immature compared to today's technology. Perhaps thankfully since 1963, atmospheric testing of nuclear weapons has been banned by international treaty called the Limited Test Ban Treaty (LTBT) signed by the United States, the United Kingdom, and the USSR [54].

After 1963, most if not all HEMP effect testing was conducted with simulators and in smaller-scale experiments as part of underground test detonations. Most of the accounts of these tests remain classified and the relevance of the data to modern society is arguable.

From 2004 until 2008, a major assessment of the effects of a high-altitude EM pulse (EMP) attack on our critical national infrastructures in the United States was executed by the U.S. EMP Commission. The Commission to Assess the Threat to the United States from Electromagnetic Pulse (EMP) Attack was established by Congress through Title XIV of Public Law 106-398. The U.S. EMP Commission reported its findings in 2008 [55].

The assessment was informed by analytic and test activities executed under Commission sponsorship. The assessment concluded that: "The electromagnetic pulse generated by a high altitude nuclear explosion is one of a small number of threats that can hold our society at risk of catastrophic consequences. The increasingly pervasive use of electronics of all forms represents the greatest source of vulnerability to attack by EMP. Electronics are used to control, communicate, compute, store, manage, and implement nearly every aspect of United States (U.S.) civilian systems."

While comprehensive effects testing and analysis of simulated HEMP were carried out on modern military and civilian systems, the U.S. EMP Commissions report falls short of actually publishing effects data.

2.3.4 HEMP Environment Summary

E1 HEMP is the primary part of the HEMP environment that is considered a threat at the system level where the size of a typical system is commensurate with the frequency content and associated wavelengths of this part of the combined HEMP waveform (see Figure 1.4). The coverage of HEMP E1 on the Earth's surface covers a wide geographical area. The standardized default HEMP E1 waveform is described in IEC 61000-2-9. The peak intensity is 50 kV/m, the rise time (10% to 90%) is 2.5 ns, and the full width at half max or 50% points (FWHM) is 23 ns. Other open-source descriptions of the HEMP E1 waveform have different parameters. The ranges of values for these three parameters are:

- Peak electric field: 50 to 65 kV/m;
- 10% to 90% rise time: 0.9 to 4.6 ns;
- FWHM: 23 to 184 ns.

Over time, as computational techniques and capabilities have improved the trend has been to prefer specifications for HEMP E1 rise times in the range of 0.9 ns to 2.5 ns and pulse widths in the range 23 ns to 24 ns.

The components of a HEMP disturbance (E1, E2, and E3) have been simulated as separate components and HEMP effects have only been assessed against these separate components in isolation. However, in reality, these components will interact with targets or victims sequentially or synergistically. It is practically extremely difficult to model or evaluate the synergistic effects of HEMP.

2.4 High-Power RF Directed Energy Environments

High-power RF (HPRF) directed energy (DE) systems have been and continue to be a focus of military research and development in many nations. HPRF DE sources intentionally generate HPEM environments to defeat (deny, disrupt, confuse, or damage) electronic systems, including hardened military systems. The accepted U.S. Department of Defense (DoD) definition of directed energy warfare is [56]: "Military action involving the use of directed-energy weapons, devices, and countermeasures to either cause direct damage or destruction of enemy equipment, facilities, and personnel, or to determine, exploit, reduce, or prevent hostile use of the electromagnetic spectrum through damage, destruction, and disruption. It also includes actions taken to protect friendly equipment, facilities, and personnel and retain friendly use of the electromagnetic spectrum."

It is not clear when the concept of HPRF DE systems and weapons was conceived, although the first discussions concerning an RF "death ray" are associated with Nikolai Tesla. Tesla attempted to build a very high-power RF transmitting station at Wardenclyffe, New York [57], for long-range radio communications. Later in 1940 when Tesla was 84 years old, he discussed the concept of an HPRF DE weapon for defeating aircraft with a journalist from the New York Times [58]. Tesla's invention of the magnifying transformer or Tesla transformer is still used in some of the HPRF generator system designs today.

After Tesla, the concept returned to the science fiction domain because no technology was available to build an HPRF DE system. In World War II, with the advent of radar and the proliferation of long-range radio communications, high-power transmitter technology started to become available for military use.

During the Cold War, there is significant evidence to suggest that the former Soviet Union researched HPRF DE concepts that apparently resulted from their recognition that they could not match the capability of Western electronics and that this could be a potential vulnerability for the West [59].

Unpredicted side effects of the 1962 Starfish high-altitude nuclear test that was discussed earlier were the disruption, damage, and power outage of electrical systems in Oahu, Hawaii, attributed to an EMP. This disruption prompted a great deal of research into EMP phenomena, especially in the areas of simulation of EMP, hardening, and protection of military equipment and systems.

In order to generate a simulated EMP event using non-nuclear methods, it was necessary to develop pulsed power sources. Pulsed power became a dedicated field of research that began around 1960 [60]. The goal of pulsed power designers was to produce efficient pulsed power machines to drive a variety of sources including EMP simulators, EM launchers, laser systems, and X-ray machines.

It may be speculated that advances in the development of pulsed power technologies together with the recorded effects of the high altitude nuclear tests and notable accidents attributed to HPEM such as the USS Forrestal disaster in 1967 [61], may have sparked the military to consider development of non-nuclear means to generate HPEM effects. Indeed, some early HPRF DE technology concepts were referred to as non-nuclear EMP or N2EMP.

2.4.1 The Status of HPRF DE Systems Today

Worldwide, HPRF DE generators continue to be researched and developed by national government defense organizations, often in partnership with industry or universities. From the open literature, it is possible to infer that countries such as Canada, China, France, Germany, India, Iran, Russia, Sweden, South Korea, United States, and United Kingdom operate specialist HPRF simulator systems [10, 14, 15] and conduct research into HPRF DE.

Examples of narrowband/hypoband simulators covering the frequency range of 0.7 GHz to 3 GHz include:

- The mobile Swedish Microwave Test Facility (MTF) in Linkoping, Sweden, consists of five microwave sources at fixed frequencies in the L, S, C, X, and Ku radar bands.
- The mobile Orion system, originally based in United Kingdom (now back in the United States), uses relativistic magnetrons and horn-fed reflector antennas [62].
- The Hyperion Hypoband simulator with frequency agility, a Super Reltron-based fixed installation at CEG, Gramat, France.
- The SUPRA Hypoband simulator with frequency agility, a Super Reltron-based fixed installation at WIS, Munster, Germany.

Photographs of the four high-power Hypoband simulators are shown in Figure 2.14.

Detailed performance specifications of these simulators are shown in Table 2.2 and can be found in IEC/TR 61000-4-35 [15], which is a compendium of HPEM simulators and includes both narrowband and wideband types.

HPRF DE systems are now increasing in maturity and may be used across a broad range of military and dual-use operations. Such operations for HPRF DE systems include but are not limited to:

- Counter improvised explosive devices (C-IED);
- Counter mobility (vehicle and boat/fast incoming attack craft (FIAC) stopping);

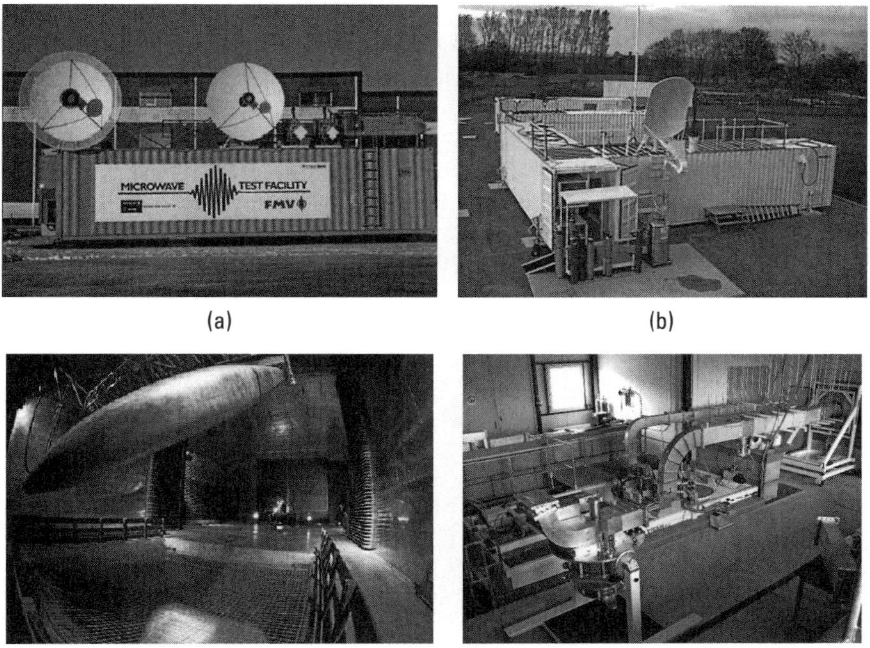

(a) (b)

(c) (d)

Figure 2.14 (a) The Swedish Microwave Test Facility (MTF), (b) the Orion Hypoband simulator, (c) the Hyperion Hypoband simulator, and (d) the SUPRA Hypoband simulator. (From: [14]. © 2004 IEEE. Reprinted with permission.)

- Ground-based air defense (GBAD), suppression of enemy air defense (SEAD), and counter unmanned aerial vehicle (C-UAV);
- Deep target attack;
- Counter personnel (usually counterpersonnel systems operate at millimeter-wave frequencies; given the application and the frequency of operation used for these systems, this operational area is considered to be out of the scope of this book).

Explosively driven HPEM generators have also been widely discussed in the literature and an excellent summary of the technology has been provided by Altgilbers et al. [63]. In popular media the term E-bomb has been used together with terms like EMP bomb, EMP/Ty bomb, and RF munitions (RFM). Articles that describe the application of the E-bomb stem from Kopp [64], and later Abrams [65].

An E-bomb is an explosively driven HPEM generator whereby explosive energy is either directly converted into RF energy or used to provide electrical

Table 2.2
Examples of Narrowband HPM Systems in Europe

Feature	MTF Sweden	ORION (formerly United Kingdom)	Hyperion France	SUPRA Germany
RF Source	Conventional tubes	Relativistic magnetrons	Relativistic magnetrons and Reltrons	Reltrons
Frequency	1.3, 2.86, 5.71, 9.3, and 1.5 GHz	1 to 3 GHz	1.3 to 1.8 GHz (magnetron); 2.4 to 3 GHz (magnetron); 0.72 to 1.44 GHz (Reltron)	0.675 to 1.44 GHz
Max. Power	25, 20, 5, 1, 0.25 MW as a function of frequency	350 MW of microwave power; 5 GW pulse power 500 kV, 50Ω	—	400 MW to 200 MW
Max. *prf*	1,000, 1,000, 1,000, 1,000 and 2,100 Hz	Single shot to 100 Hz	— (magnetron); 1 Hz (Reltron)	10 Hz
Max. pulse duration	5, 5, 5, 3, 8, and 0.53 μs	500 ns	100 ns (magnetron); 200 ns (Reltron)	>300 ns
Field level	—	—	60 kV/m (magnetron); 40 kV/m (Reltron)	70 kV/m at 15m

power to drive a microwave modulator. The energy generation mechanism is based on magneto-cumulative generators (MCG), flux compression generators (FCG), and magneto-hydrodynamic devices (MHD). Direct conversion of explosive energy to radiated RF waveforms was reported by Prishipenko [66] but later refuted.

In world terms, the United States and Russia have led the development of this technology and have been working in this field for many years [67]. The Russians have produced a mobile lightning simulator incorporating MCG technology [68] and the United States have also recently produced a portable high-power microwave (HPM) system [69] based on the E-bomb concept. The Russian work has concentrated on the injection of MCG-generated waveforms into the low-impedance earthing grids of facilities [70], allegedly to simulate positive strike lightning. It is speculated that the E-bomb could also be used to inductively inject very high currents into power distribution cables at remote locations.

Fundamentally, while the generation of the pulsed power may be explosive rather than more conventional electrical/electronic-based pulsed power systems, the environment produced by explosively driven sources tends to be similar to conventional systems. This is primarily because the largest influence on the parameters of the environment generated is related to the antenna geometry; the design of the antenna is complementary to both pulse power generation types.

Several articles have suggested that HPRF DE technologies have moved out of the government research laboratories and into defense industry products. Examples are the Northrop Grumman NUCAS platform [71], the Raytheon Vigilant Eagle system [72], and the Boeing CHAMP system [73]. There are also a few commercial suppliers of HPRF DE systems for applications such as vehicle and vessel stopping. Examples of this are the German Diehl companies HPEM CarStop [74] and the Teledyne e-2v Safe-Stop systems [75].

HPRF DE simulators for testing the effects of HPRF DE systems are also becoming commercially available [76, 77].

HPRF DE Technology is considered to be mature enough by the U.S. Department of Defense and NATO to warrant inclusion in defense EMC standards [78, 79]. Definitions of HPRF DE environments have begun to emerge in unclassified and open source standards. The narrowband HPM EME from MIL-STD-464C is shown in Table 2.3, and the wideband environment is shown in Table 2.4.

The wideband HPM environment specified in MIL-STD-464C is presented as a spectral magnitude and is actually composed of different wideband sources [80]. The detailed technical work deriving the parameters for these environments is contained within classified reports that are not broadly available. This means that the environments require significant tailoring or adaptation to a particular scenario of interest by a security-cleared specialist.

2.5 Intentional EM Interference Environments

A formal definition for intentional EM interference (IEMI) was formed by the scientific community in 1999 [81, 82]: "Intentional malicious generation of electromagnetic energy introducing noise or signals into electric and electronic systems, thus disrupting, confusing or damaging these systems for terrorist or criminal purposes."

Table 2.3
Narrowband HPM Environment from MIL-STD-464C

Frequency Range (MHz)	Electric Field (kV/m at 1 km)
2,000–2,700	18.0
3,600–4,000	22.0
4,000–5,400	35.0
8,500–11,000	69.0
14,000–18,000	12.0
28,000–40,000	7.5

Table 2.4
Wideband HPM Environment
from MIL-STD-464C

Frequency Range (MHz)	Broadband Electric Field Distribution (mV/m/MHz at 100m)
30–150	33,000
150–225	7,000
225–400	7,000
400–700	1,330
700–790	1,140
790–1,000	1,050
1,000–2,000	840
2,000–2,700	240
2,700–3,000	80

Various other terms have been used in published literature to describe IEMI or at least facets of it and include but are not limited to:

- EM terrorism;
- Side channel attacks/electronic exploits (EE), a term used by the cyber community;
- Electronic attack;
- High-energy radiated fields (HERF);
- Non-nuclear EMP (NNEMP or N^2EMP).

However, the term IEMI has become the preferred term at least by the technical community. The term EM terrorism has largely been dropped from the lexicon as it has since been realized that it is difficult to terrorize with an effect that does not directly affect people and does not have a "media-genic" manifestation.

The primary difference between HPRF DE and IEMI domains relates not so much to the technology employed to produce the HPEM environment or the characteristics of the resulting environment but to the type and motivation of the threat actor who may use the HPEM effector. HPRF DE generators are sophisticated military systems for military applications, whereas IEMI encompasses a range of generator types of varying capability that may be used for nefarious or malicious purposes.

IEMI generators perhaps of lower technical complexity or capability can produce RF signal levels just above the general RF background environment and can be effective at jamming RF communications at close range. As the intensity or energy of the RF signal that can be produced increases, the IEMI generator can be effective at jamming at longer range or can create temporary disturbances in electrical and electronic equipment at close range. The sophistication of the design of the IEMI generator has a large bearing on the efficiency and the effective range. It is therefore important to understand the relationship between technical sophistication and the technical capability of the perpetrator.

2.5.1 IEMI Technical Capability Groups

The technical capability of those individuals or groups interested in producing IEMI has been defined in several different ways. The IEC considers three technical sophistication or capability levels that are described in [83] and summarized here:

1. Novice: Individuals or small groups with minimal technical or financial support.
2. Skilled: Moderately well-funded adversaries with training and expertise in relevant technology.
3. Specialist: Well-funded adversaries with postgraduate-level training and access to substantial research capabilities, resources, and funding.

In terms of a risk analysis type of approach, five capability levels can be defined [84]:

1. Novice: Individuals or small groups with general knowledge without any particular technical knowledge or training.
2. Skilled: Individuals or small groups with basic technical understanding and some fundamental technical training.
3. Specialist: Individuals or groups with well-grounded knowledge and training in relevant technology (e.g., qualified skilled worker, technician).
4. Graduate: Individuals or groups with postgraduate-level academic knowledge and expertise in relevant technology.
5. Expert: Individuals or groups with high-level academic expert knowledge, profound expertise in relevant technology, and access to substantial research capabilities and resources.

In terms of the characteristics of the IEMI generator, the scheme low, medium, and high-tech has been used [2] and is summarized here:

1. Low-tech systems are characterized by:
   (a) Marginal performance;
   (b) Minimal technical capabilities;
   (c) Easily assembled and deployed for example while hiding behind walls or within the back of a truck or similar vehicles.
2. Medium-tech systems:
   (a) Require the skills of a qualified electrical engineer;
   (b) Incorporate relatively more sophisticated components, for example, a commercially available radar system that can be modified to become a weapon system.
3. High-tech IEMI systems:
   (a) Would require specialized and sophisticated technologies;
   (b) Could perhaps even be specifically tuned to cause severe damage to specific targets.

2.5.1.1 Novice/Low-Tech Capability Group

Technologies that block or jam RF or wireless communications traffic, jammers, are openly advertised on the Internet. Minimal expertise would be required to purchase and use jamming devices, although the effective range may be very short (perhaps less than 10m) [85] and the target set would be smaller than for other IEMI generators. Expertise is required to build an effective long-range jammer from basic principles. The use of jammers is illegal in many parts of the world, in the United Kingdom [86], for example, and in other nations. Recently laws have been passed in the United States to ban the sale of such devices [87], but the prosecution of a jamming or interference incident requires evidence to be collected by the affected party to prove that jamming has taken place.

Plans, descriptions, and demonstrations of IEMI generators are available on the Internet. The generators discussed in these open forums have a range of credibility. A cursory review of the designs shows that many are unsophisticated and use well-known voltage multiplication circuits or ideas borrowed from other high-voltage generation systems (such as car ignition systems).

The most convincing examples of novice systems use Tesla coils and microwave oven magnetrons as the basis for the design. Some academic studies have shown that undergraduates from nonengineering disciplines can successfully build IEMI generators of this type [88]. Another advantage of microwave oven technology is that it shares the same frequency band as many Wi-Fi/wireless networks (2.4 GHz Industrial Scientific and Medical (ISM) band). Potentially, this generator could jam wireless networks and may even be capable of damaging wireless network access points.

Other academic studies have examined Tasers or stun guns to create high-voltage arcs directly onto exposed connectors of information communications technology (ICT) equipment [89].

The Schriner Transient Electromagnetic Device (TED) [90] system was manufactured from commercially available components based on an automotive ignition system and an oil-cooled spark gap switch that appears to have been the primary output switch. The radiating antenna structure was a homemade horn antenna. This IEMI generator was publicly demonstrated on U.S. national television.

2.5.1.2 Skilled/Medium-Tech Capability Group

All electrical and electronic products placed on the market are required to have some level of immunity to EM disturbance; this is known as EM compatibility (EMC). Equipment for EMC immunity testing is available to purchase from specialist suppliers. Such equipment includes traveling-wave tube (TWT) amplifiers. These amplifiers are sold as test equipment for use in shielded EMC test laboratories; however, it is reasonable to assume that the same amplifiers could be used outside of the laboratory environment for IEMI purposes. In particular, TWT amplifiers designed to meet military or defense requirements are likely to be effective as an IEMI generator without significant modification.

Marine radar equipment can be procured from many sources. This equipment would require minimal adaptation to operate as an IEMI generator. Magnetron and solid-state versions are available.

Complete fixed and mobile ex-military RF broadcast and search radar systems are available on the internet. Some military surplus equipment internet sites have been known to offer complete mobile radar systems. These sites also offer radar equipment and radar or radio components including amplifiers, pulse sources, tubes, and antennas. Very high-power modulators appear to be available for sale, although prices are not given. This equipment would require minimal adaptation to operate as an IEMI generator.

Several HPRF DE sources developed for the defense and security market for test purposes or for dual-use applications such as vehicle, vessel, and counter drone-stopping have been openly advertised. While the capability to build these generators clearly belongs to the specialist capability group, some of these generators are fully available to purchase and therefore could arguably be acquired, misappropriated, and used by the skilled capability group.

Some examples of these systems are:

- The HPEMcase and HPEMcar-stop generators developed by Diehl & Rhinemettal, Germany;
- The Dutch TNO briefcase generator [91];
- The Applied Physical Electronics (APElc) Footlocker system;

- The Russian Academy of Sciences (RAS) Compact Autonomous Pulse-Repetitive Oscillator [92];
- Teledyne-e2v SafeStop;
- The Replex DS and UWB HPEM simulators.

In May 2012, a suitcase-based HPEM simulator was shipped to the United Kingdom by a U.S. exhibitor and put on display inside Portcullis House, Houses of Parliament Offices in London, as part of a summit on electrical infrastructure security [93]. fe shows a photograph of the unit as displayed at the event.

A European study that reviewed the implications of HPEM on European infrastructure, the HIPOW project [94], pointed out that the availability and proliferation of this specialist technology were particularly worrisome.

2.5.1.3 Specialist/High-Tech Capability Group

From a technical capability perspective, the specialist IEMI group is indivisible from the HPRF DE capability. The primary difference is the motivation, risk appetite, and rules of engagement, or lack of them, of the user.

2.5.2 IEMI Environment Summary

Table 2.5, which is adapted from IEC 61000-4-36 [83], provides a summary of IEMI threat source environments defined by capability group. Other defi-

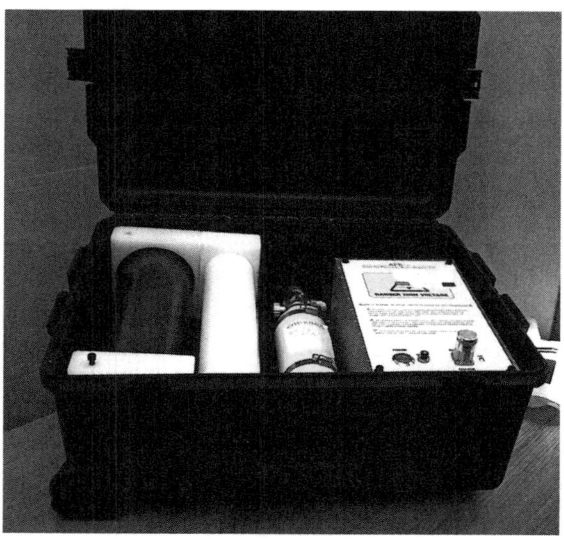

Figure 2.15 HPEM Suitcase as displayed inside the U.K. House of Parliament in 2012.

Table 2.5
Summary of Radiated IEMI Source Output (rEfar) by Capability Group

Category	IEMI Source Waveform Class	Example IEMI Source Name or Technology Type	rE_{far} (V)	Near/Far Distance (Approximate)
Novice	Hyperband	ESD gun	5,000	1m
	Hypoband	For example, microwave oven magnetron	2,000	1m
Skilled	Hyperband	For example, commercially available solid state pulser	60,000	1m
	Mesoband	For example, commercially available HPEM test source	120,000	2m
	Hypoband	Typical radar	450,000	5m
Specialist	Hyperband	Military demonstrator HPRF DE system	5,300,000	50m
	Mesoband	Military demonstrator HPRF DE system	500,000	5m
	Hypoband	Military demonstrator HPRF DE system	30,000,000	50m

nitions of the IEMI environment exist and Table 2.5 is included for example purposes only. The IEMI environment is expected to evolve as the technology matures.

The term rE_{far} describes the electric field normalized at a distance of 1m from the antenna as derived from an E-field measurement at a given distance in the far field. It is equivalent to equivalent isotropic radiated power (EIRP), which is derived from the power density at a given distance in the far field. The terms Mesoband, Hypoband, and Hyperband are fully described in the next section.

2.5.3 Open-Source Accounts of HPRF DE and IEMI Action

Qualified accounts in the open literature describing the effective use of HPRF DE/IEMI generators on electronic systems in realistic scenarios are very few in number. However, as discussed in Chapter 1, there are many factual accounts of effects to electronic systems from what could be termed as unintentional interference. Unintentional interference sources can include HPEM systems, such as radar, RF broadcast transmitters, and natural phenomena such as lightning.

Many alleged accounts of the use of HPRF DE in a military setting appear to be incorrect on inspection. One account highlights the alleged use of an HPM bomb by the U.S. military to defeat an Iraqi radar during Operation Desert Storm in January 1991. This account is perhaps misinterpreted from reports concerning the widespread use of high-speed, antiradiation missiles (HARM) such as the AGM-88, which can lock onto radar and other RF emissions [95]. The warhead is a conventional explosive device and does not produce an HPEM environment.

Another account relates to U.S. and NATO use of an EMP or HERF weapon during the Serbian conflict of 1999. This alleged weapon was apparently used to disable Serbian communications and create power outages. Again, it is probable that this story is a misinterpretation of reports referring to the first use of special soft bombs. These are based on the BLU-114/B, which simply dispensed carbon filaments shorting out overhead power lines [96]. To add to the confusion, the military has termed both of these forms of weapons as electronic attack where this relates more to the target of the attack rather than the weapon type.

Schriener, a contributor to the U.S. Senate Joint Economic Committee hearings on RF weapons, publicly demonstrated his low-tech TED IEMI source on U.S. national television [97], and other demonstrations have also apparently taken place at INFOWARCON'99. During these demonstrations, a variety of systems have been targeted. These demonstrations on the whole are fairly unconvincing because they are clearly conducted at short ranges (a few meters) in line of sight of the IEMI generator and it is possible that some victim preselection may have also taken place.

There are several unconfirmed accounts of instances where IEMI generators have been used against civilian systems [98, 99]. The Times article indicated that an IEMI generator was allegedly used to blackmail a U.K. financial institution. However, the wording in this article was obscure and could perhaps have referred to a more conventional cyber threat.

Another account alludes to the use of an IEMI generator to affect gambling machines (called pachinko machines) by a Japanese criminal group. Allegedly, a suitcase-based microwave generator (possibly a microwave oven magnetron) was placed next to the pachinko machine, which caused the machine to pay out. Apparently, the perpetrator was eventually caught [100]. Most of the evidence appears to be anecdotal, and even if the story is factual, the perpetrator was in proximity (possibly in contact) with the victim system.

In St. Petersburg, a criminal used an IEMI generator to disable a security system of a jewelry store. The reports mentioned that building the IEMI generator posed a technological challenge similar to assembling a home microwave oven [101].

In multiple European cities (e.g., Berlin), criminals used GSM jammers to disable the security system of limousines [102].

In the Netherlands, an individual disrupted a local bank computer network because he was refused a loan. He constructed a briefcase-size IEMI generator, which he learned how to build from the internet. Bank officials did not realize that they had been attacked or what had caused the disruption until the assailant was caught [103].

In 1991, 1997, and 1998, the Joint Economic Committee (JEC) of the U.S. Senate held hearings concerning low-tech RF weapons and their potential effect on critical national infrastructure [104]. In 1997 and twice in 1999, the U.S. Congress held hearings concerning the issue of EMP threats to U.S. infrastructure [105].

HPRF DE/IEMI has been associated with hacker, criminal, terrorist, and information warfare goals [106–110]. Some of the quoted advantages of IEMI over other forms of cyber and physical attacks are:

- Very low risk of detection (covert);
- No fingerprint, footprint, or evidence;
- Relatively easy to produce or procure generator technology;
- Low logistics requirement;
- Not directly lethal to people;
- Unpredictability, random effects;
- Plausible deniability by the perpetrator.

Over the past few years, the availability of technologies that could be used as an IEMI generator has increased significantly and there is evidence that these generator technologies are proliferating. A large body of clear, convincing, and documented evidence of HPRF DE/IEMI generator effects in real-world scenarios has not been obtained to date. It is speculated that this is likely to be in part due to a lack of awareness of this form of threat and a lack of deployed detection systems. However, a recent enquiry by the U.K. House of Commons Defence Select Committee [111] warned that: "While existing non-nuclear EMP (IEMI) devices may be crude and limited, the fact that viable devices could be produced by non-state actors is a cause for concern. Even localised damage could have the potential to disrupt activity, especially if combined with other forms of attack."

In 2015, three European Union (EU) Seventh Framework research projects under the security theme conducted research assessing the risk to European CNI from IEMI. The three projects high-power effects on critical infrastructure (HIPOW) [112], strategies for the improvement of critical infrastructure resilience to electromagnetic attacks (STRUCTURES) [113], and security of the railway network against EM attacks (SECRET) [114] identified cause for concern due to modern societies' reliance on electronics-based infrastructure and the potential ease with which IEMI generators could be procured or manufactured. HIPOW and STRUCTURES addressed the IEMI risk to all CNI sectors, whereas SECRET specifically addressed the IEMI risk to modern rail transportation networks and in particular the European Rail Traffic Management System (ERTMS). The ERTMS requires rail infrastructure operators to include wireless signalling and communication between the train and the trackside. The SECRET project identified that this wireless link will be vulnerable to IEMI and that this could, if not mitigated, cause the stoppage of trains and possible significant disruption of rail traffic.

There has been some press reporting and internet-based discussion that terrorist groups and, in particular, Al Qaida, have been interested in using IEMI [115, 116] for nearly a decade.

At the time of this writing, the threat HPEM environments and in particular HPRF DE and IEMI environments could pose to infrastructure are gaining interest from government and legislators in the United States and Europe, where the debate is whether to make protection from HPEM environments a legal requirement [117, 118]. In the United States, the Critical Infrastructure Protection Act (CIPA) has been passed into law and is managed by the U.S. Department of Homeland Security (DHS). The CIPA specifically mentions HPEM threats to infrastructure. In Europe, the network information security directive is ambiguous as to whether HPEM is in scope; indeed, the directive does not specifically mention threats at all. Instead, the directive mandates the requirement for operators of essential services and digital service providers to report and take action to prevent issues that may cause significant disruption.

2.5.3.1 Documented Accounts of Accidentally Generated EM Effects

Despite the very limited number of open-source accounts credited to HPRF DE or the malicious use of IEMI, there are a very large number of incidents that have been attributed to accidental or unintentional exposure to relatively benign EM environments. A very useful collection of over 500 incidents of this type was compiled by Armstrong [119]. Two very interesting and widely discussed examples are the USS Forrestal disaster and the TWA 800 aircraft crash [120]. Both accounts allude to the fact that EM disturbance cannot be proven as EM interference leaves no record or evidence.

Shahar [121] believed that these accounts of nonmalicious EM interference may encourage adversarial groups to exploit this perceived weakness. An example cited in the literature is of the warnings given by airlines to turn off personal electronic devices (PED) such as mobile phones and IT equipment on commercial flights [122]. It is believed that this, together with open-source evidence of PED disruption to aircraft systems [123] may encourage attack.

Shahar also infers that military demonstrations of HPRF DE technology, such as those conducted at China Lake [124] and the widely publicized demonstrations of military-grade IEMI technology [125, 126], could encourage adversarial groups to try out HPRF DE/IEMI technology.

2.6 Classification of HPRF DE and IEMI Environments

HPRF DE and IEMI systems are best classified by their spectral bandwidth as shown in Table 2.6. In some other documents, the terms HPM and ultrawideband (UWB) have also been used to describe the same or similar phenomena (i.e., to classify HPRF DE/IEMI systems, but these definitions are generally less well defined). Most HPRF DE/IEMI systems produce a pulsed or transient output. Note that each HPRF DE/IEMI system will have unique characteristics, but statistical descriptions of the parameters have been derived.

The band ratio (br) is the ratio of the high and low frequencies between which there is 90% of the energy and percent bandwidth (pbw) is the same factor described as a percentage. Note that a maximum pbw of 200% is possible. These factors are described by (2.6) and (2.7):

$$br = {}^{f_h}\!/\!_{f_l} \tag{2.6}$$

and

$$pbw = 200\frac{(br-1)}{(br+1)} \tag{2.7}$$

The spectral magnitudes of the different classifications of HPRF DE/IEMI waveforms are described in more detail next.

Table 2.6
Band Classification of HPRF DE Systems

Band Classification	Typical Pulse Rise Time	Typical Pulse Width	Typical Pulse Repetition Frequency	Percent Bandwidth (pbw)	Band Ratio (br)
Hypoband (or narrowband)–short pulse	>2 ns	<200 ns	100 Hz	pbw ≤ 1 %	br ≤ 1.01
Hypoband (or narrowband) – long pulse	>50 ns	>200 ns to CW	100 Hz	pbw ≤ 1%	br ≤ 1.01
Mesoband	—	<50 ns	1 kHz	1% < pbw ≤ 100%	1.01 < br ≤ 3
Sub-Hyperband	<1 ns	<10 ns	1 kHz	100% < pbw ≤ 163.4%	3 < br ≤ 10
Hyperband	< 150 ps	<1 ns	1 kHz	163.4% < pbw ≤ 200%	br > 10

2.6.1 Hypoband

The Hypoband class is also known as narrowband and HPM. Perhaps the most complete recent reference describing this technology area in full is by Benford et al. [127].

Technologically speaking, the generation of Hypoband waveforms can lead to moderate peak power (hundreds of kilowatts) and moderate average power (tens of kilowatts) for long-pulse systems or high peak power (a few gigawatts) and low average power (a few kilowatts) for short-pulse systems. This is fundamentally due to the way in which the high-power RF waveform is developed. A differentiation between long pulse and short pulse is necessary because the effect mechanism of the two different types of systems can be very different. Hypoband long-pulse systems typically operate in a regime of pulse widths from a few hundred nanoseconds to continuous wave (CW), whereas Hypoband short-pulse systems typically operate with pulse widths up to a maximum of a few hundred nanoseconds.

An analytical expression for a Hypoband pulse is given in (2.8) and an analytical example of a Hypoband short-pulse waveform is shown in Figure 2.16.

$$
\begin{aligned}
&A(t) = \sin(\omega_o t)u(t-\tau) \text{ for } 0 \le t \le (NT) \\
&\left|\tilde{A}(\omega)\right| = \left(\frac{2\omega_o \sin(N\omega T/2)}{\left(\omega_o^2 - \omega^2\right)} \right)
\end{aligned}
\tag{2.8}
$$

The waveform in Figure 2.16 is for the case of $\omega_o = 2\pi f_o$; f_o = 1GHz; T= 1 *ns*; N= 20; τ = 1 ns. This is an illustrative example of an undamped 20-cycle sine wave.

The electric field in a waveguide carrying a hypoband signal is a Gaussian modulated sine wave as shown in Figure 2.17.

2.6.1.1 Hypoband Short-Pulse Systems

Hypoband short-pulse systems typically employ relativistic or Cerenkov technologies such as backward wave oscillators (BWO), magnetically insulated line oscillators (MILO)/relativistic magnetron (RELTRON) and virtual cathode oscillators (VIRCATORS) to achieve very high peak radiated power. The efficiency of these systems is typically less than 25% and therefore the average power input requirement for such systems can be high for long bursts. Hypoband short-pulse systems typically operate in discrete bands. Tuning the frequency of operation of these systems is difficult; a tuning range of much less than 10% is typical and often requires changing out the primary tube or oscillator. The maximum pulse repetition frequency (*prf*) of such systems is typically 100 Hz.

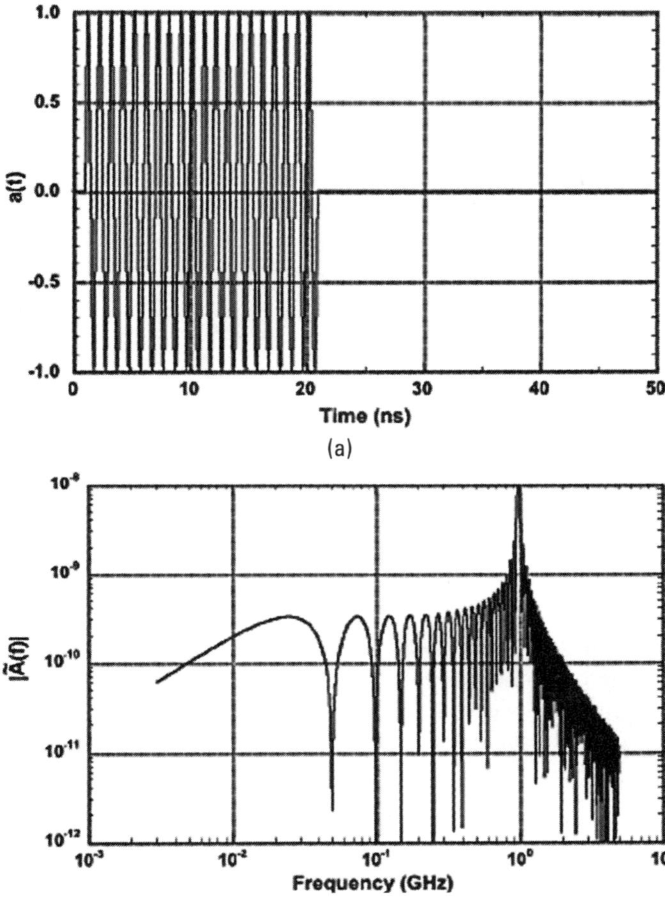

Figure 2.16 Analytical example of a Hypoband short-pulse waveform: (a) time domain and (b) frequency domain.

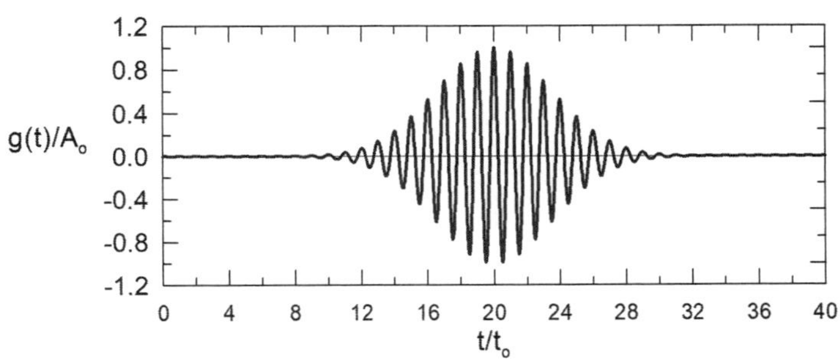

Figure 2.17 Example of a Hypoband short-pulse waveform in the time domain.

Example systems are known to employ high-gain antennas such as horn antennas, horn-fed parabolic reflector antennas, and slot arrays. The peak radiated power produced by these systems can approach or even exceed the dielectric breakdown of air at the antenna aperture (a few gigawatts) unless special design considerations are given.

Where the burst lengths of Hyperband short-pulse systems are required to be long (several tens of seconds) or bursts are required to be repeated, for example, in a test or simulation environment, then the generator can be very large. Where just a single burst is required, then the generator can be made reasonably compact and the mass is dominated by the energy storage or prime power systems.

2.6.1.2 Hypoband Long-Pulse Systems

Hypoband long-pulse systems, also referred to as ultra-hypoband systems, typically employ traveling-wave-tube amplifiers (TWTA), solid-state power amplifiers (SSPA), and magnetrons to achieve moderately high peak and average radiated power. Radar, medical radiotherapy, and radio transmitter technologies may be adapted, perhaps by utilizing a different antenna design for an HPRF DE/IEMI application.

The efficiency of these systems is typically less than 25% and therefore the average power requirement for such systems can be high for long bursts. Hypoband long-pulse systems typically operate in discrete bands. Tuning the frequency of operation of these systems is dependent on the technology employed. TWTA and SSPA can have a tuning range of an octave or more, whereas the tuning range of magnetrons is typically less than 10%. The maximum *prf* of such systems is typically 1,000 Hz or otherwise duty cycle limited. A maximum duty cycle of 6% is typical.

Example systems, for example, the Teledyne-e2v Safe-Stop, are known to employ high-gain antennas such as horn antennas and slot arrays. The peak power can be a few megawatts. The mass of systems that are required to produce many bursts can be reasonably large requiring, for example, a platform-type installation such as a vehicle for mounting and providing prime power for the generator.

2.6.2 Mesoband

The Mesoband class is also sometimes referred to as moderate band, damped sinusoid, or non-nuclear EMP. Mesoband systems typically employ Marx generators (and derivatives thereof), nonlinear transmission lines (NLTLs), and other resonant technologies (for example, Tesla transformers) to achieve moderately high peak and average radiated power.

Mesoband sources typically employ a dipole (or a derivative thereof) as the radiating element. The antenna design strongly influences the dominant radiating frequency of the source. The dominant radiating frequency can be defined as the frequency with the highest spectral magnitude contribution identified from a Fourier transformation of the measured time-domain waveform.

An analytical expression for a Mesoband pulse is given in (2.9), and (2.10) and (2.11) (magnitude) illustrate how the spectral content of the waveform can be derived. Equations (2.12) and (2.13) can be used to derive important parameters related to the Mesoband waveform.

$$E(t) = E_0 e^{-\alpha t} \sin(\omega t) u(t) \tag{2.9}$$

$$\tilde{E}(f) = \frac{\omega_0 E_0}{\left(\alpha^2 + \omega_0^2 - \omega^2\right) + 2j\alpha\omega} \tag{2.10}$$

$$\tilde{E}(f) = \frac{\omega_0 E_0}{\sqrt{\left(\alpha^2 + \omega_0^2 - \omega^2\right)^2 + 4\alpha^2\omega^2}} \tag{2.11}$$

$$\left|\tilde{E}(0)\right| = \frac{\omega_0 E_0}{\alpha^2 + \omega_0^2} \tag{2.12}$$

$$\textit{Spectral peak} = \frac{E_0}{2\alpha} \tag{2.13}$$

An analytical example of a Mesoband waveform is shown in Figure 2.18.

An actual measured Mesoband waveform produced by the QinetiQ TRI-GR Mesoband immunity test source [128] is shown in Figure 2.19.

Mesoband sources are often tunable, such that the dominant frequency (f_d) of the source waveform can be altered. However, the alteration of the dominant frequency has to be affected by changing:

- The pulsed power supply;
- The gas pressure, gas type, or gap spacing of the primary switch, if a Marx or Tesla or derivative is employed;
- The dipole antenna length or geometry;

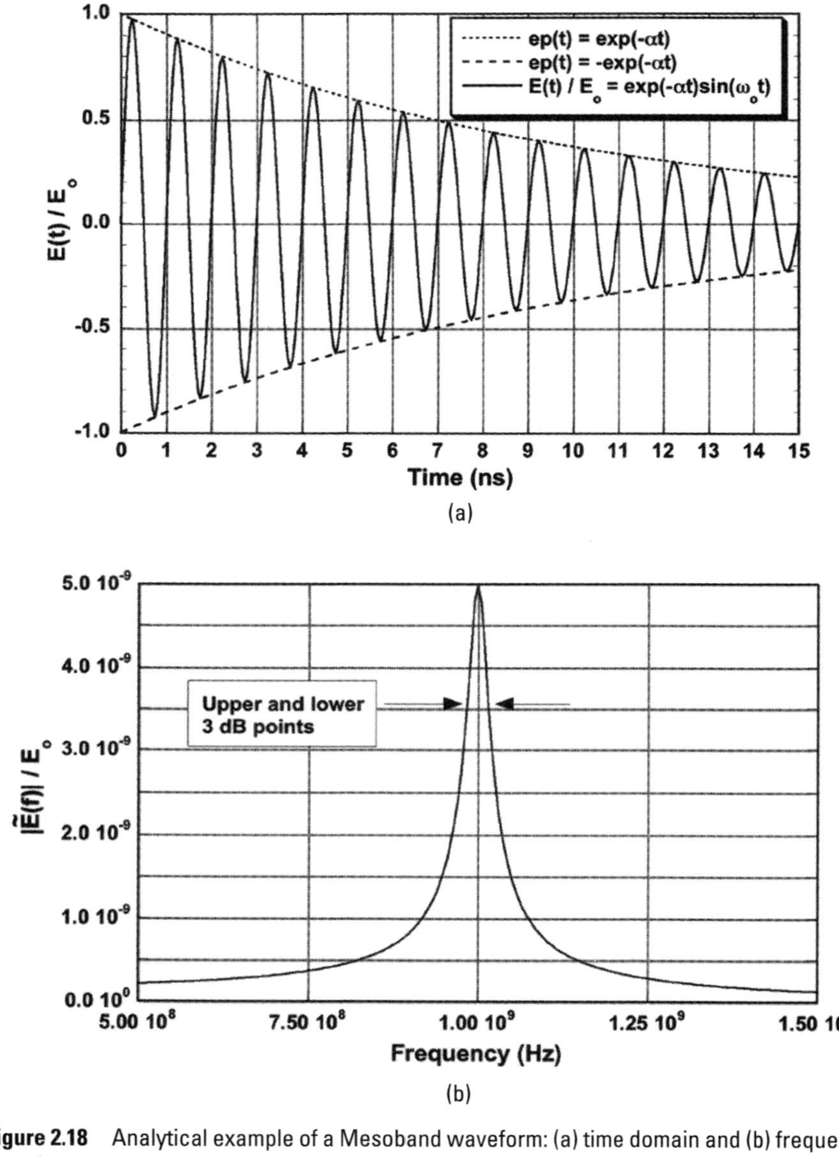

Figure 2.18 Analytical example of a Mesoband waveform: (a) time domain and (b) frequency domain.

- Any combination of the above.

Therefore, tuning of mesoband sources cannot typically be achieved dynamically (during use) and requires the user to adjust some of the key parame-

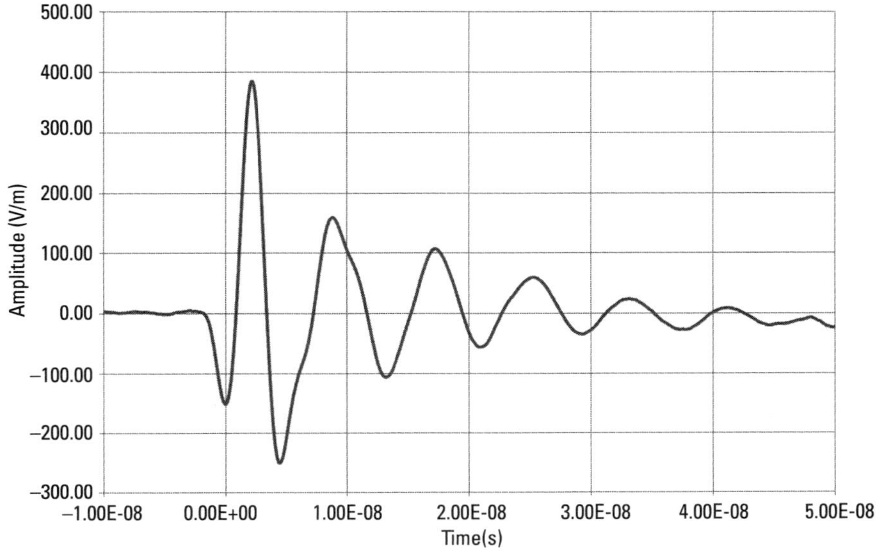

Figure 2.19 Measured example of a Mesoband waveform from TRIGR. (Courtesy of QinetiQ Ltd.)

ters described above in order to tune the source to a different radiated dominant frequency.

The typical maximum prf is 1 kHz as a spark gap switch is typically employed as the output switch.

The Diehl HPEM CarStop is an example of this technology. Mesoband systems can produce radiated peak power approaching a gigawatt. The systems can be relatively compact and briefcase-based examples have been discussed. The largest component is often the radiating antenna that must be reasonably large (>1m) to be efficient at typical operating frequencies (several hundred megahertz).

2.6.3 Hyperband

The Hyperband class is also known as ultrawideband. Sub-Hyperband and Hyperband systems typically employ Marx generators (and derivatives thereof), and solid-state devices such as avalanche diode stacks to achieve very high peak radiated power.

Hyperband sources typically employ specialized antennas such as transverse electromagnetic mode (TEM) horn antennas (including arrays of TEM horns) and impulse radiating antennas (IRA). The primary aim in Hyperband systems is to achieve a very high rate of rise, thus achieving a very wide instantaneous bandwidth.

An analytical expression for a Hyperband pulse is given in [129] and an illustrative example of a Hyperband waveform is shown in Figure 2.20.

A photograph of arguably the most powerful Hyperband simulator built to date, JOLT, is shown in Figure 2.21(a). Figure 2.21(b) is the time-domain

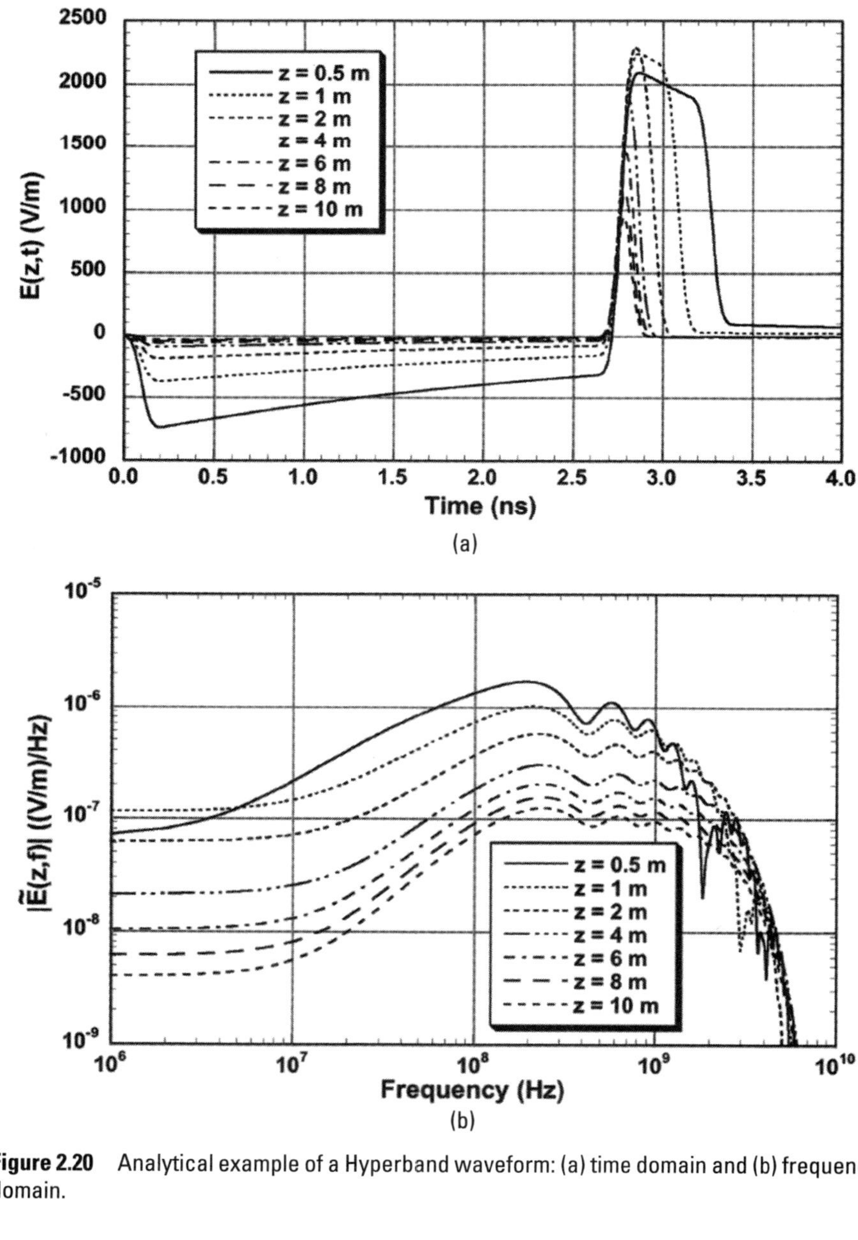

Figure 2.20 Analytical example of a Hyperband waveform: (a) time domain and (b) frequency domain.

(a)

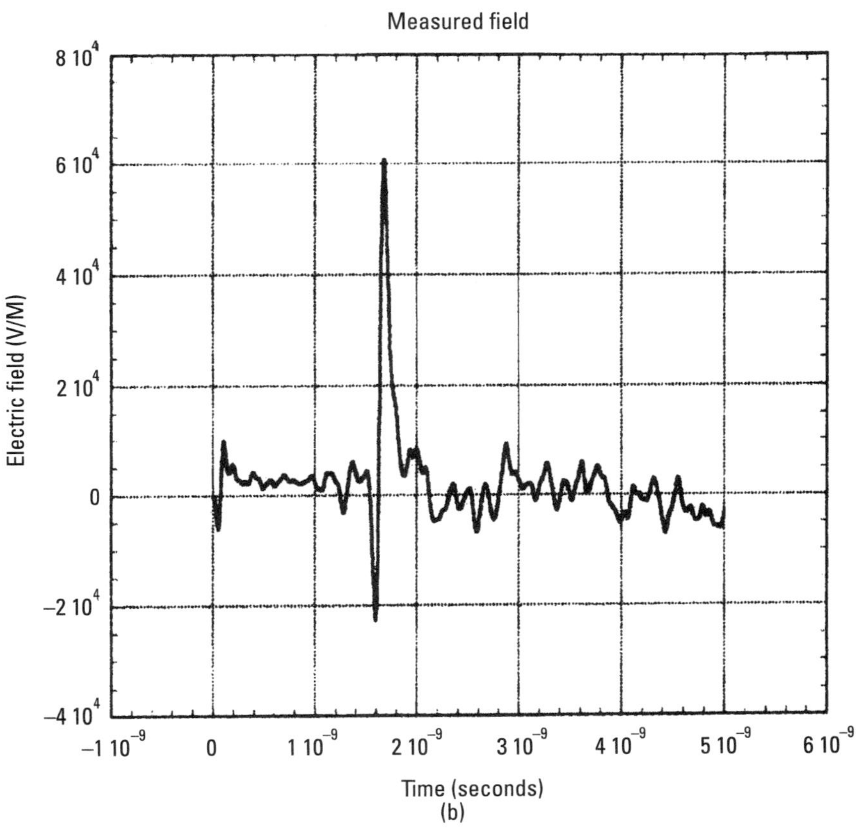

(b)

Figure 2.21 (a) Photograph of the JOLT system on a trailer and (b) measured electric field at a boresight distance of r = 85m.

radiated waveform measured at 85m from the generator. The amplitude on the figure has not been normalized.

Hyperband sources are *prf*-limited. The maximum *prf* of Hyperband sources is typically up to 1 kHz for the majority of high peak power systems where spark gap technology is often used as the output switch. There are example systems based on solid-state technology that are capable of *prfs* of several megahertz at lower peak radiated power.

Hyperband systems can potentially pose a risk of jamming of communications or sensors because they radiate energy across the wanted RF channel of these systems.

2.7 Summary

The characterization and definition of parameters related to HPEM environments and waveforms are vitally important. It provides scientists and engineers a common formal language so that effects results can be compared and contrasted and so that hardening and protection can be applied consistently. A formal definition of the lightning environment has been with us for some time yet there is continued interest in refining our understanding of this, the oldest HPEM environment. Likewise, the HEMP environment is continuously being analyzed to improve the fidelity of the environment definition. HPRF DE and IEMI environments have begun to be categorized and defined, but this is a rapidly evolving domain and the definition of the EME will no doubt have to evolve.

References

[1] IEC 61000-2-13, "Electromagnetic Compatibility (EMC) – Part 2-13: Environment-High-Power Electromagnetic (HPEM) Environments – Radiated and Conducted," March 9, 2005.

[2] Giri, D. V., and F. M. Tesche, "Classification of Intentional Electromagnetic Environments (IEME)," IEEE Transactions on Electromagnetic Compatibility, Vol. 46, No. 3, August 2004.

[3] Uman, M., The Lightning Discharge, New York: Academic, 1987, p. 118.

[4] IEC 61000-2-9, "Electromagnetic Compatibility (EMC) – Part 2-9: Description of HEMP Environment — Radiated Disturbance," 1996.

[5] Giri, D. V., C. E. Baum, and W. D. Prather, "The Relationship Between NEMP Standards and Simulator Performance Specifications," Sensor and Simulation, Note 538, 2009.

[6] Baum, C. E., "Some Considerations Concerning Analytic EMP Criteria Waveforms," Theoretical Note 285, Air Force Weapons Laboratory, Kirtland AFB NM, October 1976.

[7] Hoad, R., and W. A. Radasky, "Progress in High Altitude Electromagnetic Pulse (HEMP) Standardization," IEEE Special Issue on High Altitude Electromagnetic Pulse, Vol. 55, No. 3, Jun 2013.

[8] Giri, D. V., F. Tesche, and M. Nyffeler, "Switched Oscillators and Their Integration into Helical Antennas," IEEE Transactions in Plasma Science, 2010.

[9] Baum, C. E., et al., "JOLT: A Highly Directive, Very Intensive, Impulse Radiator," Proceedings of the IEEE, Vol. 92, No. 7, July 2004.

[10] Prather, W. D., et al., "Survey of Worldwide High-Power Wideband Capabilities," IEEE Transactions on Electromagnetic Compatibility, Vol. 46, No. 3, 2004, pp. 335–344.

[11] Giri, D. V., et al., "Design, Fabrication, and Testing of a Paraboloidal Reflector Antenna and Pulser System for Impulse-Like Waveforms," IEEE Transactions on Plasma Science, Vol. 25, No. 2, April 1997, pp. 318–326.

[12] Baum, C., "Temporal and Spectral Radiation on Boresight of a Reflector Type of Impulse Radiating Antenna (IRA)," in Ultra-Wideband Short-Pulse Electromagnetics 3, New York: Plenum Press, 1996.

[13] Giri, D. V., et al., "Intermediate and Far Fields of a Reflector Antenna Energized by a Hydrogen Spark-Gap Switched Oscillator," IEEE Transactions on Plasma Science, Vol. 28, No. 5, October 2000, pp. 1631–1636.

[14] Sabath, F., et al., "Overview of Four European High Power Microwave Narrow-Band Test Facilities," IEEE Transactions on Electromagnetic Compatibility, Vol. 46, No. 3, August 2004.

[15] IEC/TR 61000-4-35 Ed. 1, "Electromagnetic Compatibility (EMC) -- Part 4-35: Testing and Measurement Techniques -- High Power Electromagnetic (HPEM) Simulator Compendium," 2009.

[16] Giri, D. V., et al., "Switched Oscillators and Their Integration into Helical Antennas," IEEE Transactions on Plasma Science, Vol. 38, No. 6, 2010, pp. 1411–1426.

[17] Gardner, R., Lightning Electromagnetics, Boca Raton, FL: CRC Press, 1990.

[18] Uman, M. A., The Lightning Discharge, New York: Academic Press, 1987.

[19] Giri, D. V., "History of High-Power Electromagnetics (HPEM) with Illustrative Examples of HPEM Systems," Proceedings of the EUROEM Symposium, London, U.K., July 13, 2016.

[20] Baum, C. E., et al., "The Measurement of Lightning Environmental Parameters Related to Electronic Systems," IEEE Transactions on Electromagnetic Compatibility, Vol. EMC-24, No.2, May 1982, pp. 123–137.

[21] Fisher, R. J., and M. A. Uman, "Measured Electric Field Risetimes for First and Subsequent Lightning Return Strokes," J. of Geophysical Research, Vol. 77, January 20, 1972, pp. 399–406.

[22] Krider, E. P., C. D. Weidman, and R. C. Noggle, "The Electric Fields Produced by Lightning Stepped Leaders," J. Geophysical Research, Vol. 82, No. 6, February 20, 1977, pp. 951–960.

[23] Baum, C. E., et al., "Measurement of Electromagnetic Properties of Lightning with 10 Nanosecond Resolution," Proc. Lightning Tech, NASA Conference, Pub. 2128 and FAA-RD-80-30, April 1980, pp 39–82.

[24] Baker, L., et al., "Calculation of Lightning Channel Characteristics: Comparison of Existing Models," Lightning Phenomenology Notes, LPN 9, April 20, 1983.

[25] Hussein, A. M., et al., "Simultaneous Measurement of Lightning Parameters for Strokes to the Toronto Canadian National Tower," Journal of Geophysical Research, Vol. 100, 1995, pp. 8853–8861.

[26] Cai, L., et al., "The Foshan Total Lightning Location System in China and Its Initial Operation Results," Atmosphere, Vol. 10, 2019, p. 149.

[27] Romero, C., et al., "A System for the Measurements of Lightning Currents at the Santis Tower," Electric Power Systems Research, Vol. 82, 2012, pp. 34–43.

[28] Rachidi, R., and R. Thottappillil, "Determination of Lightning Currents from Far Electromagnetic Fields," Journal of Geophysical Research, Vol. 98, 1993, pp. 18315–18320.

[29] Romero, C., et al., "A Statistical Analysis on the Risetime of Lightning Current Pulses in Negative Upward Flashes Measured at Säntis Tower," International Conference on Lightning Protection (ICLP), Vienna, Austria, 2012.

[30] Pavlone, M., et al., "Measurement of Lightning Currents at the Santis Tower in Switzerland," AMEREM 2014, Paper ID 028, University of New Mexico, Albuquerque, NM, 2014.

[31] Rachidi, F., and M. Rubinstein, "Lightning Measurement Station at the Santis Tower: An Update on Recent Instrumentation and Data," AMEREM 2018, Santa Barbara, CA, 2018.

[32] Rachidi, F., and M. Rubinstein, "Modelling Lightning Strikes to Tall Towers," Chapter 25, in Lightning Electromagnetics, V. Cooray (ed.), London: IET, 2012, pp. 873–916.

[33] Giri, D. V., and C. E. Baum, "Airborne Platform for Measurement of Transient or Broadband CW Electromagnetic Fields," Sensor and Simulation, Note 284, May 22, 1984. ece-research.unm.edu/summa/notes.

[34] Pitts, F. L., and M. E. Thomas, 1980 Direct Strike Lightning Data, NASA Conference Pub. 2128 and FAA-RD-80-30, April 1980, pp. 283–299.

[35] Pitts, F. L., "Electromagnetic Measurement of Lightning Strikes to Aircraft," AIAA Paper 81-0083, February 1981.

[36] Lee, K. S. H., EMP Interaction: Principles, Techniques, and Reference Data: A Handbook of Technology from the EMP Interaction Notes, Hemisphere Pub. Corp., 1986.

[37] EP 1110-3-2, "Engineering and Design - Electromagnetic Pulse (EMP) and Tempest Protection for Facilities Proponent," December 31, 1990. http://publications.usace.army.mil/publications/eng-pamphlets/EP_1110-3-2_pfl/toc.htm.

[38] NATO AECTP 250, "Electrical and Electromagnetic Environmental Conditions - Leaflet 256 Nuclear Electromagnetic Pulse (NEMP/EMP)," Edition 2, November 2010.

[39] Glasstone, S., and D. J. Dolan, The Effects of Nuclear Weapons, 3rd ed., Washington, D.C.: U.S. Department of Defense and the Energy Research and Development Administration, 1977.

[40] IEC 61000-1-3, "Electromagnetic Compatibility (EMC) – Part 1-3: General - The Effects of High-Altitude EMP (HEMP) on Civil Equipment and Systems," 2002.

[41] Loborev, V., "Up to Date State of the NEMP Problems and Topical Research Directions," Proceedings of the European Electromagnetics International Symposium–EUROEM 94, June 1994, pp. 15–21.

[42] Bell Laboratories, Electrical Protection Department, Loop Transmission Division, *EMP Engineering and Design Principles,* Whippany, NJ: Bell Telephone Labs, 1975.

[43] Baum, C. E., "Some Considerations Concerning Analytic EMP Criteria Waveforms," Theoretical Note 285, Air Force Weapons Laboratory, Kirtland AFB NM, October 1976.

[44] DIN VG95371-10, "Bundesamt für Wehrtechnik und Beschaffung," Germany, 1995, NA-140-00-19 AA-Nuclear electromagnetic pulse (NEMP) and lightning protection, October 2014.

[45] Leuthäuser, K. D., "A Complete EMP Environment Generated by High-Altitude Nuclear Bursts: Data and Standardization," Theoretical Note 364, Air Force Phillips Laboratory, February 1994.

[46] Giri, D. V., and W. D. Prather, "High-Altitude Electromagnetic Pulse (HEMP) Risetime Evolution of Technology and Standards Exclusively for E1 Environment," IEEE Transactions on Electromagnetic Compatibility, Vol. 55, No. 3, 2013, pp. 484–491.

[47] Giri, D. V., W. D. Prather, and C. E. Baum, "The Relationship Between NEMP Standards and Simulator Performance Specifications," Sensor and Simulation Notes, Note 538, 2009.

[48] Baum, C. E., "Some Considerations Concerning Analytic EMP Criteria Waveforms," Theoretical Note 285, October 1976.

[49] U.S. MIL-STD-464A, "Electromagnetic Environmental Effects Requirements for Systems," December 19, 2002.

[50] Savage, E., J. Gilbert, and W. Radasky, "The Early-Time (E1) High-Altitude Electromagnetic Pulse (HEMP) and Its Impact on the U.S. Power Grid," Meta-R-320, January 2010, http://www.eiscouncil.com/App_Data/Upload/9b03e596-19c8-49bd-8d4e-a8863b6ff9a0.pdf. Accessed March 2018.

[51] Tesche, F. M., "Comparison of the Transmission Line and Scattering Models for Computing the HEMP Response of Overhead Cables," IEEE Transactions on Electromagnetic Compatibility, Vol. 34, No. 2, May 1992, pp. 93–99.

[52] Tesche, F. M., and P. R. Barnes, "The HEMP Response of an Overhead Power Distribution Line," IEEE Transactions on Power Delivery, Vol. 4, No. 3, July 1989, pp. 1937–1944.

[53] IEC 61000-2-10, "Electromagnetic Compatibility (EMC) — Part 2-10: Environment — Description of HEMP Environment — Conducted Disturbance," 1998.

[54] International treaty, "Treaty Banning Nuclear Weapon Tests in the Atmosphere, in Outer Space and Under Water," signed at Moscow, August 5, 1963, entered into force October 10, 1963, https://www.state.gov/t/isn/4797.htm. Accessed April 4, 2018.

[55] Graham, W. R., "Report of the Commission to Assess the Threat to the United States from Electromagnetic Pulse (EMP) Attack," April 2008.

[56] U.S. Department of Defense (2007) Joint Publication 1–02: Department of Defense Dictionary of Military and Associated Terms 12 April 2001 (As Amended Through 13 June 2007). Washington, D.C.: U.S. Department of Defense.

[57] Seifer, M., Wizard: Life and Times of Nikola Tesla, New York: Citadel Press, 1998.

[58] "Death Ray for Planes," The New York Times, September 22, 1940.

[59] Shweitzer, R. L., "Radio Frequency Weapons and Infrastructure," Statement before the Joint Economic Committee of the United States Congress, June 17, 1997, www.house.gov/jec/hearings/espionag/schweitz.htm.

[60] Smith, I. D., and H. Aslin, "Pulsed Power for EMP Simulators," IEEE Transactions on Antennas and Propagation, Vol. 26, No. 1, January 1978.

[61] Leach, P. O., and M. B. Alexander, Electronic Systems Failures and Anomalies Attributed to Electromagnetic Interference, NASA Report 1374, National Aeronautics and Space Administration, Washington, D.C., July 1995.

[62] Price, D., J. Levine, and J. Benford, "ORION--A Frequency Agile HPM Field Test System," Seventh National Conference on High-Power Microwave Technology, Laurel, MD, 1997.

[63] Altgilbers, L. L., J. Baird, and B. L. Freeman, Explosive Pulsed Power, Imperial College Press, February 17, 2011.

[64] Kopp, C., "An Introduction to the Technical and Operational Aspects of the Electromagnetic Bomb," Australian Air Power Studies Centre, Paper No. 50, November 1996.

[65] Abrams, M., "The Dawn of the E-Bomb," IEEE Spectrum Magazine, September 2003.

[66] Kristiansen, M., Fundamental Limits to Compact, Expendable Pulsed Power and Microwave Sources, Report Number: A281383, September 2000.

[67] Goforth, J. H., et al., "US/Russian Collaboration: Experiments with Explosive Pulsed Power," Pulsed Power 2001, IEE Symposium, May 1–2, 2001 pp. 1/1–1/4.

[68] Terekhin, V. A., et al., "Transportable Simulators of Electromagnetic Pulses Based on Magnetocumulative Generators," 12th IEEE International Pulsed Power Conference, 1999. Digest of Technical Papers, Vol. 2, June 27–30, 1999.

[69] Adee, S., "Portable E-Bomb to Be Tested," IEEE Spectrum Online, April 2009.

[70] Siniy, L., V. E. Fortov, and Y. Parfenov, "Russian Research of Intentional EMI Disturbances over the Past 10 Years," AMEREM 2006, Albuquerque, NM, July 2006.

[71] Fulghum, D. A., "Northrop Crafts Multimission N-UCAS," Aviation Week, Aerospace Daily & Defense Report, May 21, 2008, http://www.aviationweek.com/aw/generic/story_channel.jsp?channel=defense&id=news/NUCAS032108.xml.

[72] HSDailyWire.com, "Raytheon's Vigilant Eagle," February 21, 2008, http://hsdailywire.com/single.php?id=5591.

[73] Jackson, R., "Boeing Feature Story - CHAMP – Lights Out," October 2012, http://www.boeing.com/Features/2012/10/bds_champ_10_22_12.html.

[74] Diehl BGT Defence, "White Paper on HPEM Technology," June 2013.

[75] Teledyne-e2v, "RF Safe-StopTM – Air, Land, Sea," https://www.e2v.com/resources/account/download-literature/81.

[76] APELC Datasheet, "APELC Footlocker MIL-STD 464C Source, Model Number EMPF-464-D4-A," http://www.apelc.com/product/footlocker-system-150-kvm-hprf-source/, January 10, 2017.

[77] REPLEX, "DS and UWB HPEM Simulators," http://www.replex.co.kr/.

[78] MIL-STD 464C, "Electromagnetic Environmental Effects, Requirements for Systems," December 1, 2010.

[79] NATO AECTP 250, Leaflet 257, "High Power Microwaves," Final draft April 16, 2013.

[80] Sebacher, K., "Directed Energy and High Power Microwave Test Requirements," 2009 IEEE EMC Symposium Workshop, Austin, TX, August 17, 2009.

[81] "Workshop on Electromagnetic Terrorism and Adverse Effects of High Power Electromagnetic (HPE) Environments," Proceedings of the 13th International Zurich Symposium and Technical Exhibition on Electromagnetic Compatibility, February 16–18, 1999.

[82] IEC 61000-2-13 Ed. 1.0, "Electromagnetic Compatibility (EMC) – Part 2-13: High-Power Electromagnetic (HPEM) Environments - Radiated and Conducted," March 2005.

[83] IEC 61000-4-36, "Electromagnetic Compatibility (EMC) – Part 4-36: Testing and Measurement Techniques – IEMI Immunity Test Methods for Equipment and Systems," 2014.

[84] Sabath, F., "A Systematic Approach for Electromagnetic Interference Risk Management," IEEE Electromagnetic Compatibility Magazine, Vol. 6, No. 4, 2017, pp. 99–106.

[85] van de Beek, S., "Vulnerability Analysis of the Wireless Infrastructure to Intentional Electromagnetic Interference," Ph.D. Thesis Series No. 16-413, Centre for Telematics and Information Technology, Enschede, The Netherlands, 2016.

[86] UK Ofcom, "Radio Frequency Jammers," https://www.ofcom.org.uk/spectrum/interference-enforcement/spectrum-offences/jammers.

[87] U.S. Federal Communications Commission (FCC), "Communications Act of 1934, as amended, 47 U.S.C," Section 301, 302a(b), 333, 2009.

[88] Baker, G., et al., "The Feasibility and Effectiveness of a Common Consumer Device as an Electromagnetic Interference Source," American Conference on Electromagnetics (AMEREM), Albuquerque, NM, July 9–14, 2006.

[89] Baker, G., R. Tuttle, and J. Rudmin, "Investigation of Stun Gun Effectiveness as Intentional Electromagnetic Interference Sources," European Conference on Electromagnetics (EUROEM), Lausanne, Switzerland, July 21–25, 2008.

[90] Schriner, D., "The Design and Fabrication of a Damage Inflicting RF Weapon by 'Back Yard' Methods," Statement before the Joint Economic Committee of the United States Congress, February 25, 1998, www.house.gov/jec/hearings/02-25-8h.htm.

[91] Clingendael Centre for Strategic Studies (CCSS), "Directed Energy Weapons, a New Way of Warfare," Report No. 5, July 1, 2004, http://www.ccss.nl.

[92] Bezrukov, M. J., et al., "Compact Autonomous Pulse-Repetitive Oscillators," 7th Int. Crimean Conference Microwave Telecommunication Technology (CriMiCo'2007), Sevastopol, Crimea, Ukraine, September 10–14, 2007.

[93] EISS, "Electrical Infrastructure Security Summit – Report III Conclusions and Recommendations," The Houses of Parliament, London, U.K., May 14–15, 2012, http://www.eissummit.com/images/upload/conf/media/EISS%20III%20London%20Report.pdf.

[94] The EU 7th Framework Project, HIPOW, www.hipow-project.eu/.

[95] Isby, D. C., "Cruise Missiles Flew Half the Desert Fox Strike Missions," Jane's Missiles and Rockets, Vol. 3, No. 2, February 1999.

[96] Bender, B., "US Soft Bombs Prove NATO's Point," Jane's Defence Weekly, Vol. 31, No. 19, May 1999.

[97] Sawyer, D., "20/20 Segment on Non-Lethal Weapons," American Broadcasting Company (ABC), February 1999.

[98] Hodge, C. C., *Redefining European Security (Contemporary Issues in European Politics),* Abingdon, UK: Routledge Press, 1999.

[99] Rosenberg, E., "New Face of Terrorism: Radio-Frequency Weapons," The New York Times, June 23, 1997.

[100] U.S. Department of Homeland Security, "The Threat of Radio Frequency Weapons to Critical Infrastructure Facilities," TSWG and DTEO Publication, August 2003. http://www.dtic.mil/dtic/tr/fulltext/u2/a593293.pdf.

[101] Hoad, R., and I. Sutherland, "The Forensic Utility on Detecting Disruptive Electromagnetic Interference," Proceedings of the 6th European Conference on Information Warfare and Security (ECIW 2007), July 2007.

[102] Sabath, F., "What Can Be Learned from Documented Intentional Electromagnetic Interference (IEMI) Attacks?" Proceedings of the 2011 XXXth URSI General Assembly and Scientific Symposium, 2011.

[103] Saxton, J., "Hearing of the Joint Economic Committee 'Radio Frequency Weapons and Proliferation: Potential Impact on the Economy," February 25, 1998, http://www.house.gov/jec/hearings/02-25-8h.htm.

[104] Weldon, C., "US Congress Hearing 'Threat Posed by Electromagnetic Pulse (EMP) to US Military Systems and Civil Infrastructure,'" June 1997, http://commdocs.house.gov/committees/security/has197010.000/has197010_0.HTM.

[105] Bartlett, R. G., "106th Congress House Hearing 'Electromagnetic Pulse (EMP): Should This Be a Problem of National Concern to Private Enterprise, Businesses Small and Large, as Well as Government?'" June 1, 1999, http://frwebgate.access.gpo.gov/cgi-bin/getdoc.cgi?dbname=106_house_hearings&docid=f:59747.wais.

[106] Schwartau, W., Cybershock: Surviving Hackers, Phreakers, Identity Thieves, Internet Terrorists and Weapons of Mass Disruption, New York: Thunder's Mouth Press, 2000.

[107] Schwartau, W., Information Warfare: Chaos on the Electronic Superhighway, 1st ed., New York: Thunder's Mouth Press, 1994.

[108] Wik, M. W., R. L. Gardner, and W. A. Radasky, "Electromagnetic Terrorism and Adverse Effects of High Power Electromagnetic Environments," Workshop W4, Proceedings of the 13th International Zurich Symposium and Technical Exhibition on EMC, February 1999.

[109] Levien, F., "The Role of Directed Energy Weapons in Information Warfare," IQPC, London, U.K., January 21, 2004.

[110] Van Keuren, E., D. J. Wilkenfeld, and J. Knighton, "Utilisation of High Power Microwave Sources in Electronic Sabotage and Terrorism," Proceedings 25th Annual IEEE International Carnahan Conference on Security Technology, October 1–3, 1991, pp. 16–20.

[111] HCDC, House of Commons Defence Select Committee Report, "Developing Threats: Electro-Magnetic Pulses (EMP)," Tenth Report of Session 2010–12, February 8, 2012.

[112] HIPOW project website, www.hipow-project.eu/.

[113] SECRET project website, www.secret-project.eu/.

[114] STRUCTURES project website, www.structures-project.eu/.

[115] Shahzad, S. S., 'The Changing Face of Resistance," Asia Times Online, June 23, 2006, http://atimes.com/atimes/Middle_East/HF23Ak02.html.

[116] Vital Perspective.com, "Report: Al-Qaeda Working on Electromagnetic Bomb to Put U.S. in Stone Age," June 22, 2006, http://www.typepad.com/services/trackback/6a00d8341da99d53ef00d8352eca0653ef.

[117] 114th United States Congress (2015-2016), H.R.1073, "Critical Infrastructure Protection Act (CIPA)," https://www.congress.gov/bill/114th-congress/house-bill/1073.

[118] "Directive (EU) 2016/1148 of the European Parliament and of the Council of 6 July 2016 Concerning Measures for a High Common Level of Security of Network and Information Systems Across the Union," July 6, 2016, http://eur-lex.europa.eu/legal-content/EN/TXT/?uri=uriserv:OJ.L_.2016.194.01.0001.01.ENG&toc=OJ:L:2016:194:TOC. Accessed March 2018.

[119] Armstrong, K., "The First Five Hundred 'Banana Skins' – A Collection of Anecdotes on EMI," October 2007, www.theemcjournal.com.

[120] Pevler, A. E., "Security Implications of High-Power Microwave Technology," Proceedings of the IEEE International Symposium on Technology and Society at a Time of Sweeping Change, June 20–21, 1997, pp. 107–111.

[121] Shahar, Y., "Directed Energy Hazards to Civil Aviation," Workshop B, IQPC Directed Energy Weapons (DEW) 2009, London, U.K., February 25, 2009.

[122] Civil Aviation Authority (CAA), "Advice to Passengers – On Board the Aircraft," http://www.caa.co.uk/default.aspx?catid=1770&pagetype=90&pageid=9853.

[123] The Federal Aviation Authority (FAA), Advisory Circular 91.21.1B, "The Potential for Portable Electronic Devices (PED) to Interfere with Aircraft Communications and Navigation Equipment," August 25, 2006.

[124] Henderson, W. M., and D. A. Schriner, "Radio Frequency Weapons - 21st Century Threat Live Fire Testing of Radio Frequency Weapons," Aircraft Survivability, Summer 1998, http://www.nawcwpns.navy.mil/~pacrange/s1/news/1998/RFWeap.htm.

[125] Jackson, R., "Boeing Feature Story - CHAMP – Lights Out," October 2012, http://www.boeing.com/Features/2012/10/bds_champ_10_22_12.html.

[126] natochannel.tv, "HPEM NATO Test 2013," 2013, https://www.youtube.com/watch?v=VnDeq3Y0_FI.

[127] Benford, J., J. A. Swegle, and E. Schamiloglu, High Power Microwaves, 3rd ed., Boca Raton, FL: CRC Press, 2015.

[128] Hoad, R., et al., "Development of a Mesoband Immunity Test Method," ASIAEM 2017, Bangalore, India, July 2017.

[129] Baum, C., "Temporal and Spectral Radiation on Boresight of a Reflector Type of Impulse Radiating Antenna (IRA)," in Ultra-Wideband Short-Pulse Electromagnetics 3, C. Baum, L. Carin, and A. P. Stone, (eds.), New York: Plenum Press, 1996.

3

HPEM Coupling and Interaction

3.1 EM Interaction Coupling Model

In engineering, it is common to describe the EM interaction between an RF source and a target or victim system via the source-victim model (Figure 3.1). In this model, the source could be a lightning strike, a high-altitude electromagnetic pulse (HEMP) burst, an HPRF DE, or an IEMI generator. For lightning, the antenna can be considered as the strike itself; indeed, the lightning strike is sometimes modeled as a dipole antenna. For HEMP, the antenna could be considered to be the Earth's magnetosphere. For HPRF DE and IEMI, the antennas are actual physical structures that are connected to the generator or emitter.

The source gives rise to voltages, currents, or EM fields that can interact with the victim (the device, circuit, etc., which may be affected by the interaction). Upon reaching the device, circuit, and equipment victim the EM disturbance causes an effect, degradation of the victim's performance, malfunction, or destruction.

The abstract source-victim model, as illustrated in Figure 3.1, might give a superficial observer the impression that emitter and source are connected via a single coupling path. Unfortunately, in reality, the scenarios are rather more complex, such that the source interacts via several paths with the device, circuit, equipment, or system. For HPEM environments, the emitter generates EM fields, which can propagate directly to the victim or can be picked up by a cable that is running to it. In some cases, the source and the victim are connected to the same cable or conducting surface. In other cases, cable or, more generally,

Figure 3.1 Source-victim model.

conductors that are connected to either the RF emitter or the victim device or system interact with each other via electric (capacitive) or magnetic (inductive) fields.

As one can imagine from Figure 3.2, analyzing the actual coupling between the emitter and victim is a challenging task. As in most cases an exact theoretical or experimental analysis is not required and a suitable estimation might be sufficient. A common approach to derive an estimation model is to partition the coupling network into a set of simplified coupling paths. The partitioning process usually starts by considering the HPEM environment and identifying which coupling paths are likely to reach into the victim [1]:

- Conducted coupling, if the HPEM environment reaches the victim conducted along a cable or conducting surface;
- Radiated coupling, if the HPEM environment reaches the victim as a radiated EM field.

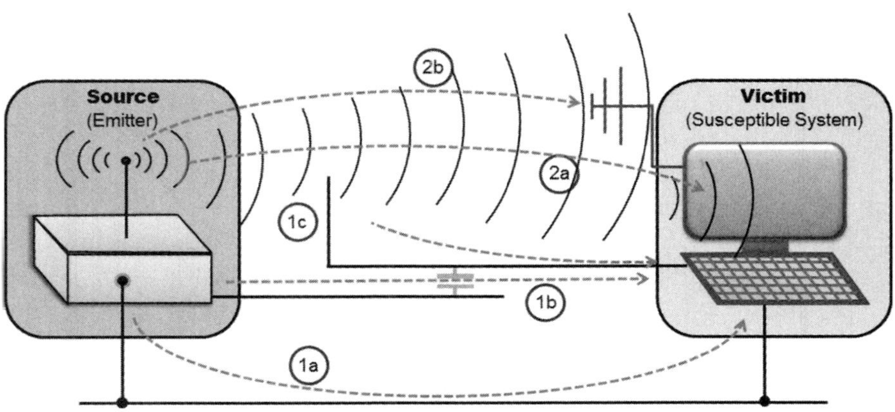

Figure 3.2 Detailed coupling model.

1. Conducted coupling:

   (a) A purely conducted (galvanic) coupling, where the emission propagates from the emitter to the victim through interconnecting cable (e.g., power line or network connection) or conductor (e.g., a copper coolant pipe).

   (b) Conducted coupling via two or multiple cables or conductors that are coupled via electric field (capacitive coupling) or magnetic field (inductive coupled) and connected to either the emitter or the victim.

   (c) Radiation from the emitter to the cables, conductors, or antennas of the victim. Antennas (e.g., Wi-Fi antennas), improperly shielded wires or cables, or metallic pipes are capable of picking up disturbances from surrounding EM fields. Cases in which the emission is picked up by nonantenna elements such as apertures, enclosures, or cables are often labeled as backdoor coupling. This differentiates it from the coupling to actual antennas or electronic sensors, which is known as front-door coupling.

2. Radiated coupling:

   (a) Direct radiation from the emitter to the victim through the air. Usually, the emitted field propagates from the source via antenna structures, penetrates the system enclosure, and enters the susceptible device via apertures.

   (b) Radiation from antennas to antennas (e.g., the source and receptors are specifically transmitting/receiving devices). This path is representative of normal radio communication (e.g., wireless local area networks (WLAN), radio broadcast/reception).

The analyses of the physical coupling mechanism that belong to these coupling paths are best approached from an understanding of Maxwell's equations, which govern all aspects of EM phenomena. However, because the application of the EM theory on real configurations is complex, only a brief overview of the pertinent aspects of the phenomena is considered necessary.

With regard to protection or hardening of the victim system, it is particularly important to consider whether the HPEM disturbance couples through an intended interface or an unintended imperfection for example an aperture in an enclosure. Due to the kind of interface, the coupling is often classified as:

- Front-door coupling, if the HPEM disturbance couples through an interface intended to communicate or interact with the external environment. Shielding of those interfaces against HPEM environments causes severe degradation or loss of its functionality. Typical examples are coupling via antennas or sensors. Front-door coupling can be subdivided into first and second orders [2]:

- *Front-door coupling, first order (in-band):* The frequency of the HPEM disturbance coincides, at least partly, with the working frequency of the equipment. First-order front-door coupling is often called in-band front-door coupling. An example is a telecom base station irradiated in its passband.
- *Front-door coupling, second order (out-of-band):* The frequency of the HPEM disturbance does not coincide with the working frequency of the equipment. Therefore, second-order front-door coupling is also called out-of-band front-door coupling. An example is the coupling of an HPEM waveform with a dominant frequency of a few gigahertz via a megahertz radio antenna. Another example is a digital television camera where the sensor is designed to collect optical wavelengths of a scene to produce an image yet RF can couple through the optical aperture at much lower frequencies.

- Antennas or sensors are typically connected with receiver front-end electronics that collects the RF energy coming from the antenna or sensor (front-door) and processes it (see Figure 3.3). Often the antenna or sensor provides some degree of frequency selectivity, but the in-band frequency selectivity of the front-door coupling is mainly managed at the receiver front-end electronics by filters, amplifier, and mixer.
- Figure 3.4 shows the idealized front-door response and the actual response to front door-coupling. In this figure, it can be clearly observed that the first-order front-door coupling (case 1) has a much higher coupling efficiency than any kind of second-order front-door coupling

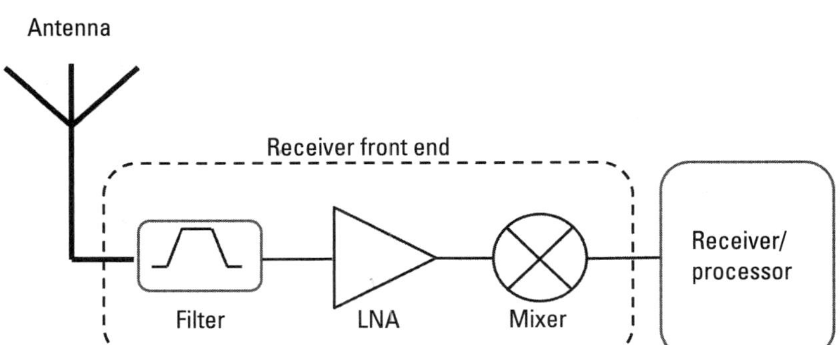

Figure 3.3 Typical receiver front-end configuration.

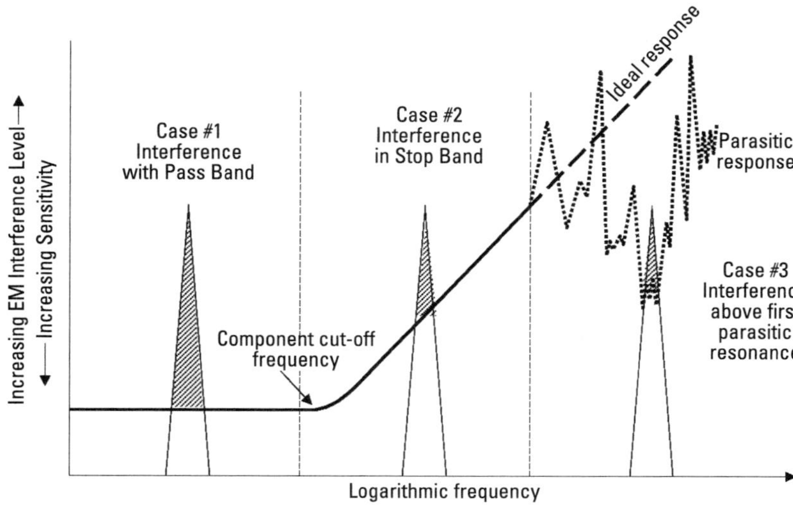

Figure 3.4 First-order and second-order front-door coupling (solid line: idealized front-door response; and dotted line: actual response).

(cases 2 and 3). The reason for this is that the antenna or sensor as well as the connected input filter are tuned to collect EM radiation in the passband and are therefore extremely sensitive in-band. Figure 3.4 also shows that parasitic resonances can enable coupling in the stopband, that is, out-of-band (case 3). In general, the coupling efficiency to first-order and second-order front-door coupling is complex and dependent on many factors.

- Backdoor coupling, if the HPEM disturbance couples through cables, apertures, or imperfections of the enclosure. Usually, backdoor-coupling points of entry can be shielded without degradation of its function. Examples of backdoor coupling are the coupling through ventilation, display, or door slit. The HPEM disturbance may also couple directly to an external wire or cable connected to a device or circuit. The reason to define such a wire as backdoor coupling and not as a second-order front-door coupling is due to the fact that the wire or cable could be shielded without degrading its function [1].

3.2 Topological Concept

A structured approach to visualize and analyze the interaction of an EM environment with complex systems is the topological concept [3–5]. It permits subdividing a complex chain of interactions into a number of simpler parts. The topological analysis starts with the source EM environment and runs through a sequence of simpler steps propagating the EM environment (radiated and conducted) until the device level victim is reached. Within each step the propagation of the EM field is characterized by measurements or described by appropriate analytical and numerical models. Often, the EM environment is characterized by norms/quantities of interest, like the peak voltages, currents, electric, and/or magnetic fields, the total energy, peak power, peak time integral, and rate of change of the aforementioned quantities.

As an example, the topological concept is applied to the radiated interaction of an emitter with a victim device or circuit. Figure 3.5 illustrates the resulting interaction sequence diagram and the corresponding factorization. The RF source (emitter) generates via antenna structures an EM field (source EM environment) that propagates as an EM wave to the exterior structure (surface) of the victim device. In Figure 3.5, the gray box marks the victim device, circuit, or equipment. The external EM environment can create a charge or current

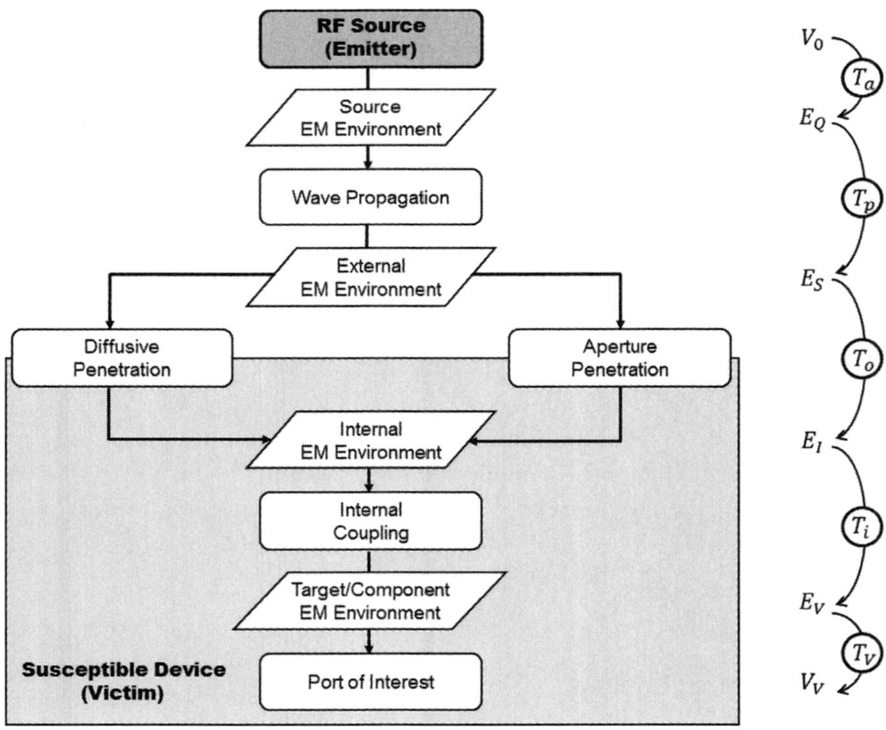

Figure 3.5 Interaction sequence diagram (left) and factorization (right) for radiated coupling.

responses on the victim's surface. The surface response than penetrates through the enclosure of the victim by diffusion mechanism. An alternative path for the EM energy to penetrate the exterior structure of the system (e.g., shielding) is propagation through apertures (e.g., windows, ventilation opening). After the RF has reached the inside of the victim, it can be redistributed from the point of entry (POE) to an internal location by internal coupling (radiated or conducted). At RF, the energy redistribution mechanism typically includes transmission-line propagation (conducted propagation), direct field radiation, and resonant fields. Most real-life systems have more complex design characterized by more than one shielding surface or level. In these cases, the process of coupling, penetration, and propagation will be repeated until the component or port of interest is reached.

The entire process of radiation, penetration, propagation, and coupling is referred to as EM interaction. The topological analysis results in a sequence of analysis steps shown in an interaction sequence diagram Figure 3.5. The interaction sequence diagram illustrates the flow of analysis, EM environments (e.g., source environment, external environment, internal environment, target environment), and the models needed for any specific part (e.g., wave propagation, diffusion coupling).

The mechanisms by which the incident EM field couples to the victim govern the efficiency with which the energy in the source waveform is transferred to the system. How the transferred energy affects the normal operation of the system, either through temporary disruption or permanent damage, is determined through the study of effects.

It has been shown that, for backdoor coupling to computer networks, the coupling efficiency at frequencies below 1 GHz will be dominated by the transfer of energy from the incident radiated field to cabling and other network interconnections. At frequencies greater than a few hundred megahertz, the coupling is dominated by ingress, or leakage, of energy through seams and apertures of individual equipment casings. In general, the target dimensions, which determine the resonant frequencies, govern the total amount of power that can be extracted from an incident field over any given frequency range. Baum [6] referred to the size of the hand rule when approximating the typical resonant frequency of equipment. This crude measure puts enclosure resonances for typical portable systems that have a human interface (e.g., laptop computer) at approximately 1 GHz.

An example of an interaction sequence diagram for conducted coupling is depicted in Figure 3.6. In this diagram, the galvanic, capacitive, and inductive coupling is shown at the boundary of the victim system. In real-life situations, these couplings can occur at any location between the emitter and the port of interest.

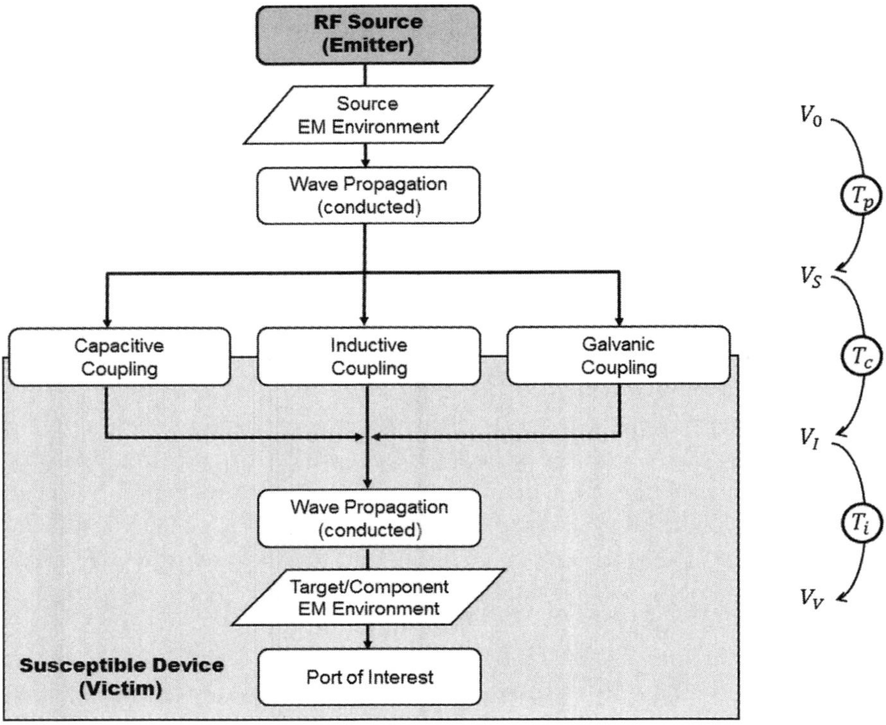

Figure 3.6 Interaction sequence diagram (left) and factorization (right) for conducted coupling.

3.3 Transfer Functions

3.3.1 Antenna Transfer Function

For the antenna transfer function (T_a), the HPEM disturbance from the source is radiated through an antenna or a conducting structure that acts as an antenna. Generally, we can assume that the RF disturbance is radiated employing an antenna that radiates the EM field with some degree of directivity. The maximum available directivity of an aperture with area A_{ant} is given by [7]

$$D_0 = \frac{S_{max}}{S_{iso}} = \frac{4\pi \cdot A_{ant}}{\lambda^2} \tag{3.1}$$

which results together with the antenna efficiency η_{re} into the maximum antenna gain [8]

$$G_0 = \eta_{re} \cdot D_0 \tag{3.2}$$

These formulas are usually derived from the aperture efficiency integrals. The antenna efficiency is the ratio of power actually radiated to the power absorbed by the antenna terminals due to the conductivity in the antenna's conductors. For the purpose of a worse-case approximation the antenna efficiency can be assumed to be $\eta_{re} = 1$. As a result, the directivity becomes identical to the maximum gain of the antenna. Kildal et al. [9] presented an equation that can be used to estimate the directivity for all antenna sizes in a heuristic manner (Figure 3.8), that is,

$$D_{0,\max} = \left(2\pi \frac{r_{\min}}{\lambda}\right)^2 + 3 \tag{3.3}$$

In this equation, r_{min} denotes the radius of the smallest sphere that can enclose the antenna (Figure 3.7) and λ is the wavelength of the radiated RF field.

Figure 3.8 illustrates the directivity function of the diameter of the smallest sphere that can enclose the antenna, including its ground plane. The solid line is the heuristic curve defined by (3.4) [9]. With this heuristic equation, the radiated power density along the boresight of the antenna is

$$S_{max} = \frac{P_{in} \cdot G_0}{4\pi R^2} = \frac{P_{in}}{4\pi R^2}\left(\left(2\pi \frac{r_{min}}{\lambda}\right)^2 + 3\right) \tag{3.4}$$

with R the distance from the antenna and P_{in} the power supplied to the antenna terminals. Using the lumped-element equivalent-circuit approach for micro-

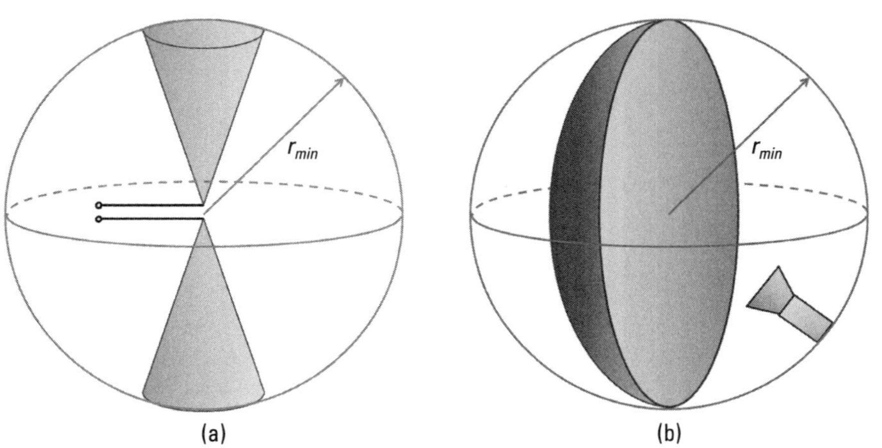

Figure 3.7 Two illustrations of the definition of the radius of the smallest sphere that can enclose an antenna.

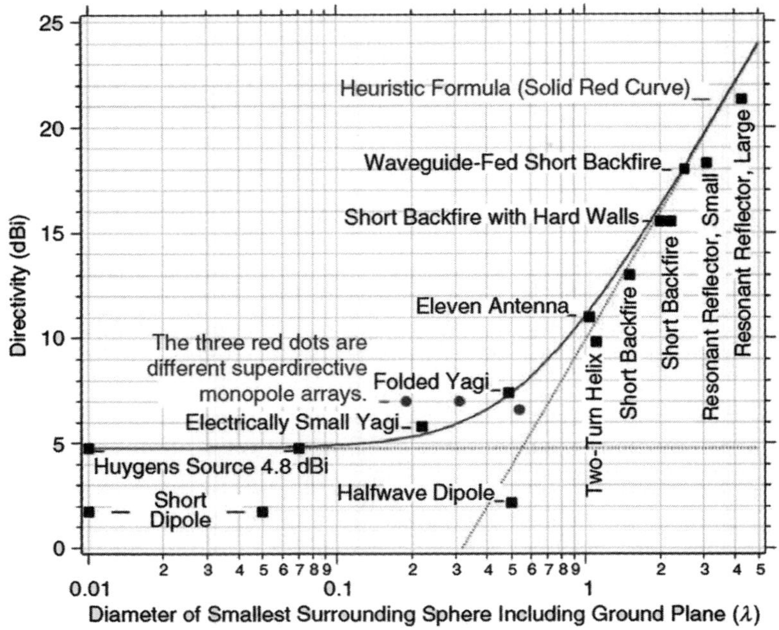

Figure 3.8 The fundamental directivity limitations of antennas (© 2017 IEEE. From: [9]. Reprinted with permission.)

wave networks [10], an RF source characterized by a source voltage V_0 feeds into an antenna input admittance Y_{s_0} the power

$$P_{in} = V_0^2 \cdot \mathrm{Re}\{Y_{s_0}\} \tag{3.5}$$

Combining (3.4) and (3.5) yields, for a plane wave,

$$E_{\max} = \sqrt{S_{\max} \cdot \zeta_0} = \frac{V_0}{R} \sqrt{\frac{\zeta_0 \cdot \mathrm{Re}\{Y_{S_0}\}}{4\pi} \left(\left(2\pi \frac{r_{\min}}{\lambda} \right)^2 + 3 \right)} \tag{3.6}$$

where $\zeta_0 = \sqrt{\frac{\mu_0}{\varepsilon_0}}$ is the characteristic impedance of free space. From (3.6), the antenna transfer function can be derived as

$$T_a = \frac{E_{max}}{V_0} = \frac{1}{R} \sqrt{\frac{\zeta_0 \cdot \mathrm{Re}\{Y_{s_0}\}}{4\pi} \left(\left(2\pi \frac{r_{min}}{\lambda} \right)^2 + 3 \right)} \tag{3.7}$$

This function increases approximately proportional to the ratio of the antenna size to the wavelength of the radiated field. At this point, we should note that a higher transfer function results from a higher directivity of the antenna (e.g., the RF energy is radiated into a smaller sector).

3.3.2 Free-Space Wave Propagation

The free-space wave propagation transfer function ($T_{p,w}$) should be considered in addition to the geometric dispersion of the spherical wave radiated from a finite size antenna, which is already included in the antenna transfer function, the EM wave propagating from the antenna to the victim is the subject of additional loss mechanisms. Chemical elements of the air, air pollution, rain, and vegetation absorb energy from the propagating EM wave. The attenuation of an EM wave propagating through atmospheric gases has been investigated by the International Telecommunication Union (ITU). The results were published in [11, 12]. The propagation of EM waves through atmospheric gas is nonlinear and highly complex. In general, the attenuation can be considered by assigning a transfer function to it. As the specific atmospheric attenuation is less than 0.01 dB/km at frequencies below 10 GHz, for typical HPEM environments in free space, the transfer function for the free space propagation can be approximated to:

$$T_{p,w} \approx 1 \tag{3.8}$$

3.3.3 Coupling/Radiation Efficiency

The ability of a distributed target to extract energy from an incident plane wave can be quantified in terms of its effective area. The effective area is analogous to the frequency-dependent effective aperture of antennas. At any given frequency, the maximum power received into a matched load, provided that the receiving antenna is orientated correctly to the polarization of an incident plane wave, is given by the product of the antenna effective receiving area and the power density of the incident plane wave. The concept of antenna effective area is used to derive the Friis transmission formula. This relation can be used to compute the maximum power received by an optimally orientated receiving antenna at a distance r_R from a transmitting antenna. The Friis formula states that the received power P_R into a matched load is given by:

$$P_R = \lambda^2 \frac{G_S \cdot G_R}{\left(4\pi r_R\right)^2} P_S \tag{3.9}$$

where λ is the wavelength of the incoming plane wave produced by the transmitting antenna with an input power of P_S. The parameters G_S and G_R are the power gains of the transmitting (source) and receiving antennas, respectively.

The concept of effective area can be extended to cables where a parameter known as effective length is sometimes used. The effective length L is a function of parameters such as:

- Cable orientation with respect to the incident field;
- Relative proximity and orientation of a cable with respect to ground planes (such as earth mats or enclosure surfaces);
- Terminating impedance of the cable.

A generic plot [6] of coupling efficiency against frequency is shown in Figure 3.9. It illustrates that the coupling efficiency falls off at a constant rate outside of the resonance region of the distributed system and that the resonance region is a function of the dimensions of the system and the wavelength of the impinging HPEM environment.

It should be noted that the effective length (L) is a complex quantity. For a cable, the effective length (L) can be affected by many factors including but not limited to:

- The orientation of the cable with respect to the direction of propagation of the HPEM waveform;

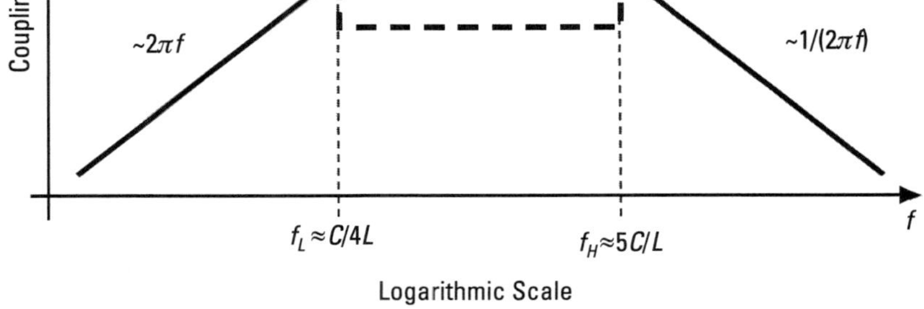

Figure 3.9 Canonical coupling efficiency model.

- The orientation of the cable with respect to other conductive and non-conductive elements (e.g., other cables in a bundle);
- The cable loop area, which can include the loop area in ground image plane;
- The proximity of the cable with respect to other conductive and non-conductive elements (e.g., the ground);
- The position of a cable within a conductive or high Q cavity.

3.3.3.1 Examples of Coupling/Radiation Efficiency

The response of a simple generic system to an illumination by a plane wave is shown in Figure 3.10. The curve was measured by injecting an RF waveform from one port of a network analyzer into a test cell and connecting the other port of the network analyzer to a connector on a victim cable located inside the test cell. At the low-frequency end of the diagram, the curve shows the expected 20 dB/decade rise. Approximately at 20 MHz, the coupling increases in magnitude. The first peak or resonance is at 90 MHz, followed by two strong resonances between 200 MHz and 300 MHz. Beyond those resonances, the response function declines and passes over into the – 20 dB/decade slope in the high-frequency integration region.

In reality, systems are far more complex and therefore possess much more complex transfer functions. Wraight et al. [13] measured aircraft cable bundle transfer functions employing the Low-Level Swept Current (LLSC) technique. This technique is a common technique for measuring coupling to aircraft and other similar sized systems and is described in several standards [14, 15]. It involves illuminating the system under test (SUT) with a uniform electric field.

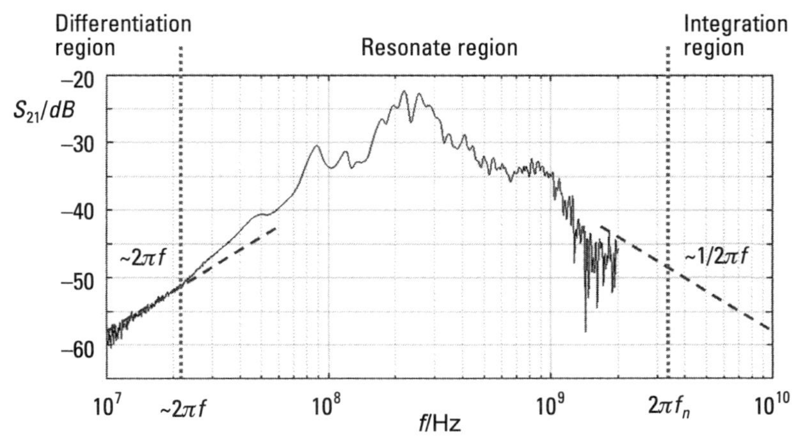

Figure 3.10 Response of a generic system (solid line) and canonical system response (dashed line) as a function of frequency.

The SUT is illuminated from four orientations in turn and completed for two polarizations of the transmit antenna. Initially, the data are frequency tied such that eight files remain (two polarizations and four orientations). The transfer data is summed to produce a generic transfer function which is an amalgamation of the highest coupling measured. These generic transfer functions are polarization and orientation-independent. Typical LLSC transfer functions for an aircraft system are shown in Figure 3.11(a) and an extrapolated generic is shown in Figure 3.11(b). In Figure 3.11(a), it can clearly be observed that the transfer function depends on both polarization and angle of incidence (orientation).

At most frequencies, the value of the transfer function varies up to 25 dB. However, all curves show the common characteristic behavior as demonstrated by the canonical resonance model (i.e., a resonance region with a fall-off in coupling efficiency on either side of the resonance region). The specific location of the resonance region and resonant peaks within the resonance region will vary depending on the size/geometry of the system.

However, it should be noted that the coupling efficiency can be extremely variable and not only dependent on polarization, orientation, and angle of incidence. In order to show this, Dr. Nigel Carter carried out an experiment to see how variable the coupling efficiency can be in regard to small perturbations of the environment. During an outdoor experiment, he observed hour by hour very large variations in measured coupling factors. For example, he noted that the coupling changed when a cloud passed in front of the Sun. Essentially, the Sun was causing thermal expansion of all of the many slot apertures and bonding points on the aircraft and these changes in condition resulted in the observed change in coupling. In order to verify his observations, he moved the test into a more controlled environment of a reverberation chamber and excited the fuselage with a wooden stick (Figure 3.12(b)). It can be seen from Figure 3.12 (a) that the maximum variation observed was of the order of 20 dB for a seemingly very small perturbation of the aircraft skin.

3.3.4 Diffusion Penetration

The transfer function associated with penetration by diffusion ($T_{o,d}$) occurs when the radiated RF wave passes through a conducting or dielectric barrier, it is attenuated by several phenomena (Figure 3.13). Due to the change of the conductivity σ, the permeability μ, and/or the permittivity ε, the incident electric field $\vec{E}_{\text{inc}}$ is reflected at the surface between outside and the barrier. The reflected portion $\vec{E}_{\text{refl}}$ of the incident electric field is given by the reflection coefficient for that surface. The nonreflected portion of the incident field is transmitted into the barrier ($\vec{E}_{\text{trans,b}}$). In the case of a conductive barrier, the amplitude is attenuated as it passes through according to the factor $e^{-\alpha z}$ where α is the attenuation constant of the material. For barrier materials that constitute

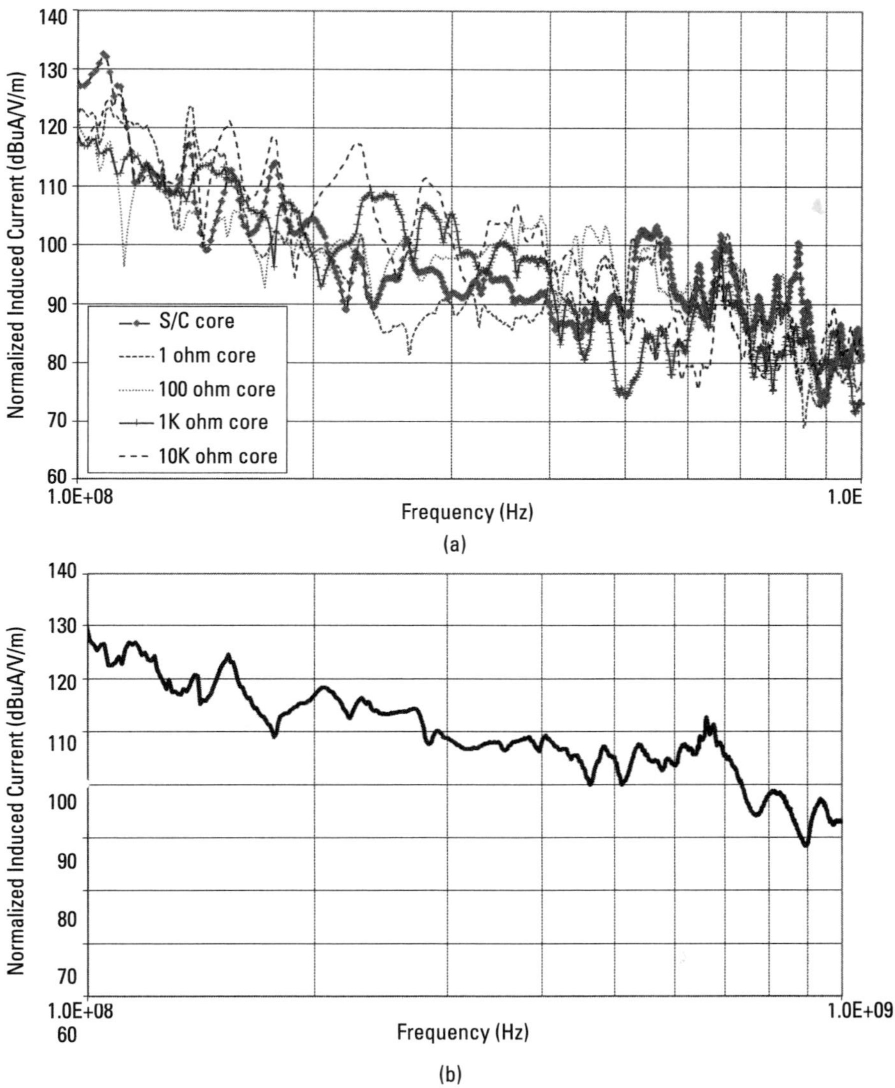

Figure 3.11 (a) Typical LLSC magnitude-only transfer functions and (b) typical magnitude-only generic transfer function. (Courtesy of Wright et al. [19].)

good conductors (as is usually the case), the attenuation constant α is related to the skin depth δ of the material

$$\alpha = \frac{1}{\delta} = \sqrt{\pi f \mu_b \sigma_b} \tag{3.10}$$

(a)

(b)

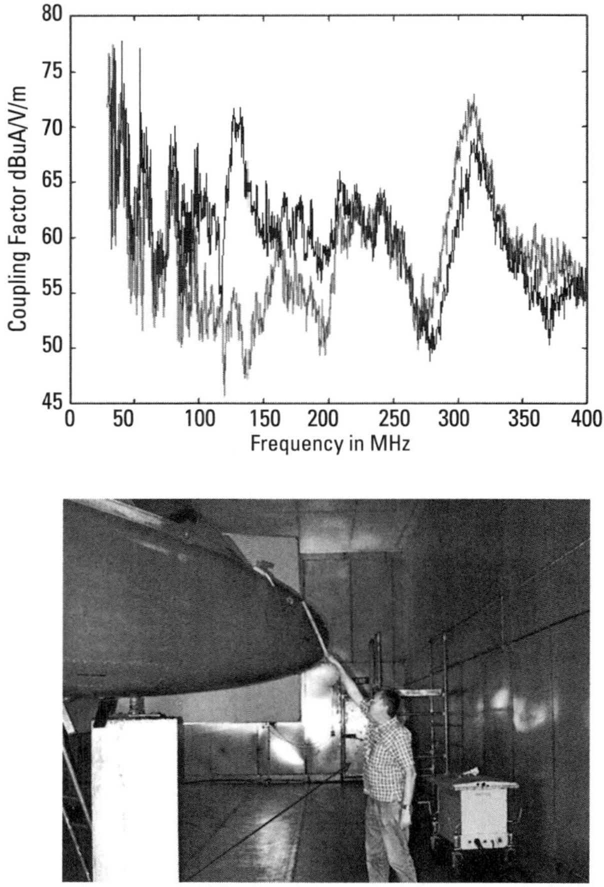

Figure 3.12 Variation in coupling to a cable inside an aircraft: (a) maximum coupling variation, and (b) Dr. Nigel Carter exercising the fuselage. (Courtesy of QinetiQ Ltd.)

where μ_b is the permeability of the barrier and σ_b is its conductivity.

If the barrier thickness d is greater than the skin depth δ of the barrier material at the frequency f of the incident wave, the internal transmitted field ($\vec{E}_{\text{trans,b}}$) is greatly attenuated when it strikes the interface between the barrier and the inside. A portion of $\vec{E}_{\text{trans,b}}$ is reflected back into the barrier ($\vec{E}_{\text{refl,b}}$), and a portion ($\vec{E}_{\text{trans}}$) is transmitted across the barrier into the medium behind the barrier (inside). The portion $\vec{E}_{\text{refl,b}}$ propagates back through the barrier and strikes the first interface. Once again, a portion of this wave is transmitted through the interface into the outside and adds to the total reflected field in outside medium, and a portion is reflected at the interface and proceeds to the right. This portion is again attenuated as it passes through the barrier.

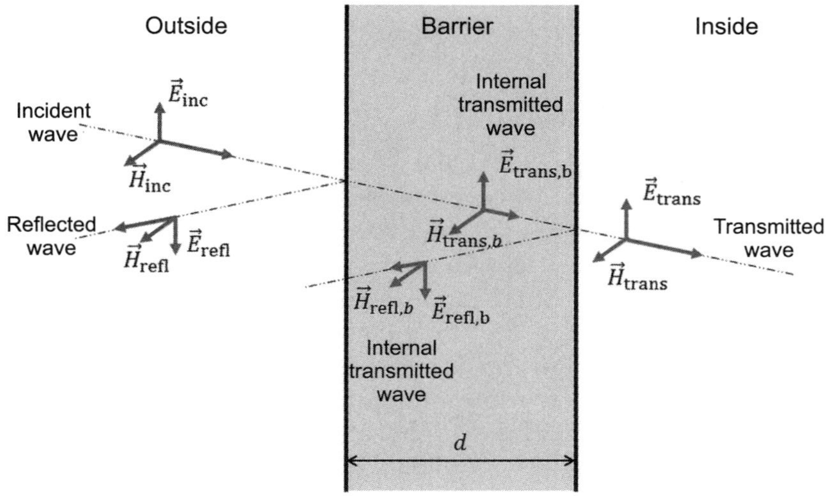

Figure 3.13 Shielding phenomena of a barrier to a plane wave [16].

Once it has passed through the barrier and strikes the right interface, a portion is reflected and a portion is transmitted through the right interface. The portion transmitted through the right interface adds to the total field that is transmitted through the shield. The process continues in like fashion, but the additional reflected and transmitted fields are progressively attenuated by their travel through the conductive material of the barrier. If a shield is designed to have a thickness that is much greater than the skin depth of the material at the frequency of the anticipated incident field, there is little consequence to the continued reflections at the interior surfaces of the barrier. These multiple reflections and transmissions can therefore generally be disregarded for shield thicknesses that are much greater than a skin depth, and only the initial reflection and transmission at the interfaces need be considered [16–19].

Summing up the described process of diffusion of an EM wave through a barrier is defined as:

$$T_{o,d} = \frac{\left|\vec{E}_{\text{inc}}\right|}{\left|\vec{E}_{\text{trans}}\right|} \tag{3.11}$$

Since barriers (e.g., system enclosures) are often used as a shield, the attenuation of the EM field is usually characterized by the shielding effectiveness (SE_E). The shielding effectiveness is defined as the ratio between the field strength at a given distance from the source without the shield interposed and the field strength with the shield interposed.

$$SE_E = 20\log_{10}\left(\frac{\left|\vec{E}_{\text{inc}}\right|}{\left|\vec{E}_{\text{trans}}\right|}\right) = 20\log_{10}\left(T_{o,d}\right) \tag{3.12}$$

While the shielding effectiveness of metal is essentially caused by its conductivity and thickness, the shielding effectiveness of dielectric material is mainly caused by its wave impedance mismatch and the absorption loss. Evidently, the absorption loss rises proportionally to the shield's thickness.

Several studies have been carried out to measure or model the SE_E of typical construction materials, in isolation from other contributing factors such as apertures and conductive penetrations, in particular [17–19]. In these studies the measurements were very well controlled using standardized measurement approaches. The graph in Figure 3.14 shows the attenuation provided by concrete with reinforcing bars (ferro-concrete) and high conductive plasterboard.

Concrete with reinforcing bars (rebar) has been of particular interest because it is a fairly common construction that could offer good shielding performance. However, studies [20–22] have shown that the spacing of reinforcing bars, the numbers of mesh layers the galvanic bonding of the bars within the mesh, and the bonding of the mesh in a slab to other slabs can have a large bearing on the shielding performance. The shielding effectiveness of a plain barrier is highly dependent on the material used, particularly its conductivity, permeability, and permittivity, as well as the specific design. Generally, typical materials for buildings and system enclosures are capable to provide a shielding effectiveness up to 60 dB. Highly conductive materials might go up to 80 dB at frequencies below 1 GHz.

3.3.5 Aperture Penetration

The aperture penetration transfer function is denoted ($T_{o,Ap}$). In reality, real enclosures, such as buildings, aircraft skins, equipment cases, or housings, contain penetrations for cable or pipes or apertures for cooling, air ventilation, sensors, screens, windows, or doors. Those penetrations or apertures into an enclosure or installation offer an unattenuated path for the RF disturbance to propagate through. However, the modeling and analysis of aperture penetration are complex because the passband of an aperture will be dependent on the size and shape of the aperture and on the wavelength of the HPEM environment.

Apertures will pass RF energy without attenuation when the wavelength of the RF waveform is a factor smaller than the aperture dimensions. It is generally accepted that Ott's Equation (3.13) [23] for aperture penetration gives an approximate measure of the attenuation provided by a regular rectangular-shaped aperture (i.e., acceptable for windows and doors).

(a)

(b)

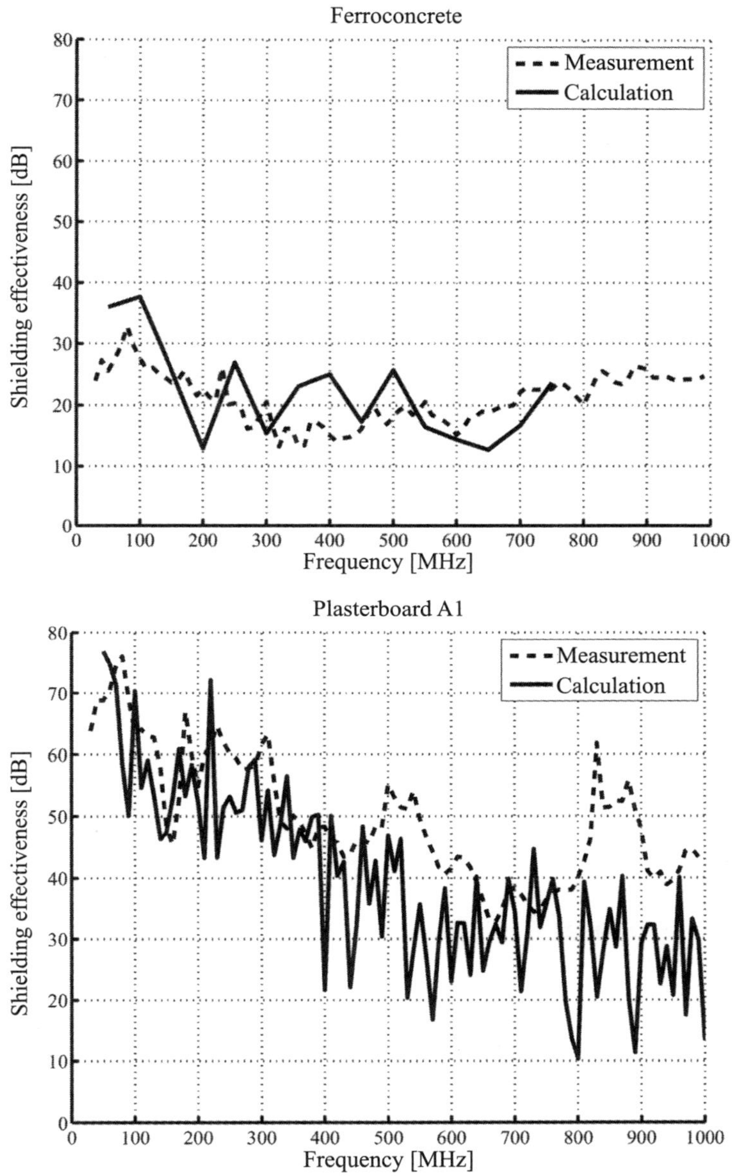

Figure 3.14 Shielding effectiveness of ferro-concrete (top) and conductive plasterboard (bottom) (From: [18]. © 2008 IEEE. Reprinted with permission.)

$$SE_{Ap} = 20\log_{10}\left(\frac{\lambda}{2\cdot l}\right) = 20\log\left(T_{o,Ap}\right) \tag{3.13}$$

In this equation, SE_{Ap} denotes the shielding effectiveness in decibels, l is the longest dimension of the aperture, and λ is the wavelength of the HPEM environment. Effectively, (3.13) is only relevant for an aperture in a conductive sheet such as steel; in practice, materials such as concrete used for installation will cause the frequency peak to shift. However, the amount of shift is difficult to predict for a generic case. The angle of incidence of the impinging field will also skew the effective aperture size.

3.3.6 Conducted Propagation

The propagation of EM waves along a conductor or cable is subject of several loss mechanisms. The transfer function associated with conducted propagation can be dnoted as $T_{p,g}$. The finite conductivity of the conductor or cable, the surrounding material (dielectric isolator), and variations of the cable geometry and radiation from the cable absorb energy from the propagating EM wave. In general, the attenuation of conducted propagation can be considered by assigning a frequency-dependent transfer function to it. Typical unshielded power lines and other cables similar in type are usually not designed to transmit or otherwise contain RF energy. High-frequency disturbances that might have been coupled to those cables will face a higher attenuation as large portions of it will be radiated into the surroundings. For typical HPEM-radiated environments below 5 GHz, the transfer function for the conducted propagation on a cable of less than 10m can be approximated to

$$T_{p,g} \approx 1 \tag{3.14}$$

3.3.7 Galvanic, Capacitive, and Magnetic Coupling

The conducted HPEM disturbance must not always propagate along a direct interconnection from source to the victim. In reality, cables that are connected to the source might be coupled to cables that are connected to the victim by shared impedance (galvanic coupling), electric fields (capacitive coupling), or magnetic fields (magnetic coupling). In general, the modeling of cable coupling is a difficult task. However, in most cases, it is possible to obtain a solution by assuming that the cables are weakly coupled [16]. The current and voltage of the cable connected to the source (source line) will induce voltages and currents in the cable connected to the victim (receptor line) through the common impedance, mutual capacitance, or mutual inductance. In general, these induced currents and voltages on the receptor line will induce currents and voltages back into the source line. By assuming weak coupling, the back-coupling from the receptor line to the source line may be ignored, that is, the induction of currents and voltages from one cable to another is a one-way effect (from the source line to the receptor line).

Finite conductivity and imperfections of cables can generate coupling between circuits, which share common impedance (e.g., common return cable or reference conductor). The situation of a common impedance coupling is depicted in Figure 3.15. Here the source feeds the source conductor with a voltage U_G. The connection of the source to the source line is characterized by the source output impedance Z_G and the source line is terminated by a load impedance Z_L. On the right side, the receptor line is connected to the victim and therefore terminated by the victim input impedance Z_V. The remaining end of the receptor line is terminated by termination impedance Z_T. For usual loads and cable conductivities, the impedance of the reference conductor is usually much smaller than the connected load impedances, so that the main part of the source conductor current returns through the reference conductor. This produces a voltage drop U_0 across that conductor. If the distributed per-unit-length impedance of the reference conductor is lumped into a single impedance Z_0, the voltage drop across the reference conductor is given by:

$$U_0 = Z_0 I_G = \frac{Z_0}{Z_G + Z_L} U_G \tag{3.15}$$

This voltage appears directly in the receptor circuit, producing contributions to the disturbing voltage received by the victim. The transfer function ($T_{p,R}$) of the common impedance coupling is:

$$T_{p,R} = \frac{U_V}{U_G} = \frac{Z_0}{Z_G + Z_L} \cdot \frac{Z_V}{Z_V + Z_T} \tag{3.16}$$

3.3.8 Capacitive Coupling

In many actual situations, source and victim systems do not share a common return conductor; instead, cables or conductors connected to them are running

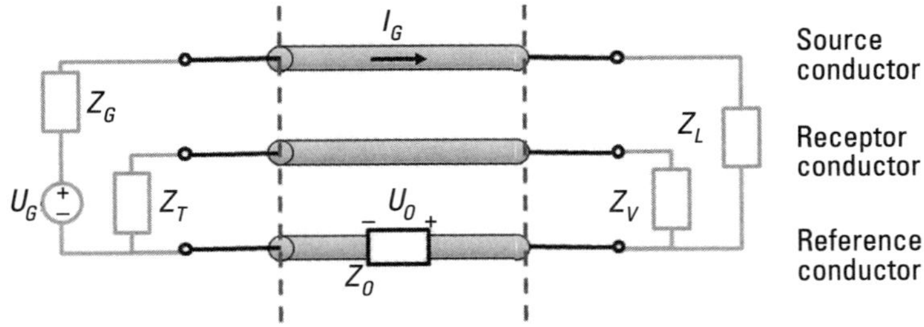

Figure 3.15 Explanation of common impedance coupling between two lines.

partially parallel to each other (e.g., they are installed in the same cable tray). This configuration enables coupling via electric and magnetic fields. The general setup of a coupling via electric fields (capacitive coupling) is depicted in Figure 3.16. The parallel running lines can be seen as a multiconductor transmission line. In this model, the charges on the four conductors are related to the voltages of the two lines (source line and receptor line) by the per unit-length capacitance matrix. The per-unit-length charge q_R induced on the receptor line is given by:

$$q_R = (c_R + c_m)U_R - c_m U_G \tag{3.17}$$

where c_R is the per-unit-length self-capacitance of the receptor line and c_m is the per-unit-length mutual capacitance between source and receptor line. As current is the time derivative of charge, the current on the receptor line is given by:

$$I_R = \frac{\partial q_R}{\partial t} = (c_R + c_m)\frac{\partial U_R}{\partial t} - c_m \frac{\partial U_G}{\partial t} \tag{3.18}$$

The first term is simply the current induced by the voltage on the receptor line. The second part is produced by the mutual capacitance and describes the coupling between both lines. In a circuit-based model, this part is represented by introducing a current source in the receptor line. Because this source is due to electric field coupling, it is usually referred to as capacitive coupling. If the distributed per-unit-length mutual capacitance is lumped into a single mutual capacitance $C_m = c_m\, \Delta z$ in the frequency domain, the induced current on the receptor line is determined to be:

$$I_S = -j\omega C_m U_G \tag{3.19}$$

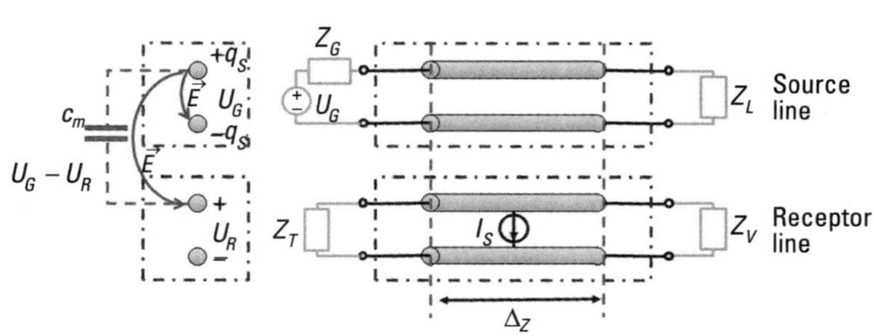

Figure 3.16 Explanation of capacitive coupling between two lines.

Figure 3.16 shows the resulting lumped circuit model. From this, the transfer function ($T_{p,C}$) of the capacitance coupling can be deduced to:

$$T_{p,C} = \frac{U_V}{U_G} = j\omega \cdot C_m \cdot \frac{Z_V \cdot Z_T}{Z_V + Z_T} \cdot \frac{Z_L}{Z_G + Z_L} \tag{3.20}$$

3.3.9 Inductive Coupling

Due to Ampère's circuital law, the current I_G on the source line generates a magnetic field. Faraday's law shows us that, if this magnetic field penetrates the loop formed by a nearby receptor line, it induces a voltage drop on the receptor line that is proportional to the time derivative of I_G.

The coupling of two lines via the magnetic field can be modeled by a per-unit-length mutual inductance l_m. Therefore, the magnetic coupling is usually referred to as inductive coupling. The induced voltage can be written as:

$$U_S = l_m \frac{\partial I_G}{\partial t} \tag{3.21}$$

If the distributed per-unit-length mutual inductance is lumped into a single mutual inductance $L_m = l_m \Delta z$ in the frequency domain, the induced voltage on the receptor line is determined to be:

$$U_S = j\omega L_m I_G \tag{3.22}$$

Figure 3.17 shows the resulting lumped-circuit model. From the transfer function ($T_{p,L}$) of the inductive coupling, it can be deduced to be:

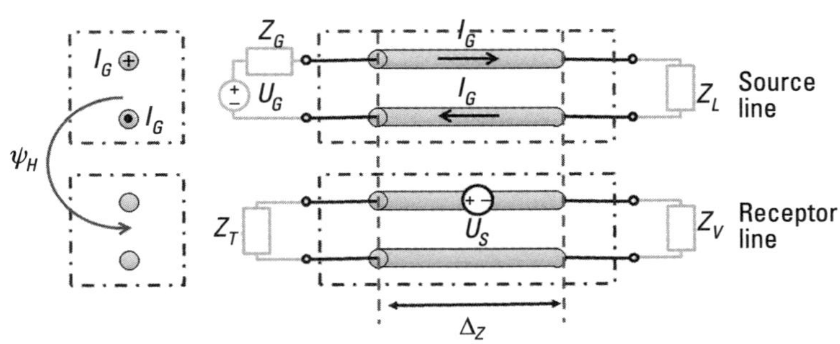

Figure 3.17 Explanation of inductive coupling between two lines.

$$T_{p,L} = \frac{U_V}{U_G} = -j\omega \cdot L_m \cdot \frac{Z_V}{Z_V + Z_T} \cdot \frac{1}{Z_G + Z_L} \tag{3.23}$$

3.4 Field Variation Inside System Enclosure

If the HPEM disturbance enters the system enclosure (e.g., shelter, building, room, or housing), it can stimulate field resonances if the enclosure cavity is even partially conductive. Those field resonances result in variations of the magnitude of the field inside the structure, which are sometimes referred to as hotspots and dead/cold spots referring to either a high-field or low-field magnitude. Indeed, this phenomenon is exploited in reverberation test chambers where the highly conductive/reflective walls of a shielded cavity are used to enhance the field and create hotspots that are then moved around the chamber using a mechanical stirrer. Ordinary rooms inside buildings and the inside of equipment enclosures also demonstrate this reverberant effect, but the magnitude is less pronounced because the enclosure walls are more lossy and the cavity is at least partially filled with lossy materials such as soft furnishings in the case of a room.

Pauli and Moldan [24], Parmentier and Junqua [25], Junqua et al. [26], and Dawson et al. [27] investigated specific aspects of the HPEM field variation inside lossy cavity structures such as building rooms. Modeling of the HPEM coupling in or through buildings has concentrated on using EM topology (EMT) principles [4, 5] and more specifically using power balance (PWB), which produces a unified model of all EM interactions and EM dissipation mechanisms. Measurements of attenuation have also been carried out to validate modeled results. Figure 3.18 shows the variation in mean attenuation over a wide-frequency range for several measurement locations inside the building.

From this, it is possible to observe that the variation of the mean attenuation from room to room within a facility is of the order of 30 dB at some frequencies. This phenomenon is well known in RF and wireless communications inside cavities such as the inside of motor vehicles and building rooms and is often termed as fading describing a cold spot where the received signal is diminished.

3.5 Overall Response

3.5.1 Devices, Equipment, Systems, Networks, and Infrastructure

Because the size and interconnectivity of a victim strongly affect the EM coupling and subsequently the effect, it is important to define different levels of

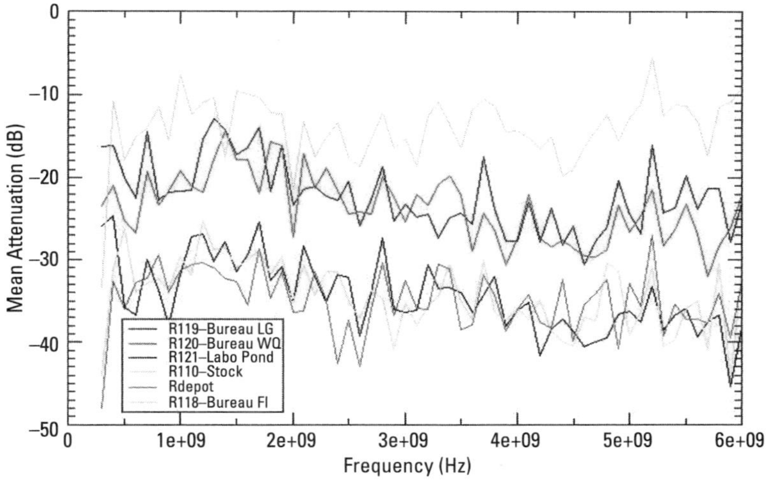

Figure 3.18 Room-to-room variation in mean shielding effectiveness (SE). (Courtesy of Junqua et al. [26].)

victim architecture based very roughly on their physical dimensions. Employing the hierarchical levels that were proposed and discussed earlier in this book, it is possible to identify the strongest radiated coupling of the different HPEM environments as a function of wavelength and victim dimension.

3.5.2 Coupling as a Function of HPEM Environment Type

Figure 1.4 shows the combined HPEM EME together with the system-of-systems hierarchy.

In the lower half of this figure, the green rectangles correspond to the typical physical size of the system-of systems element compared with the wavelength of the HPEM environment. Arrows with dashed outlines are used to show that, while coupling will dominate over certain wavelengths, the elements that make up an infrastructure, for example, are composed of all or some of the elements below it in the hierarchy and so on. The upper part of the figure shows the spectral coverage of some HPEM environments from a spectral perspective.

3.5.2.1 Lightning

As was seen in Figure 1.4, the radiated energy in a lightning stroke is mostly below 10 MHz and extends to DC. This represents a shortest wavelength of up to 30m. The radiated component of lightning therefore couples most efficiently to infrastructures, installations, networks, and systems. The effective antenna for a lightning strike is the strike itself, which can be considered as a monopole between cloud and ground, for a cloud to ground strike. This effective antenna will produce a radiated waveform radially from the strike. The radiated field therefore has very low directivity. Coupling from the radiated field will be high near to the strike, although this is also in the near field of the effective antenna, which could be a few hundred meters depending on the wavelength. In the near field, the coupling can be very variable/complex. For lightning, often the biggest concern from an effect point of view is direct attachment of the lightning strike to or through (e.g., in the case of an aircraft in flight) the victim system. In this instance then, conductors can very efficiently transfer the very high currents from a lightning strike into a system principally from magnetic field coupling. The magnetic fields can be so intense that they can bend or distort nearby conductors. Where minor discontinuities exist and there are parasitic inductive or capacitive elements, then very high-voltage potentials can be developed.

3.5.2.2 HEMP

The radiated energy in an HEMP E1 waveform is mostly below 100 MHz and extends to DC. This represents a shortest wavelength of up to 3m. The HEMP E1 has been shown to couple considerable energy into computer networks, building electrical installations, and even larger systems such as marine vessels and aircraft where the electrical installation cabling exceed 3m in length or area. The HEMP E1 also couples very well to wire-type antennas used for HF and VHF radio links and even over the horizon radar. The HEMP E2 waveform has a similar characteristic to the lightning-radiated waveform so it couples well to distributed infrastructure such as electrical transmission lines and telecommunication lines. The HEMP E3 environment can be considered as quasi-DC and couples most effectively to very long lines.

3.5.2.3 HPRF/DE and IEMI Hypoband

The coupling efficiency of a Hypoband or narrowband HPEM environment to a structure within, say, a system or a network is highly complex. While the coupling efficiency to a circuit or even a device may be high, circuits and devices are invariably housed within an equipment enclosure, which itself may be housed within a building or system enclosure. The enclosure is very likely to significantly reduce the magnitude of the E-field via attenuation, principally reflection and absorption. Furthermore, circuits and devices are often packaged in dielectric material and, because the absorption of dielectric materials is wavelength-dependent, then the magnitude of the absorbed energy above a few gigahertz rises rapidly. Much of the coupled energy is harmlessly absorbed by the lossy dielectric mass.

Hypoband waveforms also have a narrow spectral bandwidth. This means that the probability that the spectral content of the Hypoband waveform matches some critical resonance in a system can be relatively low compared with waveforms that possess a broader spectral bandwidth.

3.5.2.4 HPRF/DE and IEMI Mesoband

Mesoband sources tend to operate at VHF, UHF, up to a few gigahertz. Mesoband sources can couple very efficiently to networks within buildings, systems, and equipment. Also, because they have some spectral bandwidth, the probability of coupling some energy to a system is higher than the Hypoband example.

3.5.2.5 HPRF/DE and IEMI Hyperband

Hyperband sources have the widest spectral bandwidth of all of the HPEM environments. Coupling some energy into equipment and systems is very probable; however, because the available spectral energy in a hyperband pulse is spread so widely, the resulting magnitude of the coupled energy is generally low.

References

[1] Chatterton, P. A., and M. A. Holden, *EMC: Electromagnetic Theory to Practical Design*, New York: John Wiley & Sons, 1991.

[2] Backstrom, M. G., and K. G. Lovstrand, "Susceptibility of Electronic Systems to High-Power Microwaves: Summary of Test Experience," *IEEE Transactions on Electromagnetic Compatibility*, Vol. 46, No. 3, August 2004, pp. 396–403.

[3] Tesche, F. M., "Topological Concepts for Internal EMP Interaction," *IEEE Transactions on Electromagnetic Compatibility*, Vol. EMC-20, No. 1, February 1978, pp. 60–64.

[4] Baum, C. E., "Electromagnetic Topology for the Analysis and Design of Complex Electromagnetic Systems," *Fast Electrical and Optical Measurements*, Vol. I, January 1968, pp. 467–547.

[5] Baum, C. E., "The Theory of Electromagnetic Interference Control," *Interactions Notes*, Note 478, December 1989.

[6] Baum, C. E., "A Time-Domain View of the Choice of Transient Excitation Waveforms for Enhanced Response of Electronic Systems," *Interaction Notes*, Note 560, Summa Foundation, September 2000, http://ece-research.unm.edu/summa/notes/In/0560.pdf.

[7] Kildal, P. -S., *Foundations of Antenna Engineering: A Unified Approach for Line-of-Sight and Multipath*, Norwood, MA: Artech House, 2015.

[8] Balanis, C. A., *Antenna Theory: Analysis and Design*, 3rd ed., New York: John Wiley & Sons, 2005.

[9] Kildal, P. S., E. Martini, and S. Maci, "Degrees of Freedom and Maximum Directivity of Antennas: A Bound on Maximum Directivity of Nonsuperreactive Antennas," *IEEE Antennas and Propagation Magazine*, Vol. 59, No. 4, August 2017, pp. 16–25.

[10] Pozar, D. M., *Microwave Engineering*, 4th ed., New York: John Wiley & Sons, 2005.

[11] "Attenuation by Atmospheric Gases," Recommendation ITU-R P.676-11, ITU-R, Geneva, Switzerland, February 2016.

[12] "Propagation Data and Prediction Methods Required for the Design of Terrestrial Line-of-Sight Systems," Recommendation ITU-R P.530-17, ITU-R, Geneva, Switzerland, December 2017.

[13] Wraight, A., et al., "Phase Processing Techniques for the Prediction of Induced Current," *IEEE Transactions on Electromagnetic Compatibility*, Vol. 50, No. 3, August 2008, pp. 612–618.

[14] Carter, N. J., "The Revision of EMC Specifications for Military Aircraft Equipment," Ph.D. Thesis, University of Surrey, Guildford, U.K., 1985.

[15] Wraight, A., "Improvements in Electromagnetic Assessment Methodologies: Bounding the Errors in Prediction," Ph.D. Thesis, Cranfield University, U.K., 2007.

[16] Paul, C. R., *Introduction to Electromagnetic Compatibility*, 2nd ed. New York: John Wiley & Sons, 2008.

[17] White, D. R. J., *A Handbook on Electromagnetic Shielding Materials and Performance*, 2nd ed. Gainesville, FL: Don White Consultants, Inc., 1980.

[18] Frenzel, T., and M. Koch, "Modelling Electromagnetic Properties of Typical Building Materials," *2008 International Symposium on Electromagnetic Compatibility - EMC Europe*, Hamburg, 2008, pp. 1–6.

[19] Frenzel, T., J. Rohde, and J. Opfer, "Elektromagnetische Schirmung von Gebäuden–Praktische Messungen, Band BSI-TR-03209-2," *Bundesamt für Sicherheit in der Informationstechnik*, 2008.

[20] Stone, W. C., "NIST Construction Automation Program Report No. 3, Electromagnetic Signal Attenuation in Construction Materials," Building and Fire Research Laboratory Gaithersburg, MD, NIST United States Department of Commerce Technology Administration National Institute of Standards and Technology, October 1997.

[21] Holloway, C. L., M. G. Cotton, and P. McKenna, "A Model for Predicting the Power Delay Profile Characteristics Inside a Room," *IEEE Transactions on Vehicular Technology*, Vol. 48, No. 4, July 1999.

[22] Dalke, R. A., et al., "Effects of Reinforced Concrete Structures on RF Communications," *IEEE Transactions on Electromagnetic Compatibility*, Vol. 42, No. 4, November 2000.

[23] Ott, H. W., *Noise Reduction Techniques in Electronic Systems*, 2nd ed., New York: John Wiley & Sons, 1988.

[24] Pauli, P., and D. Moldan, "Reduction and Shielding of RF and Microwaves," *Electromagnetic Environments and Health in Buildings Conference*, London, U.K., May 2002.

[25] Parmantier, J. -P., and I. Junqua, "EM Topology: From Theory to Application –In Ultra-Wideband, Short-Pulses," *Electromagnetics*, Vol. 7, October 2006, pp. 3–11.

[26] Junqua, I., J. -P. Parmantier, and F. Issac, "A Network Formulation of the PWB Method for High Frequency Coupling," *Electromagnetics*, October 2005

[27] Dawson, J. F., et al., "Power-Balance in the Time-Domain for IEMI Coupling Prediction," *Proceedings of EUROEM 2016*, London, U.K., July 2016.

4

Overview of HPEM Test Facilities and Techniques

4.1 Introduction

The quantification of an HPEM effect on a device, circuit, equipment, or system of interest most often requires a test or experiment to be carried out. As we have discussed previously, there are a huge number of independent variables involved in HPEM interaction and coupling, and this is also an issue for the quantification of effects. Many of the variables that have a large impact on the magnitude of the E-field or the HPEM environment characteristics such as those defined by waveform norms cannot simply be inferred from a cursory inspection of the system of interest. Some features require a very detailed knowledge of the system of interest and there are those that can arise momentarily (e.g., parasitic coupling paths) and those that only manifest under the influence of the HPEM stress (e.g., nonlinear features such as rectification).

These constraints mean that the prediction of effects by analytical or computational electromagnetic modeling inevitably lead to large differences between the results of tests and experiment and the results of modeling. The difference in results does not immediately infer that the experiment or the model is in error; however, the necessary assumptions and compromises made in carrying out an effects experiment are perhaps easier to quantify as they are physical features, whereas the assumptions and compromises made within a model are often not so easily accessible to an observer. Modeling is still a very important capability for the determination of HPEM effects. At the very least, modeling enables rapid evaluation of small changes to parameters associated with HPEM interaction and the system of interest such as configuration, dimensional variables, or aspect angles.

The primary objectives of an effect test are to quantify, record, analyze, and classify the system under test (SUT) response (effect) caused by the impinging HPEM environment in ideally as representative scenario (i.e., as close to the actual real-world scenario as possible).

Here the term SUT refers to any item in the system of systems hierarchy, device through to infrastructure. The practicalities of testing at different levels of the hierarchy are discussed subsequently. Figure 4.1 shows that a system of interest interacts with the HPEM environment and both interact with the scenario. In order to carry out effect testing or experimentation, assumptions and compromises are often essential. It is very rare to have a perfect representation of the HPEM environment and at the same time a fully representative system to test, operating in a realistic scenario. As the experiment moves further from the ideal, uncertainty, inaccuracy, and error in the results inevitably creep in.

4.1.1 General Considerations for the Scenario

In order of increasing abstraction from the ideal, the following options are available for the provision of a realistic scenario:

- *The actual scenario:* To date, there have been very few examples where systems are tested in their actual operational scenario. The cost and risk associated with such experiments are usually prohibitive. One prominent example where this has been achieved was the NASA Storm Hazard Program discussed in Chapter 2. For an infrastructure asset such as a nuclear power generation facility, it is largely impractical to test in the actual scenario. A particularly large barrier to achieving testing in a fully representative manner is the fear of the impact of the HPEM environment generated on nearby or colocated systems and infrastructure,

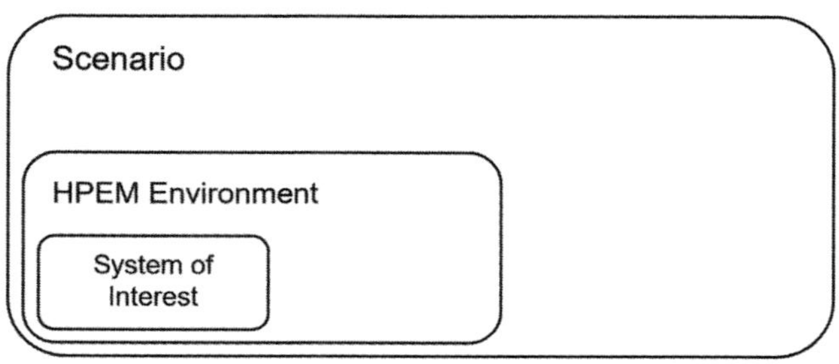

Figure 4.1 Scenario, environment, and system of interest.

for example, unintentionally affecting a data center next to the nuclear power generation facility. Figure 4.2 shows an in-situ effects test in an aircraft cockpit.

Here the objective of the test is to examine the effects of an HPEM environment on the safety critical cockpit flight displays only. However, effects may also manifest on other ancillary systems which are incidentally exposed to the HPEM environment.

- *A physical simulation of the scenario:* For the case of aircraft HPEM testing, dielectric test stands have been constructed to simulate the aircraft in flight (i.e., away from ground interaction). A good example of this approach is the ATLAS/TRESTLE HEMP simulator, which is discussed later. A hoist can also be used to suspend the aircraft again for simulation of mid-flight conditions [1].

  There continues to be great interest in countering improvised explosive devices (C-IED), basically causing predetonation of the explosive device using an HPEM generator. Often, the device is buried and several organizations fabricated sand-pits often containing sand and soil from the problem location or a close match to it in order to replicate actual conditions.

- *A controlled open air test site:* Often, government organizations have access to large open area test sites or ranges where radiated testing is

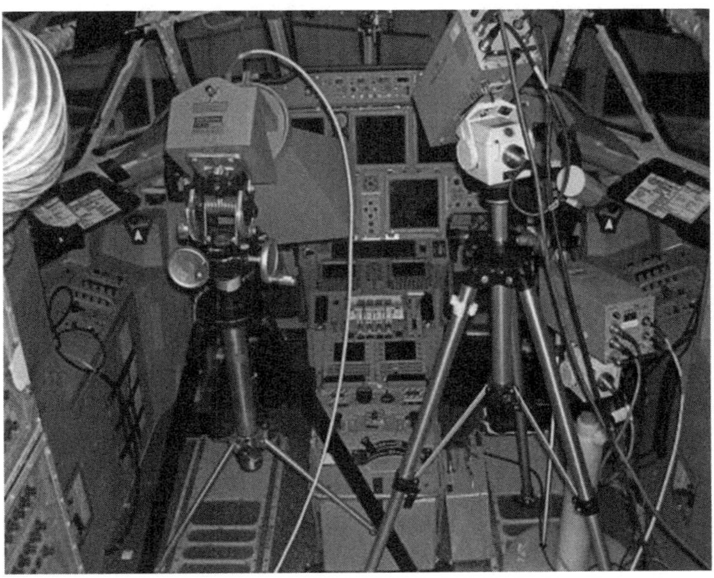

Figure 4.2 In situ testing of instruments inside an aircraft cockpit. (Courtesy of QinetiQ Ltd.).

allowed under controlled conditions. Such sites usually have a good physical security perimeter to control access and control of the airspace over the range. This enables HPEM testing to be carried out with less concern over unwanted interactions with nearby infrastructure. Often, an open area test site is constructed to have a uniform ground reflection response. Figure 4.3 shows a controlled open area test site where the ground plane has a known response.

Structures such as buildings with control rooms, which are representative of a facility or infrastructure asset of interest, can be built onto the range in order to explore interaction and carry out effect testing of systems and equipment located inside. If effects experiments are carried out outside in the open, it may also be necessary to provide measurement of E-field exposure away from the SUT for safety (E-field exposure of people) or other reasons such as to demonstrate compliance with agreed test parameters.

In practice, the majority of effects testing is carried out at the equipment or system level of the hierarchy. This is largely because HPEM effect testing typically requires radiation of high-intensity electric fields or injection of high currents, which can be a hazard to people, can cause the initiation of electro-explosive devices, can cause spontaneous ignition of flammable atmospheres, and may affect unprotected electronics,

Figure 4.3 Aircraft being tested on an open area test site, part of the RF environment generator capability located at MOD Boscombe Down, UK. (Courtesy of QinetiQ Ltd.)

which may be safety-critical, in the vicinity of the test. Managing these risks for an in situ test of, say, an element of infrastructure is often impractical. Testing must also be performed in a controlled manner, meaning that most often the item to be tested must come to the controlled test site rather than vice versa.

- *An enclosed test space:* Enclosed spaces are often the most practical test environment as the HPEM environment is entirely contained within the enclosed space. There are several types of enclosed test spaces, and these are discussed later.

4.1.2 General Considerations for HPEM Environment Simulation

In order of increasing abstraction from the ideal, the following options are available for generation of the HPEM environment for the test:

- *Testing with the actual radiated HPEM environment:* For the case of HEMP, this is totally impractical as it would require detonation of a nuclear device at altitude that is banned by international treaty. For HPRF/IEMI, it may be possible to obtain the actual environment generator. However, the objective of a test or experiment is to collect meaningful effects data, and it is very likely that many individual tests are necessary to identify the statistical accuracy of the test, for example. Often, the number of shots (number of times the test pulse burst can be delivered for a particular test setup) available from an HPRF/IEMI generator may be limited by design; this is because such a generator may be expected to produce the HPEM environment for a very short period and then may be destroyed or otherwise used up. A test of this type can be very cost-intensive and therefore it is usually reserved for a demonstration of effects to the budget holders (in the military, these kinds of test demonstrations are colloquially referred to as the Generals' test [2]) or used as a whole-system validation exercise to provide confirmatory evidence to support equipment level testing.
- *HPEM-radiated environment full-scale simulation:* A version of the actual environment generator that can perhaps provide environment levels that meet or exceed the levels expected and can be used repeatedly and reliably. Such capability exists in a few, mainly government-owned, facilities [3, 4], but it is often very expensive to procure and maintain. With some full-scale HPEM simulators, the pulse-to-pulse repeatability or fidelity can be poor. This is a result of oversimplistic modulator design or can

just be due to the output switch physics. For example, it is not uncommon for the first few pulses in a burst envelope to have a much higher peak magnitude than the remaining and majority of pulses in the burst. This can be important in terms of effects if the effect mechanism is dominated by a peak response. This type of capability was discussed in Chapter 2.

- *HPEM radiated environment subscale simulation:* A version of the actual environment generator that can be used repeatedly and reliably. This may be a legacy simulator (i.e., an earlier version or generation of a full-scale simulator). It should be noted that often for effects testing it is necessary to gradually increase the magnitude of the E-field exposure level in order to measure the onset of an effect and to protect the SUT from immediate damage. Many full-scale simulators have limited capability in this regard. Often, the waveform can suffer significant distortion from the ideal if the magnitude is dialed down. Radiated subscale test simulators for HPEM have begun to be provided on a commercial basis by several suppliers; these were discussed earlier.
- *HPEM simulation by conducted testing as a proxy for a radiated test:* The highest level of abstraction for simulating the radiated HPEM environment is a conducted or injected test. This requires an assessment or translation of a radiated HPEM waveform into a conducted waveform through an understanding (measurement or model) of coupling. The very fact that this transfer function is required leads to higher levels of abstraction from the ideal; still, sometimes conducted testing is the only practical way to carry out an effects test. This is typically the case for HEMP where the majority of the system tests will be a conducted test whereby the radiated coupling of the HEMP environment has been inferred from coupling measurements or analysis. For the lower-frequency content (i.e., less than 1 MHz) of the HEMP E1 environment, a conducted test is more meaningful as cable coupling dominates the interaction and radiated testing is also very inefficient.

4.1.3 General Considerations of the SUT

In order of increasing abstraction from the ideal, the following options are available for the SUT:

- *The actual system:* As you progress up the system hierarchy from device to infrastructure level, it is increasingly difficult to obtain access to the actual article for testing. If the interest is in a system that is embedded in infrastructure, for example, a digital protection relay used in an electrical substation, then although it is practical to have the exact or actual model, it is often impractical to configure it in exactly the way it is configured within the infrastructure asset. Testing of the protection relay, for example, within the infrastructure asset is very rarely done, as there is the inevitable risk that the infrastructure is affected.

- *An instrumented variant of the system:* Often, there is a need to instrument the SUT involving the attachment of probes or sensors in order to provide measurements of the system response to the HPEM exposure. Great care must be taken to ensure that this instrumentation does not alter the response of the SUT. Fiber optic-based solutions are now more commonly available to achieve this, but are not practical in all cases. Measurement and instrumentation are discussed later in this chapter.
- *Equipment connected as far as possible like a system:* Testing at equipment level to infer a system response is often carried out as it is the most practical level of the hierarchy to attempt effects testing. Equipment can be configured as much as possible into an approximation of the full system. An example of this is the "iron bird," also known as rig testing that is carried out for aircraft equipment [5]. Here a framework representing the aircraft fuselage is manufactured and the equipment is fitted to the rig in a representative configuration.
- *Devices or circuits of an equipment intended for a system:* The most abstract level of testing, because it is necessary to have a good understanding of the transfer function throughout the system hierarchy (i.e., system to equipment to device). In the early days of HEMP testing, a great deal of work was performed to understand the HEMP resilience of early digital devices and components [6]. The ambition was to work towards HEMP-hardened components so that equipment and systems could be assembled with a degree of intrinsic resilience. This approach attempted to mirror the concept of RadHard components developed for ionizing radiation environments. However, because the effect of HEMP varies so greatly with coupling, which depends on the configuration of the equipment, system, infrastructure, and interconnected cabling, efforts in achieving HEMP-hardened components were largely abandoned. Considerable effort is required to infer something meaningful from a device and circuit-level effect to a system-level response to an HPEM disturbance. Still, a large proportion of the published HPEM effect research is performed at the device or component level because it is relatively inexpensive to do.

4.1.4 Summary

In all cases above, the degree of abstraction from the ideal condition has a bearing on the quality of data recorded from the test and the utility of the dataset to other environments, systems, and scenarios.

4.2 Uncertainty in Effects Testing

It is very tempting to simplify effects observed for a particular item of equipment or system to a single value or range of values of the E-field. However, experience shows that the compounding of small independent variables has a significant bearing on the accuracy of the result. Factors that contribute to the uncertainty of effects test results include but are not limited to:

- Environmental conditions: temperature and humidity, the reflectivity of the ground (wet/dry), and reflections from colocated structures;
- Variation in HPEM simulator output (e.g., pulse to pulse or even burst to burst variations in amplitude, spectral content, and other important parameters);
- Variation in the coupling or other interaction of the simulated HPEM environment and the SUT (see earlier remarks in Chapter 3);
- Measuring instrumentation calibration uncertainty [7];
- Antenna or injection probe positional tolerance;
- Effect of limiting the number of illumination angles of incidence;
- Cable layout and coupling variation;
- Mode of the SUT during illumination;
- Variability between seemingly identical SUTs.

These factors must be controlled to minimize uncertainty on the effects test result. To provide an idea of the magnitude of uncertainty, one example describes the total expanded uncertainty of measurement for a well-controlled EMC immunity test that has been estimated to be 3.9 dB [8]. This uncertainty estimate excludes many of the factors listed above.

For military, space, and safety-critical applications, the concept of margins has been used and extensively researched over many years [9, 10]. The MIL-STD-464C definition of a margin is: "The difference between the subsystem and equipment electromagnetic strength level, and the subsystem and equipment stress level caused by electromagnetic coupling at the system level." The margin can therefore be considered to be related to uncertainty or for practical purposes act as indicator of the magnitude of the variability of the effect. The margin numbers include most if not all of the factors listed above derived from years of empirical observations. MIL-STD-464C applies a margin of 6 dB to safety-critical and mission-critical systems and 16.5 dB to the safety of electro-explosive devices. Similarly, the U.K. Defence Standard 59-411 [11] assumes a margin of 6 dB for mission-critical systems not tested as platforms and 12 dB for tests where a transfer function must be used (i.e., a conducted test simulating a radiated exposure). The margin can be considered to be the one-sided version of the expanded uncertainty with a rectangular probability distribution. It seems clear that the expanded uncertainty of an effect test could be as large as approximately ±15 dB at a 95% confidence level (CL) even under well-controlled conditions.

A good understanding of the individual contributions and control thereof during testing is essential and all identified uncertainties and margins should be combined through analytical treatment, numerical analyses, or empirical measurements.

4.3 HPEM Effects Test Methods and Facilities

HPEM effects testing can be done using purely empirical means without additional data collection for the purposes, for example, for a demonstration of an effect. However, for a deeper scientific understanding of effects, sensitive measurements must be made and more control must be exercised over variables. Very simply, effects testing can be divided into two broad categories:

- *Radiated effects testing:* where the SUT is exposed to a radiated field;
- *Conducted effects testing:* where currents and voltages are injected directly into or onto the SUT, the waveform that is injected is very often derived from a radiated exposure through the measurement of coupling or transfer function.

Effects testing is complex and requires suitable equipment to carry out. The equipment needed to reproduce the HPEM environment may include signal sources, HPEM waveform generators or simulators, and antennas. The equipment needed to measure the HPEM environment will likely include a field probe, attenuation, some form of isolation, and a spectrum analyzer, receiver, or oscilloscope. Equipment may also be required to operate or exercise the SUT. A suitable location or facility is also required.

For scientific purposes, it is necessary to measure parameters of the HPEM environment and correlate this with the observed effect. It may also be necessary to make measurements of other parameters such as time, temperature, humidity, and/or shielding integrity during HPEM exposure.

4.3.1 HPEM-Radiated Testing

A typical radiated effect test setup, in this case for a SUT inside a shielded chamber, is shown in Figure 4.4.

Standards associated with electromagnetic compatibility (EMC) immunity and susceptibility can be considered to provide a best practice methodology for radiated effects testing. The objective of EMC immunity testing is different from that of an HPEM effects test. The objective for an EMC immunity test is simply to test to a threshold or limit and the observation of an effect is an unwanted consequence or a test fail. However, the methods described to control the test variability and therefore, to minimize the uncertainty for EMC immunity testing, are largely applicable to an HPEM effects test. The methods included within standards IEC 61000-4-36 is a good example [12]. That are applicable to best-practice effect testing include:

- *Calibration of measuring equipment:* Any well-controlled test or experiment should take care with the calibration of instrumentation. The

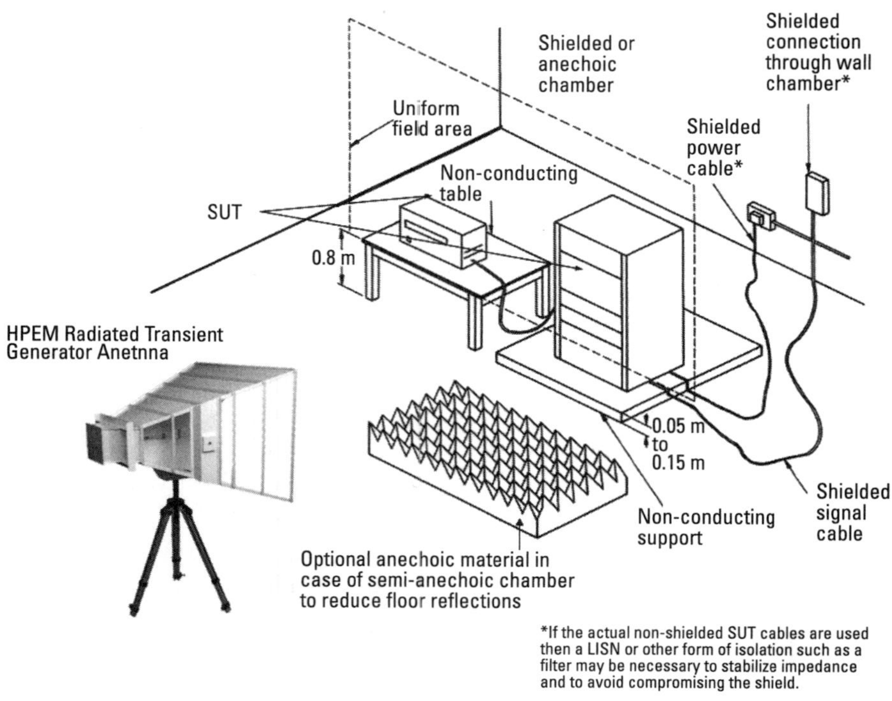

Figure 4.4 Typical HPEM effect test setup.

primary objective of calibration is to minimize measurement uncertainty and therefore overall to manage measurement errors. However, for HPEM measurements, because the magnitudes of the environments are large, there is the distinct possibility of damage to components in the measurement chain. Sometimes this damage can appear more like degradation such as a change of impedance or attenuation. A calibration during effects testing preparation can reveal these issues, which may not otherwise be immediately obvious.

- *Calibration of the E-field/E-field uniformity:* The purpose of field calibration is to ensure that the uniformity of the field over the SUT is sufficient to ensure the validity of the test results:
  - The aim is principally to achieve a far-field, plane wave at the location of the SUT. Testing in the reactive near-field zone of the HPEM simulator or antenna should be avoided as the field variation in this zone fluctuates enormously and results would not be repeatable.
  - It is important that the ground or floor between the HPEM simulator antenna and the SUT must have a uniform impedance or reflection coefficient. Options can include either a perfectly conducting ground plane (if this is expected to be the case for the actual installation) or to use absorber on the ground plane to minimize the ground reflected wave. Testing outdoors can lead to large variability depending on the climatic conditions (e.g., snow-covered versus sun-baked concrete). Metallic structures in the ground such as rebar in concrete might affect the ground reflection at certain frequencies.
  - Calibration of a uniform field area typically requires several measurements (e.g., 9, 12, or 16 grid points) to be made in a vertical plane at the location of the SUT over which variations in the magnitude of the E-field are acceptably small. For IEC 61000-4-36 [12] for example, the E-field is considered uniform if its magnitude measured at the grid points is within –0 dB to +6 dB of the nominal value for not less than 75% of all grid points.
- *Standardizing SUT setup:* All testing should be performed with the SUT in a configuration as close as possible to the actual installation conditions:
  - If the equipment is designed to be mounted in a panel, rack, or cabinet, it should be tested in this configuration.
  - If the final installation of the equipment in service means that it will be mounted on a metal plate or bulkhead, perhaps, for example, inside an aircraft bay or vehicle chassis, then it should be tested on a conducting ground plane.

- If the SUT is free-standing or tabletop-mounted, then a metallic ground plane should not be used, but if it still needs to be supported, then the supporting structure must be constructed of a nonmetallic, nonconductive material. Low dielectric constant (low permittivity) materials, such as rigid polystyrene, should be used. Tables or supports made from wood or glass-reinforced plastic can be reflective at some frequencies, so these should be avoided to minimize field perturbations, which could degrade field uniformity.
- If parts of the SUT would not typically be grounded, then nonconductive supports should be used to prevent accidental earthing of the EUT and distortion of the E-field.

- *Standardizing cable layout:* The layout of cables has a very large bearing on the coupling to the SUT as was discussed in Chapter 3, and therefore great care must be taken to replicate typical installations:
  - The manufacturer's specified wiring types and connectors for the SUT must be used.
  - Cables up to 3m long can typically be managed easily in the uniform field area of the test. If the uniform field area is small or the SUT has long cables, then longer cables (>3m) should be bundled at the midpoint to provide a low-inductance coupling path. Cables that extend out of the uniform zone require very careful treatment so that erroneous results are not produced.
  - Alternatively, for larger SUTs with long cables, a multipoint exposure can be used where there is sufficient overlap in the uniform field area, although this is clearly a compromise.
  - A minimum of 1m of cable should ideally be exposed to the EM field.
  - If a conducting ground plane is used for the SUT, then it is good practice to elevate the cables at 5 cm above the ground plane on nonconducting stand-offs. This is because the field at distances less than 5 cm from the ground plane falls off rapidly and there may otherwise be no coupling to the SUT cables. Raising the cables off of the ground plane creates a kind of worst-case condition because in a practical installation the cables are likely to be run close to a conducting chassis, but there will inevitably be points along the cable length where the cable must be routed away from the conducting surface.
- *Artificial mains network (AMN), line impedance stabilizing network (LISN), or isolation transformer use:* These can be used to control the apparent length and therefore coupling via the mains port of the SUT. These are particularly useful for situations where a longer than usual or

typical power cable is used to provide power to the SUT. Misleading results can be obtained if the cable coupling length is not controlled and the power impedance is not stabilized.

It is good practice to control the setup and take photographs of the setup so that if there is a need to repeat effects testing after some break in time, then these variables are minimized. Often, the configuration of the SUT during an effects test is a compromise between the ideal and practicalities.

4.3.2 HPEM-Radiated Test Facilities and HPEM Environment Simulation

The test facility must ideally:

1. Allow the radiated environment at the SUT to be a faithful representation of the HPEM environment.
2. As far as possible to recreate the dynamic engagement scenario between the SUT and the HPEM environment, addressing such factors as angle of incidence, polarization, and propagation considerations.
3. Allow observation of the SUT so that effects can be characterized and recorded.
4. Allow the radiated HPEM environment on the item under test to be produced without affecting the measuring equipment, observation equipment, or indeed other electronics in the vicinity of the facility.
5. Incorporate measures to reduce the hazard to people, fuels, ordnance, and colocated safety-critical electronic systems from the HPEM environment.

There are a wide variety of solutions for radiated effects testing facilities. Examples include:

- Outdoor and open area test sites;
- Shielded or screened room;
- Transverse EM mode (TEM) cells and lines, which can be used for HEMP and HPRF DE/IEMI environments;
- Anechoic chambers and reverberation chambers, which are more appropriate for HPRF DE/IEMI environments;
- Lightning and HEMP injection rigs;
- HEMP simulators;
- HPRF DE/IEMI simulators, outdoor and enclosed.

4.3.2.1 Outdoor and Open Area Test Sites

Carrying out radiated effects testing outdoors is in many ways close to the ideal situation. For lightning, HEMP and HPRF DE/IEMI environments much of the interaction with systems, installations, networks, and infrastructures will take place in the outdoor, open-air environment. However, if the outdoor environment is not controlled in terms, for example, with the weather, temperature, and ground conditions (soil type), then the variability in the SUT response can lead to very high uncertainty in the effects test result.

Unless the outdoor test site has exemptions or is very remote from population, then HPEM environment simulators or generators used for outdoor testing are likely to require a transmission license to be obtained from the spectrum regulating authorities. The license may limit the number of frequencies that can be tested, the modulation used, and the magnitude of the radiated environment.

Special considerations may be required to de-conflict testing with, for example, over-flying aircraft, marine vessels, road vehicles, or other users of the test site. In practice, it is better to standardize many features of the outdoor environment as far as possible so that results can be extrapolated to other environments.

Open Area Test Site

An open area test site incorporates certain features to help standardize certain conditions. An open area test site must be flat, must be free from overhead cables, must be located a significant distance from metallic objects, and must have a conducting or reflecting ground plane of certain standardized dimensions [13]. The ground plane must be continuous, free from gaps and voids that are large compared with the wavelength to be evaluated. The ground plane is usually rectangular and is required to extend a minimum of 1m from the SUT. Sometimes a turntable is incorporated into an open area test site to allow the SUT to be rotated so that different aspect angles can be evaluated.

An example of large open area test sites was shown in Figure 4.3.

Shielded or Screened Room

A shielded or screened room is a six-sided metallic conductive enclosure, basically a Faraday cage. Testing inside of the shielded room prevents the propagation of potentially hazardous fields outside of the room. There are several different construction techniques for shielded rooms that generally fall into three classes: welded, prefabricated, and architectural. Architectural shielding refers to the fitting of shielding materials to an existing structure. Given that the existing host structure is often not designed to accommodate shielding, the performance achieved when using this approach can be variable. Welded seam-shielded rooms typically have the best shielding performance, typically better than 120 dB over the required frequency range, but are also the most expensive solution. The most common shielded room construction method is the prefabricated clamped seam solution. For this scheme, a clamping or jointing system, perhaps using a conductive gasket or a filter/labyrinth scheme known as "hats and flats," is used to electrically and mechanically join panels of shielded material fixed to both sides of a particleboard or plywood core. A shielded room that uses this fabrication method is shown in Figure 4.5.

While the shielded room may well adequately attenuate generated HPEM environments inside the room to the outside, the situation inside the room is not ideal for plane-wave radiated effects testing. This is because the HPEM fields inside the room will interact with the shield; the most troublesome interaction is reflection. Reflection off of walls, room corners, and other surfaces will distort the field arriving at the SUT, causing poor field uniformity. Even for HPEM-transient environments where the pulse duration may be very short, unwanted interactions in the form of constructive and destructive interference can occur. The magnitude of these variations can be as much as ±30 dB at a 95% confidence level over an area equivalent to a 6-foot computer rack.

The reverberation chamber, discussed later, makes use of these reflections. The anechoic chamber, discussed later, tries to eliminate as far as possible reflections using absorbing materials fitted to the inner surface of the chamber.

If a shielded environment is to be used, then care must be taken in managing penetrations through the shield boundary. It is likely to be essential to power the SUT inside the shield room so that effects can be observed, and if this cannot be managed by a battery supply, then access to a low-voltage (LV) electrical power is likely to be required inside the shield room. The analog to the shield for a power cable or any conducting penetration is to use circumferential

Figure 4.5 Typical shielded room.

bonding of cable shields to the shield wall, decoupling, and/or to use a filter of some sort. LV power filters prevent the HPEM environment generated within the shield room from propagating out of the shield room through conduction. LV power filters are commonly available and can provide a similar degree of isolation (e.g., 80 to 100 dB) for conducted signals as the shielding analog provides for radiated signals. Another penetration that needs to be managed is air flow, which is essential to equalize air pressure inside and outside of the shield room and may be required for air conditioning or cooling. The air-flow penetration is managed by using the waveguide beyond cutoff principle described in Chapter 7 of this book. Again, solutions for managing this air flow are commonly available and the typical solution is to use a honeycomb waveguide vent, where a large aperture is divided up into many small apertures each with the required length or depth to prevent the radiated environment from propagating through. Other types of penetrations that may commonly be encountered include cables or optical fibers for monitoring, measuring, or exercising the SUT. Optical fibers can be passed through a waveguide beyond cutoff. Data cable shields must be circumferentially bonded to the shield wall on both sides of the shield room wall. It may also be necessary to use a filter to prevent the HPEM environment propagating outside of the test environment by conducted means. A data filter can be difficult to design as it is required to pass the wanted data signal, which may have frequency components comparable to the HPEM environment which must be blocked.

Increasingly modern SUTs have a wireless feature, for example, Global Navigation System (GNSS)/Global Positioning System (GPS) as an integral part of the SUT and must be functioning to allow the SUT to function properly. Inside the shielded room, the GNSS/GPS signal from a satellite or a wireless signal from a router external to the chamber should be attenuated to the extent that reception is not possible (if it is, then the shield is not working). In these cases, it may be necessary to use a fiber-optic bridge or repeater to allow the wanted wireless or GNSS/GPS signal to reach the SUT. For GNSS/GPS specifically, special exercising equipment is available to record GNSS/GPS data and play them back. In this way, a GNSS/GPS signal can be recorded outside of the test environment and played back inside the test environment. The GNSS/GPS signal generator must itself be protected from the HPEM environment.

4.3.2.2 TEM Cells and Lines

The TEM cell is also known as the Crawford cell after its inventor [14]. The original TEM cell design by Crawford had a tapered feed section, connected to a rectangular transmission line and a tapered end section. Both the feed section and the end section were fitted with a coaxial connector. Radio or radar absorbent material (RAM) is used inside the cell to reduce higher-order modes.

This double-tapered design of TEM cell is now typically used for testing small test items, devices, circuits, and small equipment.

A more popular design that subsequently evolved is the gigahertz TEM (GTEM) cell [15]. This is very similar in construction to a shielded room, except it is tapered or wedge-shaped rather than rectangular and an offset stripline is incorporated to supress higher-order modes caused by the corners and side walls of the cell. A schematic of the GTEM cell is shown in Figure 4.6(a) and a photograph is shown in Figure 4.6(b).

(a)

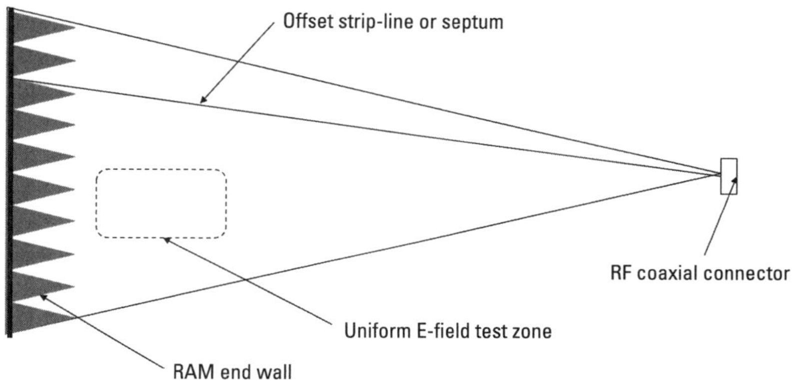

(b)

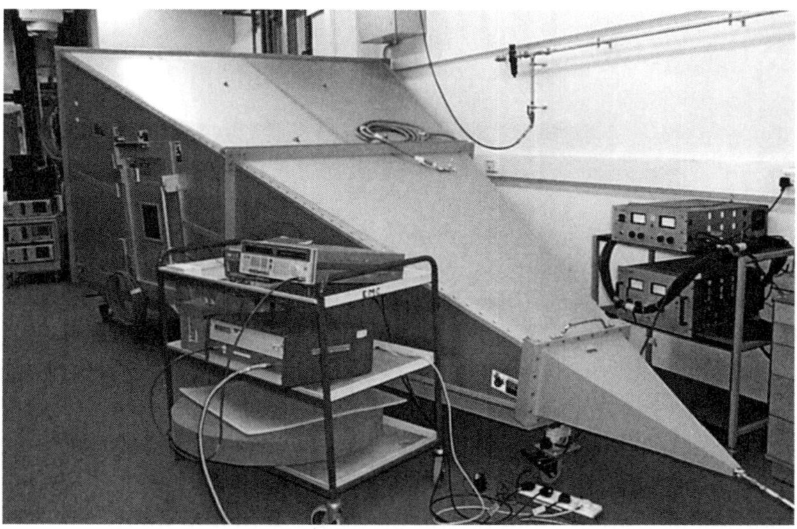

Figure 4.6 (a) Schematic drawing of the cross-section of a GTEM cell. (b) Photograph of a typical GTEM cell. (Courtesy of QinetiQ Ltd.)

The GTEM cell can also be used for HEMP effects testing if a high-voltage feed section capable of holding off 50 kV or more is incorporated.

Open-sided TEM cells are very similar in construction to HEMP-guided wave simulators and can also be used for effects testing. This type of TEM-guided cell has been used extensively by different effect research teams throughout Germany. An example of an open-sided, guided-type TEM cell is shown in Figure 4.7.

This TEM cell simulator based in WIS in Germany is capable of testing with Hyperband, Mesoband, and HEMP E1 type waveforms with a minimum rise time of 90 ps. The test volume is 2.75m high by 2.3m wide. The overall length of the TEM structure is 6m.

TEM cells have a number of advantages for system effect testing, namely:

1. There is no need for a radiating antenna;
2. For GTEM cells, the potentially hazardous E-fields required for effects testing are contained within a shielded enclosure.

TEM cells can have the following limitations:

1. For guided type open-sided TEM cells, potentially hazardous E-fields can radiate away from the cell;
2. Unless the cell is very large, it is only practical for testing small items of equipment, circuits, and devices;
3. The polarization of the E-field inside the chamber is constrained by the geometry of the cell.

Figure 4.7 Photograph of a typical TEM cell. (Picture courtesy of Research Institute for Protective Technologies and CBRN-Protection (WIS), Munster, Germany.)

4. Equipment and particularly cable layout require careful control because the ground plane is the bottom plate of the cell and in the GTEM cell the shielded walls of the cell are also at ground potential.

4.3.2.3 Anechoic and Semi-Anechoic Chambers

Anechoic and semi-anechoic chambers are shielded enclosures with RAM on the inside surfaces to reduce as far as possible reflections and the creation of higher-order modes [16]. The aim of an anechoic chamber is to create a uniform field region around the SUT that is as close as possible to a free-space environment. A photograph of a typical 3-m semi-anechoic chamber is shown in Figure 4.8. The term 3m refers to the typical separation between the transmitting antenna and the SUT for field uniformity.

The RAM used is typically the pyramidal cone type; however, the length of the cones relates to the wavelength at which it is effective. For EMC testing, typically the requirement for radiated testing starts at 30 MHz for military equipment and 80 MHz for civilian equipment [17]. At these frequencies, the RAM can be up to 1m long to be effective and this can use up a large volume of available space within the chamber.

If lower-frequency performance of the RAM is required, then ferrite tiles may be used to absorb low-frequency radiation where the pyramidal cone is less effective.

Anechoic chambers have a number of advantages for system effect testing, namely:

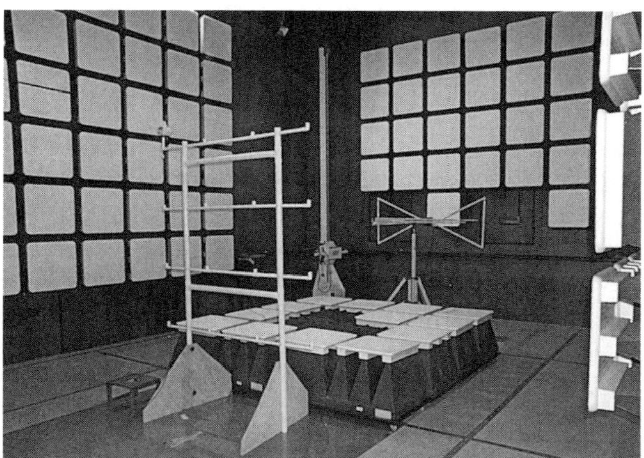

Figure 4.8 Typical 3-m semi-anechoic chamber. (Courtesy of QinetiQ Ltd.)

- They contain the HPEM environment and prevent hazardous fields propagating outside of the facility;
- They are fairly common, used for EMC testing, for example, so that they can be affordable;
- Solutions to providing necessary penetrations of air for cooling and conductors for powering the SUT and the HPEM simulator are generally available.

Anechoic chambers can have the following limitations:

- The fields generated inside the chamber are damped by the absorber, meaning that it can difficult to produce high peak fields at the SUT;
- There may be limitations on the peak and average field that is allowed because the binder materials for the RAM can be flammable;
- The physical dimensions of the chamber and sometimes the dimensions of the chamber door will constrain the size of the SUT that can be tested.

4.3.2.4 Reverberation Chambers

A reverberation chamber (RC) [18, 19] is a conductive shielded room or closed cavity similar in construction to a shielded room. This allows for the E-field from a radiating antenna to reflect around the chamber causing constructive and destructive interference. In this way, hotspots, regions of high peak intensity and nulls, regions of field cancellation, are formed within the chamber. This is the primary advantage of the RC test technique in that modest power from an amplifier can produce intense E-fields in the hotspots.

The chamber is usually equipped with a mechanical tuning or stirring device or devices whose dimensions are significant fractions of the chamber dimensions. When the chamber is excited with RF energy, the resulting multimode EM environment can be stirred by the mechanical paddle, tuner, or stirrer (i.e., the hotspots and nulls are circulated around the chamber). A schematic and a photograph of a reverberation chamber are shown in Figure 4.9.

A snapshot of the variation of the spatial field intensity is shown in Figure 4.10.

This 2-D image shows regions of varying field intensity with the highest intensity field (hotspots) shown. The image was captured by placing a heat-sensitive film (colored region of the image) within the chamber and viewing the heat induced by the RF field via a thermal camera. The images show the effect of four different stirrer positions on the location of the hotspot within the chamber.

(a)

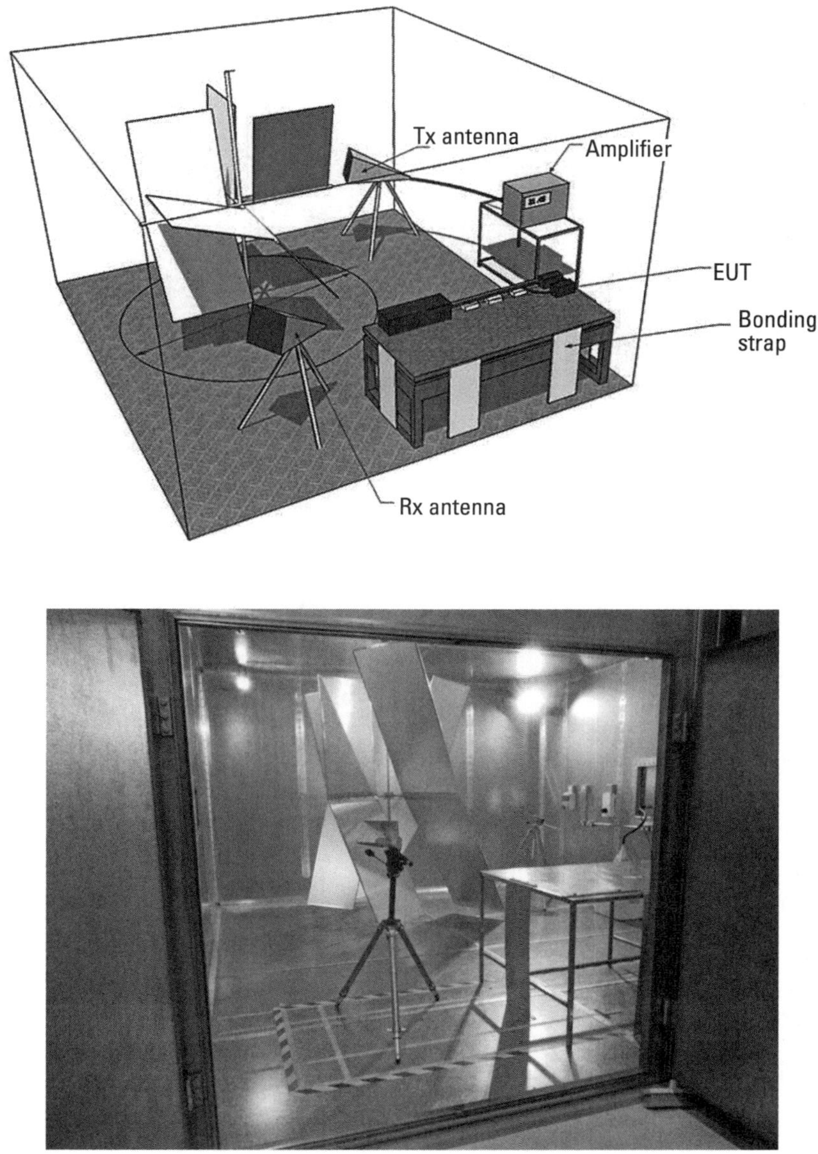

(b)

Figure 4.9 (a) Schematic and (b) photograph of the Reverberation chamber QinetiQ Farnborough. (Courtesy of QinetiQ Ltd.)

RCs have a number of advantages for system effect testing, namely:

- The field strengths available are much higher per watt of input power than a conventional anechoic room or TEM cell.

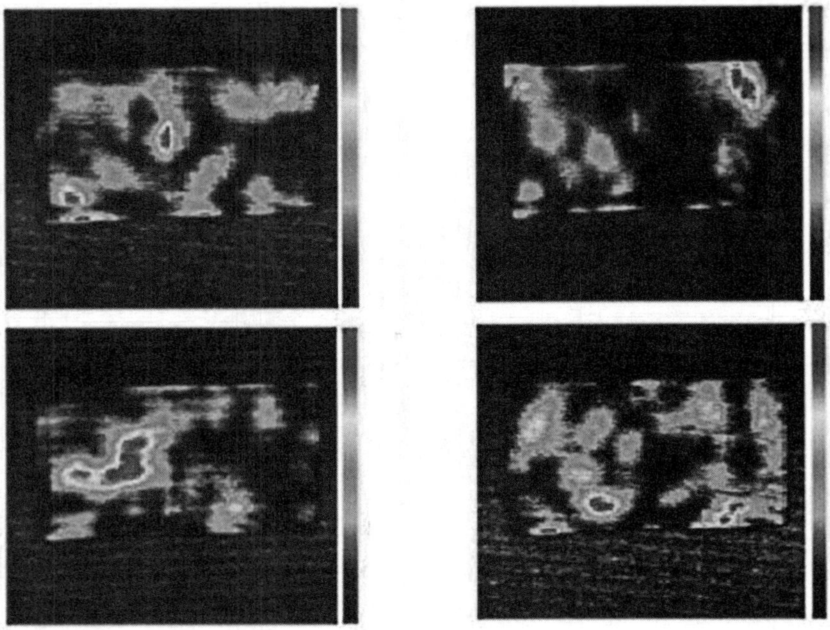

Figure 4.10 Variation of spatial field intensity inside the RC.

- The peak E-field strengths at different locations within the room are more uniform (time averaged over one paddle or tuner rotation) than is possible while employing an anechoic chamber or room or TEM cell.
- The SUT will be subjected to the peak field E-field levels in a multitude of polarizations and angles of arrival, which is prohibitively time-intensive in an anechoic room, in a TEM, or with an HPEM simulator. The RC can be considered to provide a worst-case illumination, and it has been argued that this is more representative of the true EM environment when the system is within its intended operating environment [20].
- All parts or faces of the system will experience the peak E-field levels over the time for one paddle or tuner rotation. Therefore, the particular orientation of the SUT cables has less bearing on the effect threshold.
- The repeatability is better than conventional anechoic and TEM techniques due to time averaging, which produces a more consistent test method.
- Most importantly, the measurement uncertainty is lower than that of anechoic, TEM, or HPEM simulator techniques. Some [21, 22] have quoted the expanded uncertainty of the reverberation technique to be of the order of $\pm$ 1.5 to $\pm$1.8 dB at a 95% confidence level.

The disadvantages of the reverberation chamber technique are:

- For mode stirring (i.e., continuous stirrer rotation), the duration that the peak field dwells on the SUT can become an important factor if the cycle time of the SUT is long.
- Some authors find it conceptually difficult to relate the effect levels recorded using the reverberation method to other radiated techniques; this inevitably leads to a higher expanded uncertainty.
- The size of the smallest dimension of the chamber dictates the lowest usable frequency (LUF) at which the chamber can be used.
- Crucially, the pulse rise time that can be used is limited by the *Q*-factor of the chamber; this restricts the pulse modulations that can be used. Reverberation chambers are therefore only appropriate for Hypoband (narrowband) long-pulse HPRF DE and IEMI environments where the pulse length exceeds the *Q* of the chamber, typically a few microseconds.
- The directivity characteristics of the SUT are not preserved. This means that the critical illumination angle cannot be revealed and any enhancement in SUT coupling efficiency is canceled out [23].

RCs have been shown to be a useful environment for achieving a thoroughness of exposure that is difficult to achieve in practice with any other test method.

As mentioned, the lower-frequency limit of an RC is dependent on chamber size. The chamber's lowest frequency is calculated from:

$$N = \frac{8\pi}{3} abd \frac{f^3}{c^3} \tag{4.1}$$

where a, b, and d are the chamber's internal dimensions in meters, f is the operation frequency in hertz, and c is the speed of light ($3*10^8$ m/s).

RC standards typically state that if, for a given frequency (f), N is less than 100, then the chamber should not be used at or below that frequency. An RC of approximate dimensions of $a = 8$m, $b = 5$m, and $c = 3$m, will have a LUF of approximately 100 MHz.

It is normal good practice to step, rather than sweep, through a range of frequencies. The range is normally governed by the practical limitation of the bandwidth of the amplifier used for the test. It is common and good practice to select 100 frequency steps per frequency decade, derived by

$$f_{(n+1)} = f_n * 10^{\left(\frac{1}{99}\right)} \tag{4.2}$$

where f_n is the test frequency and $n = 1$ to 100, f_1 is the start frequency, and f_{100} is the end frequency.

The paddle/stirrer or tuner can be stepped (discrete steps at a given fixed angle) or stirred (continuously rotated during RF exposure). For stirring, the speed of the mode-stirring paddle wheel is typically between 0.5 and 4 revolutions per minute (rpm). The objective is to allow sufficient time for the E-field hotspots to dwell and upsets to occur if the field strength is sufficient to do so while still enabling all propagation and coupling modes to be assessed during a single test.

At each frequency, the E-field level is set to some minimum low level well below the expected effect threshold of the SUT. The E-field is allowed to dwell on the SUT for one revolution of the paddle. The magnitude of the E-field is slowly increased until an effect on the SUT is observed. The effect level and effect type by classification are then recorded.

The HPEM exposure in terms of the E-field can be derived from (4.3). There are a variety of other equations that can be used to derive the field depending on which standardized test method is selected:

$$E_c = \frac{8\pi}{\lambda} \cdot \sqrt{\frac{5P_r}{\eta_a}} \tag{4.3}$$

where P_r is the ensemble average antenna received power and η_a represents the efficiency of the antenna.

For radiating the field inside the RC, the selection of the transmit antenna is relatively unimportant because the directivity characteristics of the antenna are not preserved. It is usually more important that the antenna is capable of surviving the very high fields generated inside the RC without thermal or dielectric breakdown. For frequencies between 100 MHz and 1 GHz, a log-periodic type antenna is typically preferred. The log-periodic antenna has a quoted nominal efficiency of 0.75.

For frequencies between 1 GHz and tens of gigahertz, a double-ridged waveguide horn is the preferred transmitting antenna. The horn antenna has a quoted nominal efficiency of 0.9 [24]. These factors allow for the calculation of the electric field within the chamber via (4.3).

4.3.2.5 HEMP Simulators

The HEMP E1-radiated EME discussed earlier is an intense single pulse of 50 kV/m or more with a unipolar double exponential form with a rise time of the order of 1 to 2 ns. At equipment level, testing is usually carried out using injection of damped sinusoidal signals of current and voltage into the equipment cables. For system-level testing and testing of installations, injection techniques can also be used, but simulators that produce a radiated HEMP waveform can be used as well.

HEMP simulators may be classified in three categories according to Baum [25, 26]: guided wave, dipole, and hybrid. A compendium or directory of HEMP simulators comprising over 40 detailed datasheets describing simulator parameters for both active and dormant simulators from many nations around the world is available [4].

All HEMP simulator designs have inherent limitations; thus, the large variety of designs that exists. Some analysis and extrapolation of results must always be done in order to infer the implications of the effects observed in the actual HEMP E1 EME.

Guided-Wave HEMP Simulators

Guided-wave simulators were originally referred to as bounded wave simulators because the HEMP waveform is primarily bounded or constrained between two conducting plates. The plates are more typically a series of parallel conductors or a wire mesh to save weight. However, it was subsequently realized that the pulse was not truly bound within the plates and that some radiation outside of the simulator was inevitable. The term guided was therefore coined and is now the preferred term.

The plates are typically driven by one or more high-voltage generators to propagate a nominally TEM wave through a region frequently called the working volume. The test object is located inside of this working volume. This class of simulator is used primarily to simulate the free-space HEMP E1. Most existing guided-wave simulators produce a vertical electric field (and horizontal magnetic field) because in this case the Earth or ground plane can be used as one of the conducting plates. The Swiss VERIFY guided-wave HEMP simulator is shown in Figure 4.11.

The high-voltage pulser is shown in the foreground. The ground plate is actually a complete conductive ground plane covering the entire floor of the facility. The triangular top plate is made of wire mesh and at its widest end it is terminated into a distributed load impedance mesh. RAM lines the far wall of the simulator to try to absorb stray radiation that is not absorbed by the termination.

This type of HEMP simulator is highly efficient. For example, a 1-MV Marx generator can provide E-fields with strengths of >100 kV/m over objects as long as 6m. Guided-wave structures can propagate pulses with subnanosecond rise times if the generator is capable of producing them and careful matching of the generator output to the plates is implemented.

Figure 4.11 The Swiss VERIFY guided wave HEMP simulator. (Courtesy of Armasuiss.)

Guided-wave simulators do not provide the ground reflection needed for assessing the HEMP E1 coupling characteristics of systems situated on the Earth's surface and are therefore the preferred simulator type for testing of systems such as aircraft in simulated in-flight configurations. For good simulation fidelity, the test object dimensions should not exceed two-thirds of the plate spacing.

The largest HEMP simulator of any type in the world is the ATLAS/TRESTLE guided-wave HEMP simulator. A photograph of ATLAS/TRESTLE is shown in Figure 4.12.

The structure was built to perform tests of aircraft with horizontally polarized waves. The structure was capable of accommodating aircraft the size of large passenger aircraft. The wooden platform on which the aircraft sits is 36m above the Earth and has a ramp over 180m long.

Dipole HEMP Simulators

Dipole simulators can radiate very fast rise-time pulses, and the fields produced include the ground reflected wave component. Dipole simulators can be mobile or fixed. Ideally, the test object is positioned far away (tens of meters) from the dipole simulator structure so that the waveform incident on the SUT is a freely propagating TEM wave.

Because these are radiating antennas, they are not as efficient at converting pulse-power energy into fields as are guided-wave simulators. Dipole simulators also suffer from a deficiency in low-frequency energy because they cannot radiate at near DC and their physical size must be held to practical limits.

The U.S. Army EMP Simulator Operation (AESOP) Horizontally Polarized Dipole (HPD) [27], was a prominent example of this type of simulator.

Figure 4.12 ATLAS/TRESTLE guided-wave HEMP E1 simulator.

This simulator was a 300-m-long dipole, which produced mostly horizontally polarized waves. Some vertically polarized components were introduced by the ground termination structures, and these components were more strongly observable off the antenna center line. The biconic section (Marx generator) radiated the high-frequency components, and the horizontal antennas (dipole transmission lines) radiate the lower-frequency components of the waveform.

The simulator produces a radiated pulse whose amplitude varies roughly as $(\sin\theta)/R$, where θ is the angle away from the conic monopole or dipole and R is the distance from it. Ground effects make amplitudes deviate from this formula and also distort the waveshapes and polarization. Close to the simulator, the radiated pulse is not a plane wave. Therefore, an SUT must be placed at a 25-m to 50-m distance to approximate a plane wave.

The output characteristics of AESOP HPD HEMP simulator are shown in Table 4.1.

Vertical polarized dipoles (VPDs) have the appearance of a monocone over a ground plane. VPD simulators produce vertically polarized fields at a

Table 4.1
Parameters for the AESOP HPD HEMP Simulator

Parameter	Value
Peak output voltage	7 MV
Peak E-field at stated range	50 kV/m at 50m
Height	20m
Length	300m
Bicone impedance	120Ω

single angle of incidence. Most of these antennas are resistively loaded to prevent reflection of the currents when they reach the top of the cone [28].

The U.S. VPD-II HEMP simulator is shown in Figure 4.13. The VPD-II simulator had a 60Ω antenna was 40m high and was driven by a 4-MV pulser. The peak field strength at the center of the test zone, 100m away, exceeded 36 kV/m with a rise time of 10 ns.

Hybrid HEMP Simulators

Hybrid simulators are used to generate the HEMP E1 EME, together with the ground reflected wave. This means that the simulated pulse only accurately represents the desired waveform at some critical point in space (distance from the simulator and height above ground).

Hybrid simulators are used for HEMP testing of systems that could be exposed to HEMP at or near the Earth's surface (ships, parked aircraft, communication centers, infrastructure assets). In this case, the Earth surface (soil or water) gives an essential contribution to the interaction of the incident HEMP environment with the system. Indeed, the condition of the soil or the surface (moisture content) has a large impact on the waveform produced. In some instances, a metallic ground plane may be used to stabilize the waveshape.

The hybrid type of EMP simulator is more complex theoretically than bounded-wave types and combines far and near-field considerations in a particular way. Although not as efficient as a guided-wave type of simulator (in terms of the peak electric field produced from equivalent pulse power generators), it can achieve the hybrid double-plane-wave field distribution [29].

The hybrid simulator combines several EM concepts:

Figure 4.13 The U.S. VPD-II HEMP simulator.

- The early-time (high-frequency) portion of the waveform is radiated from a relatively small part of the simulator compared to the major simulator dimensions.
- The low-frequency portions of the waveform are associated with currents and charges distributed over the major dimensions of the simulator structure.
- The structure is sparse so that most of the high-frequency energy radiates out of the simulator without reflecting off the simulator structure. The structure is also impedance-loaded (including resistance) to further reduce unwanted reflections in the simulator.

A hybrid simulator is therefore an electromagnetically complex structure. The fields produced by the simulator cannot be described by a simple formula characteristic of an electric or magnetic dipole.

Elliptical Hybrid Designs

Most working designs and implementations of hybrid EMP simulators are elliptical types, as typified by the WIS HPD as shown in Figure 4.14.

Elliptically shaped hybrid HPD simulator designs were first developed in the United States in the mid-1970s. This simulator has antenna arms with a diameter of 5m and contains discrete resistors uniformly distributed throughout

Figure 4.14 The German hybrid HPD EMP simulator: elliptical design. (Picture courtesy of Research Institute for Protective Technologies and CBRN-Protection (WIS), Munster, Germany.)

its length to provide the desired ratio of electric and magnetic field amplitudes at low frequencies. The output characteristics of the WIS HPD are shown in Table 4.2.

These simulators typically have a Marx generator housed within a cylindrical structure with the output of the Marx connected to a biconical feed section to generate the waveform. The taper of the bicone elements set the impedance of the line [30]. The bicone effectively feeds the dipole arms or transmission lines. For elliptical designs, the transmission lines form an arch with the Marx generator suspended at the apex of the arch and the dipole arms terminated at the ground.

4.3.3 Measuring the Radiated HPEM Environment

During effects testing, it is important not only to record the functional response, effect, or degradation of the SUT but also some other technical parameters related to the HPEM exposure [2]. These parameters can include:

- The electric (E) and/or magnetic (H) fields (e.g., incident fields or incident plus scattered fields within an SUT);
- The current (I) (e.g., induced by a transient field or within an SUT);
- The voltage (V) (e.g., induced by a transient field or within an SUT).

Particular quantities of a measured transient waveform that are important for effects testing, the *waveform norms* are more fully discussed in Chapter 5.

With some HPEM simulators, the pulse to pulse repeatability or fidelity can be poor. This can be important in terms of effects. For this reason, it is important to measure the burst envelope in addition to the parameters above. The mean amplitude of the burst can then be calculated.

Table 4.2
Parameters for the WIS HPD HEMP Simulator

Parameter	Value
Peak output voltage	1.2 MV
Peak E-field at stated range	18 to 80 kV/m
Height	8m
Length	50m
Bicone impedance	—
Rise time	1 to 3 ns
Pulse duration	5 to 20 ns

4.3.4 The Measurement Chain

Figure 4.15 shows a typical measurement chain that is used for measuring HPEM parameters.

- *Sensor:* A sensor is a device that converts the measured quantity (E-field, H-field, current, or charge) into a voltage that can be measured. For transient HPEM E-field measurements, the preferred sensor is often a d-dot sensor. The d-dot sensor was invented by C. E. Baum in the 1960s [31]. During measurements of radiated E-fields, very wide bandwidth antennas were initially used but the transient response of these sensors was not uniform over the bandwidth of the measured signal and therefore a great deal of postprocessing was required to correct for the antenna distortion. A schematic of a d-dot sensor is shown in Figure 4.16(a) and photograph is shown in Figure 4.16(b).

  A d-dot sensor measures the electric displacement vector, essentially *dD/dt*. In order to relate the electric displacement to the voltage at the output of the sensor (V_0), (4.4) and (4.5) must be used.

$$V_0 = A_{eq} R \frac{dD}{dt} \tag{4.4}$$

  where

$$D = \varepsilon_0 \varepsilon_r E \tag{4.5}$$

  where A_{eq} is the sensor equivalent area (m^2), R is the sensor output resistance (typically 100Ω for a dipole and 50Ω for a monopole or ground-plane type), ε_0 is the permittivity of free space (8.854×10^{-12} F/m), ε_r is the relative permittivity (1 in free space), D is the electric displacement (Cm^{-2}), and E = electric field vector (V/m).

  To resolve an accurate representation of the electric or magnetic field from the voltage produced by the sensor, it is necessary to carry out a

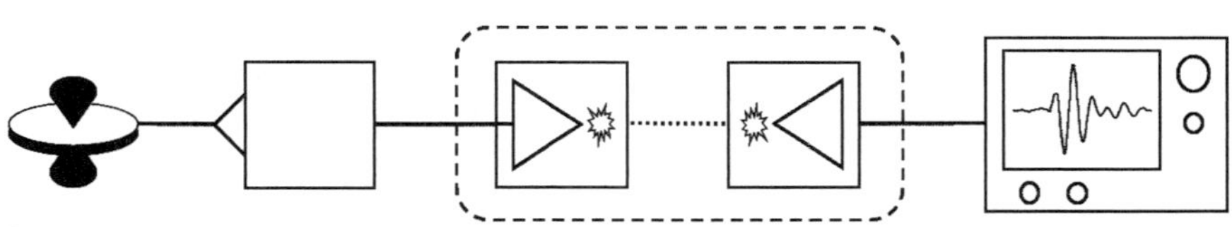

Figure 4.15 Typical HPEM transient environment measurement chain.

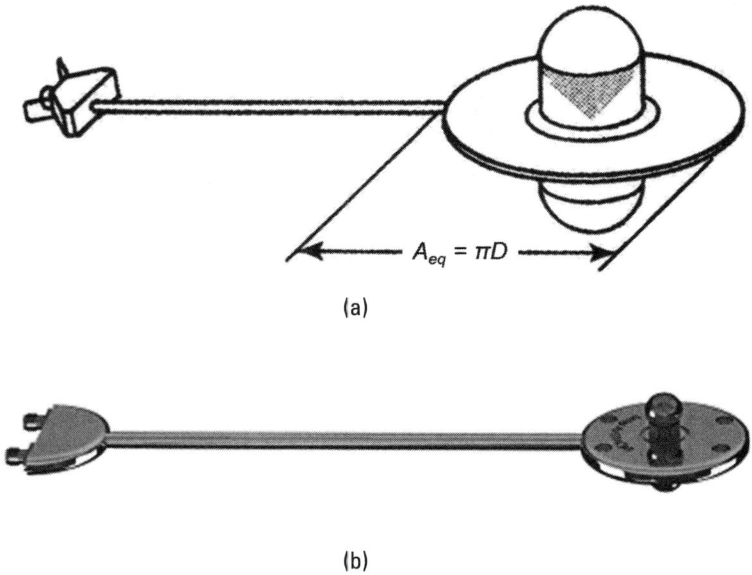

Figure 4.16 (a) A d-dot sensor, *Aeq*, is the equivalent area, where *D* indicates the diameter of the d-dot ground plane. (b) Photograph of a d-dot sensor. (Courtesy of Montena.)

mathematical integration of the measured parameter. Originally, hardware integrators were used for this task, but now it is more common to use a mathematical function on the instrument or process this using software on a computer. The integration phase can be problematic as small features such as DC offset may significantly distort the waveform.

- *Balun:* A balun is a device that operates as a matching transformer to ensure that the sensor is impedance-matched to the coaxial signal line. The balun should also suppress common-mode signals (i.e., prevent the HPEM environment coupling into the balun or the next element of the measurement chain). There are several practical details that must be taken into account when specifying a balun for use in a measurement system. These include the following: the bandwidth of the balun; the change of impedance level, if any, from the balanced to unbalanced ports; the effective attenuation of the signal level due to the insertion of the balun; and the maximum peak voltage and power rating of the device. A 100 ohm to 50 ohm balun usually has a loss factor of 8 dB. This is usually added to the calibration factor for the sensor.
- *Attenuator:* An attenuator is often required because the magnitude of the voltage or current delivered by the sensor is too large for the next

element of the measurement chain. An attenuator simply reduces the magnitude of the voltage by a fixed and calibrated amount.

- *Fiber-optic link (FOL):* A FOL comprises an electrical to optical signal converter (transmitter), a length of fiber optic cable, and an optical to electrical converter (receiver). The FOL provides a high degree of isolation of the sensitive instrumentation that is used to digitize the transient voltage that is representative of the HPEM environment. Fiber optics do not respond to HPEM environments and are the preferred solution for providing isolation. Additional isolation can be provided by having the instrumentation a considerable distance from the HPEM environment or by using a shielded enclosure around the HPEM environment or the instrument.
- *Waveform digitizer:* This is the detector in the measurement chain, which receives the sensor electrical analog signal, converts it into a stream of digital data, and then passes these data on to a recording device. Many modern digitizers have maths functions that can be used to process the measured waveform such as integration or fast Fourier transformation.
- *Data acquisition and control computer:* This may perform several functions. It may be used for controlling the test parameters for the HPEM simulator and setting up the digitizer. It may provide a recording and analysis interface for the waveform data. It may also be used to record the effects observed and collate any time referenced video or other data.

The setup described in Figure 4.15 is typical for measurement of HEMP, Hyperband, and Mesoband waveforms. For Hypoband measurements, the sensor is often an E-field measurement system or perhaps a waveguide or resistive sensor [32]. It is more typical to use a spectrum analyzer to capture the spectral shape and magnitude for these waveforms.

4.3.4.1 Common Issues with the Measurement Chain

There are several fairly common issues or problems associated with making high-power measurements. The first and most serious is breakthrough where the high-power fields find a way of coupling significant energy into the measurement system causing damage to measurement chain components. This damage may not be easily apparent; for example, it can be quite common for the fine interconnecting wire in one hemisphere of the d-dot to fuse. This means that one side of the d-dot is open circuit. The measurement chain may still provide a result, but effectively it is not possible to correct the waveform because the d-dot is now acting as an asymmetrical monopole rather than a symmetrical dipole.

Failure of attenuators is also fairly common and can be difficult to diagnose if the failure is a short circuit. An attenuator is often a simple resistive divider and therefore the reduction in magnitude is achieved by dissipating some of the waveform as heat in the resistance of the attenuator. Attenuators typically have a power limit but these are normally only quoted for continuous and not pulsed waveforms more typical of HPEM waveforms. It is good practice to put a 6-dB or smaller value attenuator as the first attenuator after the sensor or balun. This way, the amount of power or voltage that the first attenuator attempts to dissipate is more manageable.

Another common issue is compression of the front end of the FOL. The FOL has a specific dynamic range of analog input voltage. If the attenuation between the d-dot balun and the FOL is set too low, the front end can go into compression. This is manifested as distortion in the waveform on the oscilloscope or digitizer and can be hard to spot. A practical method of finding out if compression of the FOL is occurring is to increase the attenuation by, say, 6 dB and see if there is a similar corresponding change in the amplitude of the measured waveform on the next test run.

4.3.5 HPEM Conducted Testing

Conducted testing is more abstract from the ideal than radiated effects testing, but often the compromise is essential. If the conducted test waveform is required to represent a radiated HPEM environment, then it is necessary to understand the radiated coupling to the SUT and quantify the transfer function. Different types of transfer functions were discussed in Chapter 3.

4.3.5.1 Measurement of Transfer Functions

The quantification of a transfer function can be done through modeling or measurement. Modeling tools are particularly well suited to this problem, although for very large SUTs computational EM models can become highly complex and unwieldy. The concept of EM topology [30, 33, 34] has demonstrated utility. Topological models can be used to represent the many paths of coupling that can exist and are shown in a simple format that includes all possible coupling paths.

The transfer function can be modeled or measured in the time domain (such that phase contribution is inherent in the quantification) [35] or in the frequency domain. Frequency-domain solutions are generally preferred over time domain due to the relative ease of the test.

For a frequency-domain transfer function, it is necessary in the strictest sense to include magnitude and phase components. However, it has been shown [36] that it is practically very difficult to measure phase with sufficient fidelity to be applied consistently. It is therefore quite typical for the measurement of a transfer function to be magnitude only and an assumption is made with respect to the phase, usually by applying a minimum phase approach [37]. The Minimum Phase Algorithm (MPA) is used to add phase information to a magnitude only transfer function measurement and has been validated extensively. The MPA uses a Hilbert transform to create phase information that is related to the measured magnitude component. Once the phase is constructed, a complex function exists for the transfer function ($f_c(\omega) + \Phi_c$).

Transfer functions are typically stated as a ratio of induced current per incident E-field or H-field or a ratio of external E-field or H-field to the internal E-field or H-field. Examples of a frequency-domain transfer function measurement were shown in Chapter 3.

The measurement of transfer functions is well established and described in many standards, for example, [38–40]. Specific techniques exist for lightning and HEMP due to the lower-frequency content of these environments compared with HPRF DE and IEMI. A particular advantage of the measurement of transfer functions is that the test does not have to be carried out at high radiated power levels. The benefit of using a low-power technique to drive the illuminating antenna means that these assessment techniques are not restrictive in terms of assessment location. The use of low power also removes many of the concerns over nonionizing radiation hazards to people or damage to the system.

However, if there is a nonlinear function or system response such that the SUT coupling is different at high power to low power (perhaps through an electrical breakdown or thermal effect), then the transfer function measured at low level may be inaccurate.

Typically, as identified in the standards referenced above, low-level transfer functions are measured in the frequency domain with a swept continuous wave signal. This is primarily due to economic reasons since a very large volume of useful transfer function data can be gathered relatively quickly and relatively easily. Usually four illumination angles and two different field polarizations are chosen for expediency. Mobile or test in situ capabilities are available for measurement of transfer functions using the low-level swept (LLS) technique but fixed installations have also been made, such as the Ellipticus CW system [41, 42] shown diagrammatically in Figure 4.17.

While the radiated transfer function may be measured in the frequency domain, HPEM environments are typically best described as a radiated transient in the time domain. It is therefore necessary to employ some process to the transfer function data to enable the coupled transient waveform to be revealed. This is generally achieved by the process of convolution which is shown diagrammatically in Figure 4.18.

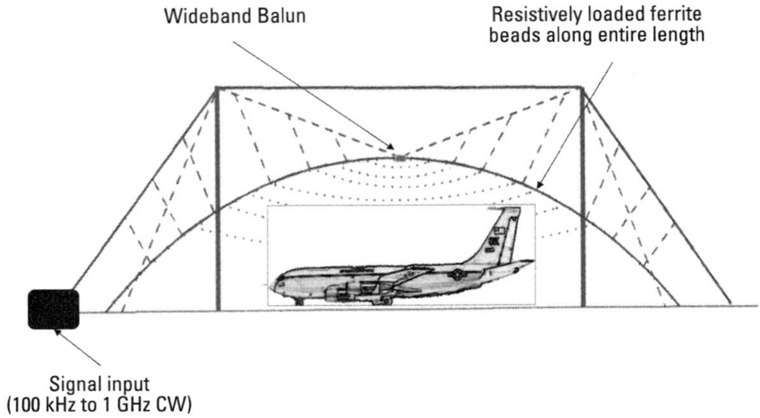

Figure 4.17 Ellipticus CW LLS test facility.

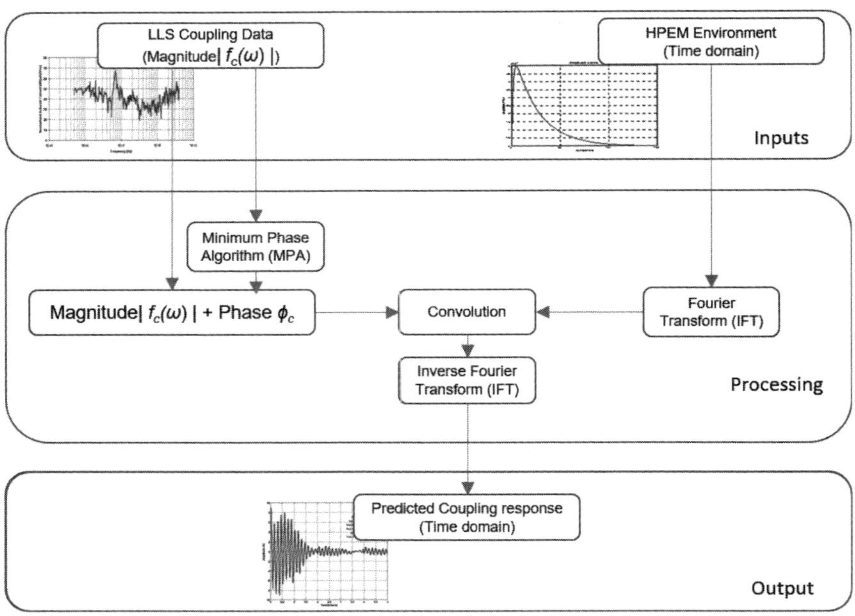

Figure 4.18 The convolution process.

In this process, the Fourier transform (FT) of the incident transient-radiated HPEM environment is computed such that a complex function now exists for that environment ($f_t(\omega) + \Phi_t$). These two complex functions are convolved (multiplication in the frequency domain), resulting in the frequency and phase information of the prediction. The final stage of the process is to calculate the inverse Fourier transform (IFT) of the complex result, which gives the resultant time-domain waveform. This resultant time-domain waveform is the waveform that is required for conducted transient testing.

4.3.5.2 Conducted Effects Test Methods

A typical conducted effects test setup is shown in Figure 4.19. Here the HPEM environment is produced by a transient waveform generator. The environment is applied to a cable bundle or loom connected to an SUT via an injection probe. The voltage V_S and current I_S must be measured.

Simplistically, any high-voltage generator or pulsed power system can be used to drive high voltages or high current into the SUT directly or into cables attached to or adjacent to the victim system. The generator impedance or source impedance should ideally match the conditions attempting to be simulated or at the very least have a higher impedance than the SUT test setup. The simulator must be able to withstand considerable reflected power as often the coupler or transducer used is mismatched to the SUT and much of the power injected is reflected back to the simulator. Care must also be taken to consider stray or parasitic paths, perhaps through the grounding or Earth circuit.

Many systems such as damped oscillatory wave, surge, and even lightning impulse generators exist and are available for EMC, lightning, and HEMP testing. The waveform produced by such simulators is invariably derived from a radiated HEMP or LEMP environment and it is necessary to assume some standardized coupling parameters in order to arrive at a standardized conducted waveform. This often leads to the situation where several conducted waveforms are necessary to represent one radiated environment and for large differences to exist between different SUT types because the transfer function is different.

An example of one type is the Electrical Fast Transient (EFT) generator test set. This generator is capable of producing a 4-kV impulse with a rise time

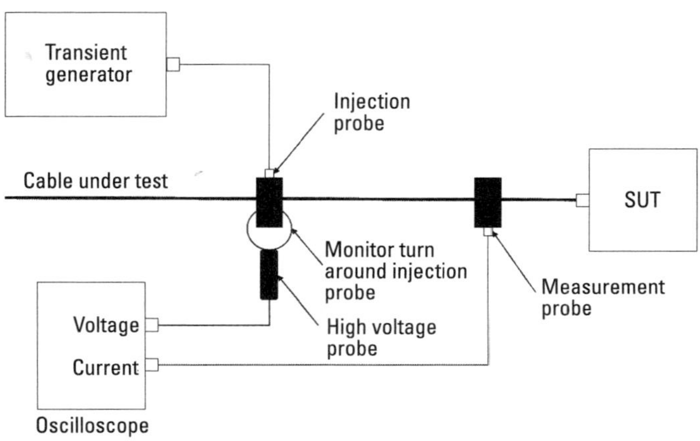

Figure 4.19 Typical conducted effects test setup.

of 0.5 ns and a pulse width of 50 ns. HEMP E1 conducted testing can require testing with an open-circuit voltage limit of 160 kV and short-circuit current limit of 3.2 kA [43].

Another example is the Repetitive Random Square wave Pulse Generator (R2SPG), developed for military conducted immunity tests in the United States [44].

Powerful mobile conducted HEMP generators have been developed in Russia [45, 46]. These mobile simulators are known as Zenit-A and Zenit-K. Figure 4.20 shows how the HEMP Simulator could be configured and Table 4.3 shows the specifications for the two simulators.

There are several transducers or couplers or injection mechanisms that can be considered.

- *Direct injection, also known as direct current injection (DCI):* This technique involves the direct attachment of a HPEM simulator to the SUT with or without using a specially designed coupler or transducer. Often, there is a significant mismatch between the simulator and the SUT and the SUT can act as an antenna and radiate the injected current potentially causing a disturbance to nearby equipment. Direct injection of HEMP and lightning transients onto aircraft fuselage and missile bodies has been employed in effects testing. The technique is considered to be useful for frequencies less than 1 MHz as it is very difficult to produce high-magnitude radiated fields at these frequencies [47, 48].

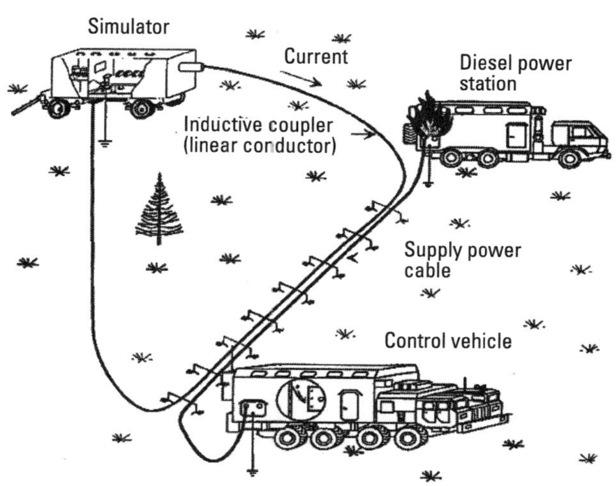

Figure 4.20 Zenit Conducted effects HEMP simulator assessing the HEMP effect on a control vehicle power cable. (From: IEC 61000-1-3 [46]. Copyright © 2003 IEC Geneva, Switzerland. www.iec.ch.)

Table 4.3
Parameters for the Zenit Conducted HEMP Simulators

Output Parameters	Zenit-A	Zenit-K
Voltage pulse amplitude	100 to 800 kV	10 to 35 kV
Voltage pulse rise time	10 to 80 ns	3 to 10 μs
Voltage pulse width	0.5 to 5 μs	Up to 100 ms
Current pulse amplitude (short circuit)	20 kA	Up to 80 kA
Current pulse width (short circuit)	Up to 1 μs	Up to 70 μs

Directly connecting a simulator to the conductor in a cable can be done but is not preferred because it is often necessary to introduce a decoupling network into the circuit to protect the simulator from the SUT line voltages and current, for example, if the SUT is connected to a medium-voltage (MV) power line, typically 415V ac/60A in Europe, and there is a desire to inject a HPEM threat environment onto that cable.

- *Point injection probe:* A bulk current injection probe specifically designed for HPEM environments is shown in Figure 4.21. The term "bulk" is used to refer to the fact that the probe injects current and voltage onto the whole cable bundle. The probe can be split in half for easy placement around the cable, which is placed through the center of the probe. If the cable is a shielded cable, then the injected current and voltage will be coupled to the shield only unless a modification of the shield is made such that the inner conductors of the cables are broken out, but this is a significant violation of the test geometry. It should be noted that

Figure 4.21 Bulk current injection probe.

the injected current and voltage will flow in both directions along the SUT cable. This type of probe uses a magnetic core and care must be taken that the magnetic core does not saturate due to the applied HPEM environment. Saturation can also occur if there is high system current already flowing in the SUT cable bundle. Saturation of the probe core can result in damage to the probe, distortion, or nonlinearity of the applied waveform.

- *Capacitive coupling:* For this type of conducted technique, an electrostatic charge is built up on a plate or similar structure and induces an opposing charge in the SUT or SUT cables. Capacitive coupling clamps are available for certain tests such as the EFT test specified in IEC 61000-4-4 [49] (see Figure 4.22).

  The cable under test is placed along the length of the clamp. The length of the clamp must be physically longer than the rise time of the applied transient waveform (i.e., a rise time of 1 ns has an effective length of 30 cm in air).

- *Planar magnetic induction:* This technique is seldom used but can be implemented using a lossy cable laid along in parallel with the SUT conducting channel as shown in Figure 4.20. The advantage of this technique is twofold; as with the capacitive clamp, the applied transient HPEM waveform can be orientated to induce current flowing towards the SUT. This technique therefore replicates end-fire coupling (i.e., the situation where the direction of propagation of the transient is along the direction of the cable). End-fire coupling can be shown to induce higher stress on the SUT for certain situations [51] when compared with other coupling geometries. Also, there is limited contact with the cable connected to the SUT so if the cable being tested is passing large voltages or current, the SUT can be tested with power on. This has proved to be important as, in tests of power-line insulators, researchers have found that damaging effects can occur due to the follow-through of the systems power [52].

Figure 4.22 Capacitive coupling clamp. (Courtesy of EMC Partner [50].)

4.3.5.3 Conducted Test Facilities

A particular advantage of conducted testing is that specialist test facilities are not generally needed. Conducted testing can be carried out in situ or in a simple laboratory environment.

Consideration should be given to any radiated fields unintentionally produced during a conducted test. It has been observed, for example, from DCI tests that the SUT acts as an antenna and can radiate significant fields, which in the extreme can effect instrumentation and colocated equipment.

4.3.6 Measuring the Conducted HPEM Environment

During conducted effects testing, it is important not only to record the functional response, effect, or degradation of the SUT but also some other technical parameters related to the HPEM exposure [2]. These parameters can include:

- The voltage and current injected into or onto the SUT;
- The voltage and current (e.g., at some particular important location in or on the SUT).

The simultaneous measurement of voltage and current enables the calculation of impedance, ideally the load impedance, of the SUT, which can be a very useful parameter.

The measurement of HPEM transient quantities such as rate of rise, peak amplitude, and pulse width requires measurement, processing, and recording equipment that must be time-correlated with the observed SUT functional effect response.

4.4 Exercising and Observing the SUT

Often, particularly for radiated effect testing, it is necessary to exclude people from the test zone because of the hazard of EM field exposure (EMF). In Europe, there are now laws to protect workers from exposure to EMF [53], and in the rest of the world voluntary guidelines exist which recommend limits of exposure to EMF [54, 55]. The exclusion zone is often large enough to render direct human exercising or observation of the manifestation of HPEM effects difficult.

The range of exercising or monitoring equipment required could include the following: real-time monitoring, post-exposure examination, and effect or disturbance measurement via direct connection (nongalvanic) or stand-off.

Exercising the SUT is very important. It has been anecdotally observed that during effects testing of a computer in an idle mode results in a higher effect threshold than when tested while the computer is being exercised (i.e., reading and writing files to the hard drive, typing on the keyboard, and moving the mouse). For many complex systems, it may be necessary to provide stimulus for sensor inputs to trick the SUT into believing that it is within its normal operating envelope. For example, many modern SUTs (e.g., autonomous platforms such as unmanned aerial vehicles (UAVs) or drones) incorporate GPS receivers for precision timing or position. This challenge was discussed earlier. SUTs may also contain inertial switches, environmental sensors for air flow that must be tricked into believing that they are operating within normal parameters.

It is very important that the exercising method does not introduce an unintended coupling path into the SUT. Figure 4.23 shows a picture of a pneumatic exercising scheme mounted on a dielectric plate over the keyboard of an SUT. Pneumatic actuators are triggered using compressed air via the pipework shown to press keys on the keyboard exercising the keyboard and providing a stimulus to the SUT. Such schemes can become very elaborate.

In order to exercise the processor with an item of equipment or system, it may be necessary to use exercising software. Such software needs to exercise the volatile and nonvolatile (RAM and hard disk) of the SUT, peripherals such as wired and wireless ports (Ethernet and Wi-Fi), and the SUT display. A screenshot of an exercising software program is shown in Figure 4.24, which illustrates that the exercising software program provided consistent 100% central processing unit (CPU) usage.

For monitoring or observing the SUT, digital television (DTV) cameras are often employed. Hardened DTV cameras that have a good degree of protection from the HPEM exposure are widely available.

Actually probing the SUT in order to have a reference measurement inside the SUT enclosure or even at the component level in a way that does not perturb the measurement is very challenging. However, some techniques using

Figure 4.23 An exercising scheme using pneumatics.

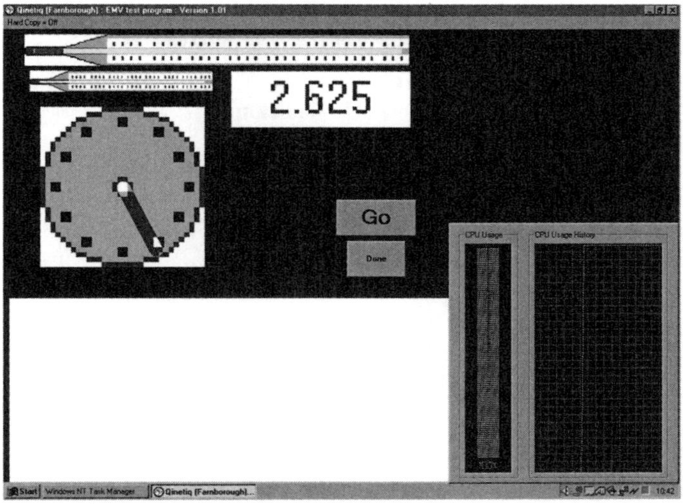

Figure 4.24 Screenshot of an exercising program.

the electro-optic Pockels effect [56] and other optically isolated schemes [57] are maturing.

4.5 Effects Data Presentation

The classification of effects is discussed in Chapter 6. However, the representation of effects data is also required. In recent times, there has been an impetus not to include the actual units on the axes of a graph. This is motivated by commerical confidentiality security concerns with some military organizations requiring that such data are security classified. The most typical forms of effect data representation is either a tabular or graphical format. An example of the tabular form, from [58], is reproduced in Table 4.4.

This table describes the effects thresholds for various SUT types from Hypoband exposure. The testing was carried out in a semi-anechoic chamber. An example of the graphical format is shown in Figure 4.25 [59].

Here the data are represented with the operating or carrier frequency of the Hypoband exposure along the *x*-axis and the measured SUT E-field exposure level on the *y*-axis. Each symbol represents the recording of an effect (class D.1 according to the classification scheme discussed later) at a specific test frequency. This testing was carried out in a reverberation chamber. Representing those data in this manner is useful to identify if there is a dependence on the effect with carrier frequency (i.e., to identify if there is a specific resonance, trend, or other response that is interesting).

Table 4.4
Minimum Radiated Effect Threshold of Computers

SUT Type	Carrier Frequency (GHz)	Upset Threshold (V/m)	Modulation Used	Effect
133-MHz Pentium	1.133	50	AM*	Reset
	1.133	50	Pulse**	Reset
	2.675	50	AM	Loss of access
	2.675	75	Pulse	Loss of access
	2.713	30	CW***	Loss of data
	2.770	50	AM	Loss of data
	2.887	75	AM	Loss of access
233-MHz Pentium II	1.070	100	Pulse	Disk write error
	1.460	100	CW	Power down
	1.460	100	AM	Power down
	1.460	100	Pulse	Power down
	1.480	100	CW	Power down
300-MHz Pentium II	1.040	45	Pulse	Power down
	1.400	100	CW	Power down
	1.43 to 1.55	50	Pulse	Power down
	1.510	100	AM	Power down
	1.510	75	Pulse	Power down
	1.515	100	AM	Reset
	1.690	85	Pulse	Power down
	1.750	75	Pulse	Power down

Source: [58].
*AM: Amplitude modulation with 80% modulation depth.
**Pulse: 217 Hz, 50% duty cycle (2.3-ms pulse width).
***CW: continuous wave (i.e., no modulation applied).

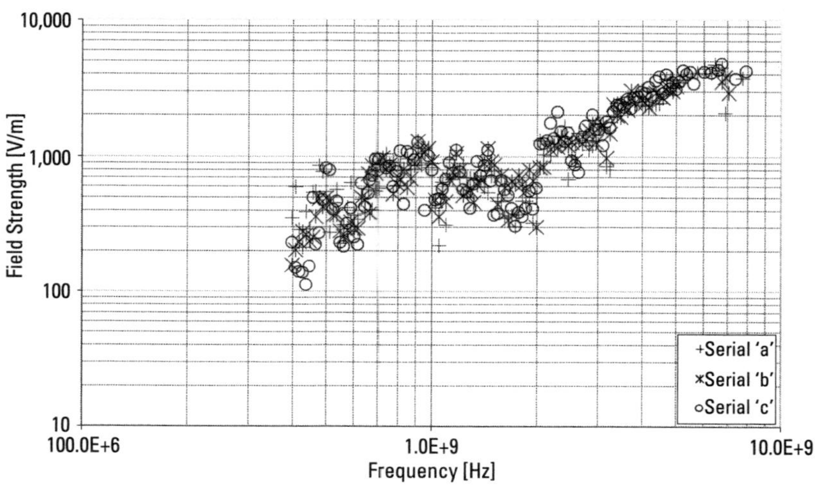

Figure 4.25 Effects threshold data for three different specification computers.

A multidimensional graphical representation of effects has been given by Nitch et al. [60] (Figure 4.26).

Here the *x*-axis represents different HEMP and Hyperband test simulators used in the test campaign (up = unipolar, bp = bipolar) the *y*-axis is the electric field strength and the *z*-axis refers to low or high-pulse repetition frequency (LPRF/HPRF). This format enables ready comparison of the parameters that dominate the effect. This testing was carried out in a TEM cell.

More recently, there has been a trend to represent the data in a more statistical representation. This is because the variability of the effects test results is fairly large, as discussed, and it is considered that a statistical or probabilistic representation of the data allows the inclusion of the variability. Camp and Garbe [61] summarized effect data in terms of breakdown failure rate (BFR), as shown in Figure 4.27.

Here the BFR is the total number of failures (specifically, upsets of a certain type) divided by the total number of pulses applied. A BFR of 0 implies that no effect was observed at a particular threshold for all pulses applied. A BFR of 1 implies that every pulse applied above a certain threshold had an effect on the SUT. The effect testing was carried out on many different computer types with HEMP and Hyperband waveforms in a TEM cell.

These examples are only really capable of representing one type of effect at a time and different graphs are needed to show different effects. Li et al. [62]

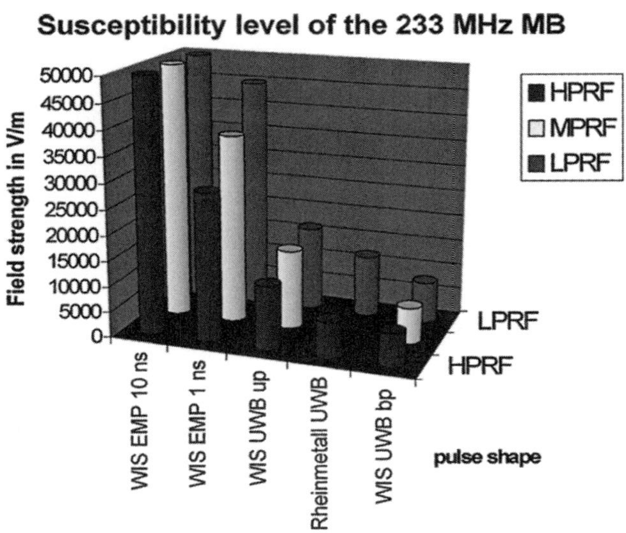

Figure 4.26 Effects data represented in three dimensions. (From: [60]. © 2004 IEEE. Reprinted with permission.)

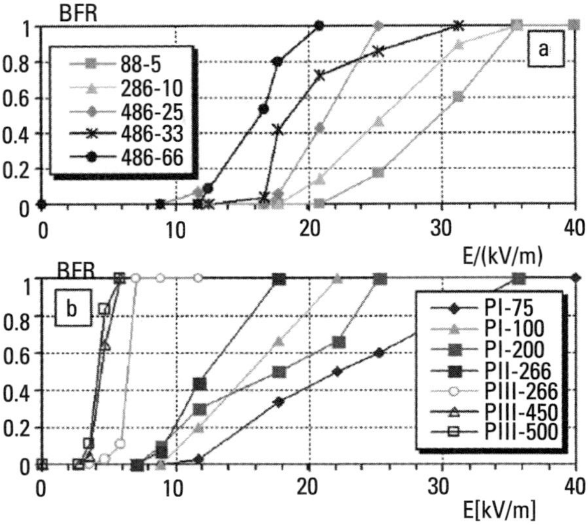

Figure 4.27 BFR effects data representation. (From: [61]. © 2006 IEEE. Reprinted with permission.)

proposed a method to represent multilevel effects, for example, no effect, upset, damage, or a single graph as a probability density function (PDF) (Figure 4.28).

Here the SUT exhibited three types of effects during exposure of an HEMP E1 pulse of varying amplitude. Two types of multinomial linear regression algorithms (Softmax and MINDA) are used to curve-fit the PDF. The

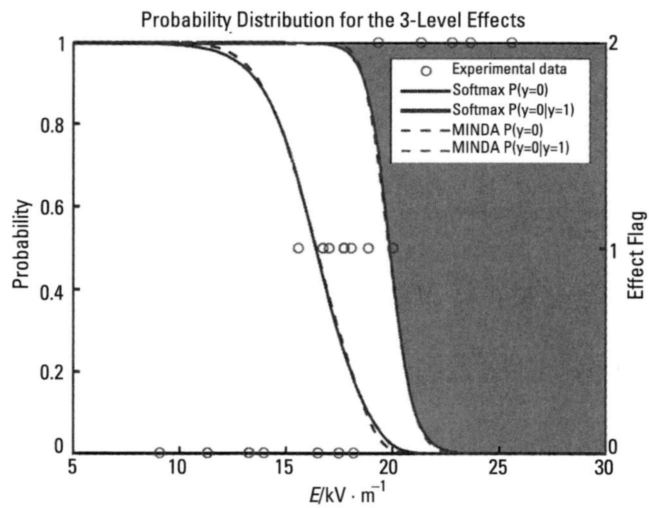

Figure 4.28 The PDF effects data representation.

testing was carried out on a computer communication system in a guided-wave EMP simulator.

4.6 Other Practical Considerations for HPEM Effects Testing

Over many years of experience of HPEM effects testing, due to the necessity of having to compromise in the way that testing is carried out, several issues and questions remain. This section elaborates on some of these issues.

- *Frequency sweeping:* When trying to evaluate whether some critical frequency is important during an effects test, it is often tempting to sweep through a range of frequencies. However, sweeping through a range of frequencies at a high rate may introduce a low-frequency modulation or chirp into the exposure. In addition, the SUT may have a response time or control cycle time (i.e., it may take hundreds of milliseconds for all sensors to be addressed by a CPU). The dwell time of the exposure can therefore become an important factor.
- *Repeat testing, cumulative effects, or aging:* During an effects test, an SUT may be exposed to an HPEM environment multiple times. It has been at least anecdotally observed that the SUT effect threshold appears to be lower after the first effect is observed when the test is repeated. This has also been experimentally observed by Esser and Smailus [63] when carrying out conducted testing on components. To date, limited evidence of this aging effect is available to enable a conclusion about the implication on effects testing. However, it is likely that if such unwanted effects are observed, they are related to thermal or charging (capacitive or inductive) response of the SUT and therefore the relaxation time between exposures will be just as important a contribution as the exposure or dwell time. This observation could also be responsible for an unwanted effect known as windowing, where it appears that the measured effect threshold may only appear at certain values of exposure and effects may cease or the threshold may increase as the exposure is increased above some level. Practically speaking, it is not cost-effective to test a new sample of the SUT for every HPEM exposure.
- *Identical SUT paradox:* Since it is now recognized that small-scale sample testing with very low numbers of samples of a particular SUT type can lead to high uncertainties in the measurement of effects, there is increasingly a drive to test many more samples of the same SUT. In an experiment carried out by Hoad [64], three nominally identical SUTs were

procured simultaneously and tested consecutively. These SUTs appeared to be identical via visual inspection of the external casing, although the serial numbers were nonconsecutive.

Postprocessed analysis of the effects data showed considerable variance, perhaps as much as 18 dB at some frequencies of the recorded effect level, compared with a measurement uncertainty of approximately ±1.8 dB (95% CL). This testing was carried out in a reverberation chamber where the field uniformity is averaged and variations in SUT layout have very little effect, still the differences are large. On closer inspection of the circuit boards inside the SUT, it was found that different components from different manufacturers had been used though the overall specifications of the three SUTs were identical.

- *Synergistic effects:* This has been observed during conducted effects testing. The majority of SUTs contain numerous cable bundles and testing them all simultaneously is not possible. It is feasible to identify those cable bundles that are associated with a particular function or subsystem but even then significant numbers of cable bundles are likely to require testing at the same time. In reality, the radiated HPEM environment will illuminate the system of interest at the same time such that synergistic excitation of all ports occurs. However, typically for a conducted effects test, only a single port is excited at any one time. Figure 4.29 shows a comparison of the effect threshold for single-port and multiport injection of a piece of electronics equipment with multiple cable bundles.

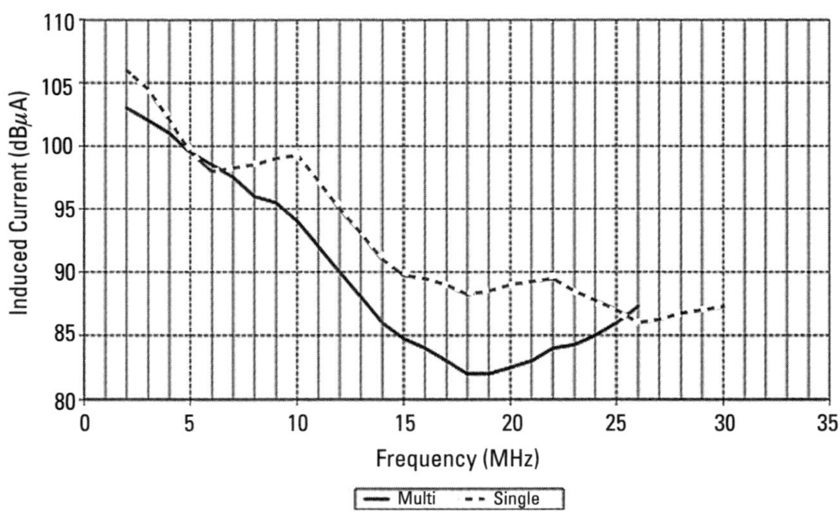

Figure 4.29 Comparison of single-port and multiport injection.

It can be observed that the multiport injection resulted in as much as 8 dB difference in the induced current in the SUT.

4.7 Summary

As has been discussed, abstraction from the ideal case is often unavoidable for HPEM effects testing. Compromises must be made with respect to the scenario, the simulation of the HPEM environment and the SUT in order to make an effects test affordable and manageable. A test or experiment is often essential though analytical and computational modeling of coupling can be very useful for exploring small deltas such as changes to the SUT orientation or the position of the SUT above the ground plane. An independent body in the United States, the National Research Council carried out a major review of HEMP effects assessment including modelling and testing [65]. This was a very thorough study which identified that, "...repeated testing of systems and sub-systems at as high a simulated threat level as possible is essential..." The Council also advocated the use of statistics in testing to "...collect, analyse, interpret and present the test data together with the associated uncertainties..."

Many different test methods have been developed and the capabilities to carry out effects testing continue to be improved.

Tight control of many variables is essential, still even with such controls in place the variability or uncertainty in an effects test result could easily still be of the order of ± 15 dB at a CL of 95%. Well controlled testing with as close as practical simulation of the environment and the scenario of very large sample sets can clearly improve this situation perhaps tending to a reduction in the effects result uncertainty to perhaps ± 6 dB at a confidence level of 95%. However, to date, there have been very few examples of large sample-set effects testing.

For now, the only practical way to minimize uncertainty or variability is to test as close as possible with the actual situation of interest (i.e., the actual scenario, the actual HPEM environment, and the actual SUT). Attempts to extrapolate the results of any given effects test to other situations should be considered with caution and a large uncertainty budget should be applied.

References

[1] Pywell, M., and M. Midgley-Davies, "Aircraft-Sized Anechoic Chambers for Electronic Warfare, Radar and Other Electromagnetic Engineering Evaluation," *The Aeronautical Journal 1*, Royal Aeronautical Society, 2017.

[2] IEC 61000-4-33: 2004, "Electromagnetic Compatibility (EMC) - Part 4-33: Testing and Measurement Techniques - Measurement Methods for High Power Transient Parameters," 2004.

[3] IEC 61000-4-35: 2009, "Testing and Measurement Techniques – HPEM Simulator Compendium," 2009.

[4] IEC 61000-4-32: 2001, "Testing and Measurement Techniques – HEMP Simulator Compendium," 2001.

[5] Roemelt, S., "Electrical Systems Engineering & Integration in AIRBUS," *ICAS Biennial Workshop 2015*, Krakow, Poland, August 31, 2015.

[6] Eichwald, L. G., *EMP Electronic Design Handbook*, AFWL-TR-74-58, April 1973.

[7] LAB34, *The Expression of Uncertainty in EMC-Testing*, Edition 1, United Kingdom Accreditation Service (UKAS), Feltham, U.K., August 2002.

[8] Van Troyen, D. L. R., and F. Nauwelaerts, "Uncertainties in EMC — Calibration and Testing," *Progress in Electromagnetics Research Symposium (PIERS-Toyama)*, 2018, pp. 220–226.

[9] MIL-STD-464C, Department of Defense Interface Standard – Electromagnetic Environmental Effects – Requirements for Systems," USA, December 1, 2010.

[10] ESA ECSS-E-ST-20-07C Rev. 1, 'Space Engineering – Electromagnetic Compatibility," February 7, 2012.

[11] DEF STAN 59-411 Part 3, "MOD, Electromagnetic Compatibility, Part 3," *Equipment Level Test Techniques*, UK, Issue 2, March 31, 2007.

[12] IEC 61000-4-36, "Electromagnetic Compatibility (EMC) – Part 4-36: Testing and Measurement Techniques – IEMI Immunity Test Methods for Equipment and Systems," 2019.

[13] ANSI C63.4, "Methods of Measurement of Radio-Noise Emissions from Low-Voltage Electrical and Electronic Equipment in the range of 9 kHz to 40 GHz," June 20, 2014.

[14] Crawford, M. L., "Generation of Standard EM Fields Using TEM Transmission Cells," *IEEE Transactions on Electromagnetic Compatibility*, Vol. EMC-16, No. 4, November 1974, pp. 189–195.

[15] Hansen, D., et al., "Device for the EMI Testing of Electronic Systems," U.S. Patent No. 4,837,581, June 1989.

[16] Hemming, L. H., *Electromagnetic Anechoic Chambers: A Fundamental Design and Specification Guide*, New York: John Wiley & Sons, 2002.

[17] Montrose, M. I., and E. M. Nakauchi, *Testing for EMC Compliance: Approaches and Techniques*, New York: John Wiley & Sons, 2004.

[18] Hill, D. A., "Electromagnetic Theory of Reverberation Chambers," NIST Technical Note 1506, December 1, 1998.

[19] International Electrotechnical Commission 61000-4-21, "Electromagnetic Compatibility (EMC) – Part 4-21: Testing and Measurement Techniques – Reverberation Chamber Test Methods," 2003.

[20] Borgstrom, E. J., "A Comparison of Methods and Results Using Semi-Anechoic and Reverberation Chamber Radiated RF Susceptibility Test Procedures in RTCA/D)-160D, Change One," *IEEE International Symposium on EMC*, Santa Clara, CA, August 2004.

[21] Arnaut, L. R., and P. D. West, *Evaluation of the NPL Untuned Stadium Reverberation Chamber Using Mechanical and Electronic Stirring Techniques*, NPL report CEM 11, August 1998.

[22] Musso, L., "Assessment of Reverberation Chamber Testing for Automotive Applications," Ph.D. Thesis, Politecnico Di Torino, February 2003.

[23] Freyer, G. J., and M. G. Backstrom, "Comparison of Anechoic and Reverberation Chamber Coupling Data as a Function of Directivity Pattern," *IEEE International Symposium on EMC*, Washington, D.C., August 2000.

[24] RTCA DO160/D Change Notice 1, Section 20.6, "Radiated Susceptibility (RS) Test; Alternate Procedure – Reverberation Chamber," December 14, 2000.

[25] Baum, C. E., "EMP Simulators for Various Types of Nuclear EMP Environments: An Interim Characterization," *IEEE Transactions on Electromagnetic Compatibility*, February 1978, pp. 35–53.

[26] Baum, C. E., "Prolog to 'From the Electromagnetic Pulse to High-Power Electromagnetics," *Proceedings of the IEEE*, Vol. 80, No. 6, June 1992.

[27] Miletta, J. R., R. J. Chase, and B. Luu, "Modeling of Army Research Laboratory EMP Simulators," *IEEE Transactions on Nuclear Science*, Vol. 40, No. 6, December 1993.

[28] Baum, C. E., "Resistively Loaded Radiating Dipole Based on a Transmission-Line Model for the Antenna," *Sensor and Simulation*, Note 81, April 1969.

[29] Baum, C. E., "Review of Hybrid and Equivalent Electric Dipole EMP Simulators," *Sensor and Simulation Notes*, Note 277, October 1982.

[30] Dong Fang, Z., et al., "The Application of Electromagnetic Topology in the Analysis of HPM Effects on Systems," *6th International Symposium on Antennas, Propagation and EM Theory*, October 2003, pp. 630–633.

[31] Baum, C. E., et al., "Sensors for Electromagnetic Pulse Measurements Both Inside and Away from Nuclear Source Regions," *IEEE Transactions on Electromagnetic Compatibility*, Vol. EMC-20, No. 1, 1978, pp. 22–35.

[32] Kancleris, Z., et al., "Recent Advances in HPM Pulse Measurement Using Resistive Sensors," *Digest of Technical Papers, 14th IEEE International Pulsed Power Conference (PPC-2003)*, Vol. 1, 2013, pp. 189–192.

[33] Baum, C. E., "Electromagnetic Topology: A Formal Approach to the Analysis and Design of Complex Electronic Systems," *AFRL Interaction Notes*, Note 400, October 1980.

[34] Parmantier, J. P., et al., "Electromagnetic Topology: Junction Characterization Methods," *AFRL Interaction Notes*, Note 489, May 1990.

[35] Tesche, F. M., et al., "Measurements of High-Power Electromagnetic Field Interaction with a Buried Facility," *Proceedings of the International Conference on Electromagnetics in Advanced Applications*, Torino, Italy, September 10–14, 2001.

[36] Audone, B., M. Audone, and I. Marziali, "On the Use of the Minimum Phase Algorithm in EMC Data Processing," *International Symposium on Electromagnetic Compatibility - EMC EUROPE*, 2012, pp. 1–6.

[37] Wraight, A., "Improvements in Electromagnetic Assessment Methodologies: Bounding the Errors in Prediction," Ph.D. Thesis, University of Cranfield, U.K., December 2007.

[38] IEC 61000-5-9 Ed. 1, "Electromagnetic Compatibility (EMC) -- Part 5-9: Installation and Mitigation Guidelines - System Level Susceptibility Assessments for HEMP and HPEM," 2009.

[39] EUROCAE ED107, "Guide for the Certification of Aircraft in a High Intensity Radiated Field (HIRF) Environment," March 2001.

[40] AEP 4, "Electromagnetic Environmental Effects (E3) Committee. ARP5583 - Guide for the Certification of Aircraft in a High Intensity Radiated Field (HIRF) Environment," January 2003.

[41] Prather, W. D., and C. E. Baum, "Elliptic CW Antenna Design," *Miscellaneous Simulator Memos*, Memo 22, March 1987.

[42] Prather, W. D., "Aircraft EMP Hardening in the 21st Century - Aircraft EMP Hardening as a Part of an Integrated E3 Design," *Proceedings of EUROEM 2016*, London, U.K., July 2016.

[43] IEC 61000-4-25: 2002, "Electromagnetic Compatibility (EMC) - Part 4-25: Testing and Measurement Techniques - HEMP Immunity Test Methods for Equipment and Systems," 2002.

[44] Hoeft, L. O., et al., "Upset Thresholds of Various Systems as Measured by the R2SPG Technique," *Symposium Record. Compatibility in the Loop, IEEE International Symposium on EMC*, 1994, pp. 264–268.

[45] Golikov, R. Y., V. M. Kondratiev, and Y. F. Chibisov, 'Simulation of Early HEMP Impact on Distribution Power Lines Under Working Voltage," *International Symposium on Electromagnetic Compatibility*, September 2002.

[46] IEC 61000-1-3: 2003, "Electromagnetic Compatibility (EMC) — Part 1-3: General — The Effects of High-Altitude EMP (HEMP) on Civil Equipment and Systems," 2003.

[47] Wellington, A. M., "Direct Current Injection as a Method of Simulating High Intensity Radiated Fields (HIRF)," *IEE Colloquium on EMC Testing for Conducted Mechanisms*, 1996.

[48] Zhang, B., and U. Jiang, "Research Progress of Direct Current Injection Technique in Aircraft EMC Test," *3rd IEEE International Symposium on Microwave, Antenna, Propagation and EMC Technologies for Wireless Communications*, 2009, pp. 843–849.

[49] IEC 61000-4-4, "Electromagnetic Compatibility Electrical Fast Transient/Burst Immunity Test," 2012.

[50] https://www.emc-partner.com/products/immunity/eft-burst-surge/eft-burst-accessories/cn-eft1000.

[51] Vukicevic, A., et al., "On the Evaluation of Antenna-Mode Currents Along Transmission Lines," *IEEE Transactions on Electromagnetic Compatibility*, Vol. 48, No. 4, November 2006, pp. 693–700.

[52] Parfenov, Y. V., et al., "Research of Flashover of Power Line Insulators Due to High-Voltage Pulses with Power ON and Power OFF," *IEEE Transactions on Electromagnetic Compatibility*, Vol. 55, No. 3, 2013, pp. 467–474.

[53] Directive 2013/35/EU of the European Parliament and of the Council on the Minimum Health and Safety Requirements Regarding the Exposure of Workers to the Risks Arising from Physical Agents (Electromagnetic Fields), June 26, 2013.

[54] International Commission on Non-Ionizing Radiation Protection, "Guidelines for Limiting Exposure to Time-Varying Electric, Magnetic, and Electromagnetic Fields (Up to 300 GHz)," *Health Physics*, Vol. 74, No. 4, 1998, pp. 494–522.

[55] IEEE Std. C95.6-2002, "IEEE Standard for Safety Levels with Respect to Human Exposure to Electromagnetic Fields, 0-3 kHz," 2002.

[56] Kohler, S., et al, "Simultaneous High Intensity Ultrashort Pulsed Electric Field and Temperature Measurements Using a Unique Electro-Optic Probe," *IEEE Microwave and Wireless Components Letters*, Vol. 22, No. 3, 2012, pp. 153–155.

[57] Yan, J., et al., "Performance Investigation of VCSEL-Based Voltage Probe and Its Applications to HPEM Effects Diagnosis of Embedded Systems," *IEEE Transactions on Electromagnetic Compatibility*, Vol. 60, No. 6, 2018, pp. 1923–1931.

[58] LoVetri, J., A. T. M. Wilbers, and A. P. M. Zwamborn, "Microwave Interaction with a Personal Computer: Experiment and Modelling," *Proceedings of the 1999 Zurich EMC Symposium*, 1999.

[59] Hoad, R., et al, "Trends in EM Susceptibility of IT Equipment," *IEEE Transactions on Electromagnetic Compatibility*, Vol. 46, No. 3, August 2004.

[60] Nitsch, D., et al., "Susceptibility of Some Electronic Equipment to HPEM Threats," *IEEE Transactions on Electromagnetic Compatibility*, Vol. 46, No. 3, August 2004.

[61] Camp, M., and H. Garbe, "Susceptibility of Personal Computer Systems to Fast Transient Electromagnetic Pulses," *IEEE Transactions on Electromagnetic Compatibility*, Vol. 48, No. 4, November 2006.

[62] Li, K. J., et al., "Multinomial Regression Model for the Evaluation of Multilevel Effects Caused by High-Power Electromagnetic Environments," *IEEE Transactions on Electromagnetic Compatibility*, Vol. 61, No. 1, February 2019.

[63] Esser, N., and B. Smailus, "Measuring the Upset of CMOS AND TTL Due to HPM-Signals," *Digest of Technical Papers, 14th IEEE International Pulsed Power Conference (PPC-2003)*, Vol. 1, 2003, pp. 471–473.

[64] Hoad, R., "The Utility of Electromagnetic Attack Detection to Information Security," Ph.D. Thesis, University of Glamorgan, December 2007.

[65] Pierce, J. R., "Evaluation of Methodologies for Estimating the Vulnerability to Electromagnetic Pulse Effects," National Research Council, SDAN 0027, 1984.

5

HPEM Effects Mechanisms

5.1 Introduction

HPEM effects on electrical or electronic systems have been and continue to be an important area of research. An understanding of HPEM effects is enhanced through an understanding of HPEM environments (Chapter 2), HPEM interaction and coupling (Chapter 3) and the constraints, limitations and uncertainties posed during effects testing (Chapter 4).

Evidence for HPEM effects range from empirical observations after an often unexpected event or system failure, for example, the Forrestal disaster discussed previously through to observations from controlled systematic effects testing or experiments. The scientific experimental research has revealed and continues to reveal the underlying effects mechanisms, but this area of research is still relatively immature. The various terms used to describe HPEM effects are not precisely defined at this stage and therefore a common understanding can be more difficult to achieve. Also in the last 25 years or so, there has been a tendency for researchers to normalize, effectively obscure, or hide important parameters of an effect when publishing data. The reasons that this normalization or obscuration of important effect parameters is applied seems to be motivated by either security concerns (i.e., the parametric data may reveal a weakness in a system that is in use) or intellectual property concerns (i.e., the parametric data may indicate that a system from one manufacturer compares unfavorably with another or the data may reveal a particular ability of a particular HPEM effector). These reasons are valid, but there is no doubt that this has hampered the development of a common language around effects mechanisms and made it more difficult to form a uniform view or at least a consensus in the understanding of effect mechanisms.

It is worth reiterating, as we have reiterated throughout this book, small changes in variables such as coupling, degree of system under test (SUT) abstraction from normal operation, and system state at the time of exposure lead to very large variability in an effects test result.

5.2 Terminology

In this book, we define an HPEM effect as: "Any temporary or permanent manifestation at the system due to the HPEM exposure of a system, where a system in this context could be any item in the system of systems hierarchy (i.e. device, circuit, equipment, system, network or infrastructure)."

The terms immunity, susceptibility, and vulnerability are also well-used terms in the context of HPEM effects. The IEC definition [1] of immunity is: "The ability of a device, equipment or system to perform without degradation in the presence of an electromagnetic disturbance."

In EMC standards, it is very possible for a system to be declared as having immunity to EM disturbance but still have experienced an effect. For example, if during HPEM exposure a display screen freezes momentarily, then this could still be judged as immune if the degradation (duration of the freeze) is not considered to significantly affect the performance of the system.

The IEC definition of susceptibility is: "The inability of a device, equipment or system to perform without degradation in the presence of an electromagnetic disturbance. NOTE: Susceptibility is a lack of immunity." As stated susceptibility is the lack or the absence of immunity. Clearly, with respect to this definition, *susceptibility* is synonymous with *effect*. However, a point of confusion emerges because the military and defense communities use the term susceptibility to mean what the civilian world calls immunity. Within the military EMC standards, the term susceptibility testing describes a test of EM exposure up to a limit. During a test to the EM exposure limit, if the system does not demonstrate an effect, then it is considered to have passed the susceptibility test.

There is no formal definition of vulnerability in the IEC or the military context; however, from the literature, one could consider the definition of vulnerability to be: "Disruption or damage of a system function or mission such that the impact on the function is considered critical."

In the context of vulnerability, the system experiences an effect and that effect has the consequence of disrupting or damaging the critical function of the system. Functions that experience effects (e.g., a warning indicator illuminates in an aircraft cockpit) that are not critical (e.g., do not cause the aircraft to crash) are not considered as vulnerable.

In this book, we have used the more generic term of HPEM *effects* because the definitions of immunity, susceptibility, and vulnerability are clearly highly subjective, in part confused between military and civilian meanings and because they are highly dependent on the system function.

5.2.1 About This Chapter

The scientific understanding of HPEM effects is best explored by utilizing specially developed test and analytical techniques. An understanding of the mechanism that drives a system to manifest an HPEM effect is of vital importance, and this chapter aims to explore and expand on some of the more well-known HPEM effect mechanisms.

In the beginning of this chapter, the main types of HPEM effects at the circuit and component level of the hierarchy are described and some examples are given. The effect mechanisms can be summarized as follows:

- Rectification;
- Noise;
- Interference;
- Saturation;
- False information;
- Transient upset;
- Shift of operation point;
- Chaotic behavior;
- Damage and destruction;
- Latch-up, punch-through, and thermal.

Due to the design and functionality of electronics, the employed HPEM environment, and the overall setup, the observed effects differ significantly from each other. For example, if a given system is exposed to different HPEM waveforms, different HPEM effects are caused. The reason for this is that the different HPEM effects mechanisms depend on different characteristics of the HPEM environment.

In many instances in EM testing of systems, it is necessary to compare several measurements to infer some important property of the SUT. For example, in a shielded electrical system, measurements of the internal electric (E) and magnetic (H) fields may be made, and an evaluation of suitable waveform characteristics made to determine if there is adequate protection installed in the facility. Typically, such measurements can be performed in the time domain, for which a transient record of the measured quantity is recorded. Conversely, continuous-wave (CW) measurements can be performed, resulting in a wideband spectrum consisting of a real and imaginary part (or magnitude and phase). Such responses can be rather complex, with rapid oscillations and large variations from one part of the response to another.

A difficulty arises in trying to compare one particular measured data record with another. Usually, the human eye can easily compare various data records and see subtle differences contained in the data. However, this approach is inefficient, especially when large data sets need to be analyzed in a consistent manner. Furthermore, the human interpretation is qualitative in nature, which leads to a difficulty in trying to make quantitative statements about how similar (or different) one response is to another.

In an attempt to describe the similarities and differences in measured responses, Baum suggested the use of mathematical norms [2]. These are mathematically well-defined numbers (i.e., scalar quantities) that may be easily calculated from the measured transient or spectral data. Other types of response characterizations are also possible, and although they are not norms in a strictly mathematical sense, they are equally useful in distinguishing one response from another.

Later in this chapter, we will consider both the mathematically well-defined norms and the other parameters that are not norms in a strict mathematical sense, but that are useful in describing a response. Here signal indicators are used to refer to both mathematical norms, as well as other waveforms or spectral scalar quantities.

Finally, we discuss the correlation between signal indicators and HPEM effects mechanisms.

5.3 Device and Circuit-Level Effects

Since at least the mid-1960s, a variety of HPEM effects observed in HPEM tests have been reported. The mechanisms that largely manifest at the device or component and circuit levels of the hierarchy are summarized here.

5.3.1 Rectification

Rectification is one of the primary mechanisms whereby energy contained in an out-of-band HPEM signal is converted to an in-band signal that can cause effects in electronic systems. Rectification, in simple terms, is the conversion of an alternating current (AC) into a direct current (DC) [3]. It can also lead to the creation of intermodulation or harmonics. Rectification is a vital process in all electronic devices without exception. In order to produce rectification, nonlinear devices are often employed and the simplest example is that of the semiconductor junction implemented in a diode. An ideal diode would be able to rectify all AC signals and create pure DC signals without distortion over an infinitely wide bandwidth; however, in practice, implementing a diode on a semiconductor wafer and making vital conductive connections to the junction lead to the addition of parasitic capacitance, which moves the diode performance far from the ideal case.

Importantly, nonlinear diode-like junctions can also be formed unintentionally or passively (e.g., rusty bolt effect) [4] or intentionally (e.g., ESD protection diodes) at various levels within a device or circuit. These unintentional nonlinear junctions produce a diode-like response to RF rectifying and demodulating the RF signal. This is shown in Figure 5.1.

Rectification is therefore one of the primary mechanisms through which HPEM signals can be converted to in-band signals that are then processed by the electronic devices.

If a biased AC voltage occurs across a PN junction such as a diode, the nonlinear current-voltage characteristic causes the current to flow in a series of nonsymmetrical pulses. The average magnitude of the pulses is higher than the DC value of the current that would exist if only the bias voltage were present. Therefore, it appears that the signal has caused a DC offset current to flow. This effect gives rise to an RF-induced diode characteristic that is different to the normal diode characteristic.

The nonlinear rectification effects of PN junctions can be observed in op-amps when exposed to high frequencies. The output voltage begins to exhibit fluctuations even when the inputs are held constant. The effects cause demodulation of out-of-band signals. For example, an unmodulated carrier produces a DC shift in the output; an AM or FM modulated carrier would produce an AM-modulated output or a pulse-modulated carrier would produce a pulsed output tracking the RF interference signal modulation. The modulated

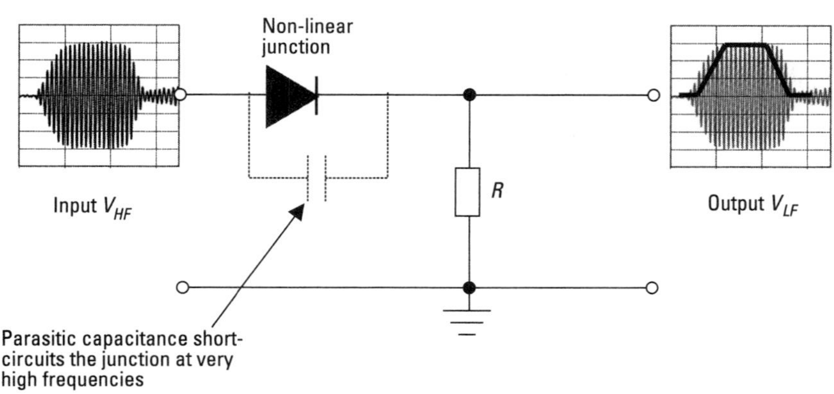

Figure 5.1 Rectification of an RF signal.

RF interference signal is demodulated. Then the demodulated envelope can be amplified or decoded because it is within the bandwidth of subsequent signal processing blocks. However, many of the parameters that drive the rectification process are difficult to quantify and or extremely variable in a complex system. Any analog amplifier or digital device can be shown to have an in-band and out-of-band response to EM interference. This was discussed in Chapter 5 and shown diagrammatically in Figure 3.4.

5.3.2 Noise

Even very low-magnitude EM environments can add additional noise to a signal line or a receiver. If the signal is weak, the receiver might be able to correctly detect bits so that bit errors can be restored. However, additional noise reduces the signal-to-noise ratio (SNR) and might therefore decrease the usable data rate. Analog circuits are more sensitive to signal noise. Therefore, an induced noise signal on signal or power lines of analog circuits might cause noisy displays or flashing of displays.

RF receivers are becoming more commonly integrated into electronic systems, for example, for wireless communication or for command control and telemetry such as an aircraft instrumented landing system (ILS) or for precision timing and navigation such as GPS/GNSS. RF receivers are designed to collect very low-level RF signals (as low as μV/m levels) within a narrowly tuned receiver bandwidth. It is very easy to overpower the signal to be received, by a decade or more of field strength (noise jamming). The user of the electronic device or equipment merely experiences noise in the receiver that lasts as long as the disturbing environment is present. The consequences of such a disturbance are unlikely to be critical to the function as there are likely to be backup or manual reversionary methods available. In the worst-case scenario, for an aircraft on approach to an airfield, which is being assisted by ILS, the pilot may abort landing and make another try or go to an alternate airport.

An increased noise level can result from both front-door and backdoor coupling.

5.3.3 Interference or Jamming

Interference or jamming occurs if the interfered signal causes a noise background that masks the wanted signal by some, usually small, margin. A jamming or interference signal only needs to be a few decibels more than the wanted signal and the wanted signal may be very small, perhaps as small as a few tens of μV/m or a few pW/m^2 [5–8].

Due to their extremely high sensitivity, GPS/GNSS receivers are particularly vulnerable to jamming. The extreme sensitivity of GPS receivers is because satellites have a limited transmission power and signals have to travel from orbit to Earth and thus are very weak. As a rule of thumb, an interference signal that exceeds the normal signal level of the receiver by 20 dB to 30 dB is sufficient to block the communication of a GPS receiver [9]. In [9–11], it was estimated that a field strength of the order of only tens of μV/m is sufficient to block or jam a GPS receiver. As a comparison, levels of in-band interference of WLAN have been reported to be around 1 V/m.

For modern digital modulation schemes, such as differential quadrature phase shift keying (DQPSK), jamming does not have to simply be a power battle. Intelligent jamming can exploit the processing or decoding function of the receiver creating the same effect, for example, by just masking the preamble or cueing bit of the wanted communications signal so that the back-end process does not receive a cue to begin processing a data packet. When the interfering signal is removed normal communications with the wanted signal is immediately restored.

As the coupling efficiency to a front-door, front-end system can be considerably higher than coupling via backdoors and because front-end components can have a much higher sensitivity to HPEM environments. Usually, interference results from the front-door coupling of an HPEM environment to front-end electronics.

5.3.4 Saturation

Saturation of receiver components, such as the low noise amplifier (LNA) or the mixer, requires much higher magnitude input signals than for interference or jamming. Saturation implies that the LNA goes into compression and results in nonlinearity and hence desensitizes the receiver and may cause production of spurious modulation and intermodulation products [12]. The compression point for an amplifier, denoted as the P1dB point, is the point where the amplifier output no longer linearly increases with a linear increase in input power. The saturation effect may persist for a short time after the interfering signal is removed. The P1dB point or saturation point for a given LNA chip is a known parameter and is invariably included within the data sheet or specification for the device.

This type of effect mechanism can also occur in backdoor coupling, particularly in analog circuit components with feedback such as op-amps.

5.3.5 Shift in Operating Point

The behavior of digital and most analog electronic circuits and components are characterized by nonlinear junctions. During the circuit design, the chosen operating point for these components are set so that they operate in a linear regime. Interference in the power distribution system of electronic components and circuits are capable of causing a change of the potential at the electronic component and into a shift of the operating point of the component. As an immediate consequence, the real behavior of the component deviates from the linear approximation. A shift of the operating point of a digital-to-analog converter results in wrong data. Other effects are compression of analog data and dc offset of data.

5.3.6 False Information

If a transient interference signal reaches the signal level of the system, the EM interference signal may feed false information to the data stream or alternate information bits of the data stream. The intentional creation of such a situation is known as spoofing.

Consequences can be critical, as the exposed system operates on corrupted or false information. During HPEM tests, the corruption of data streams resulted in a malfunction of the system or component up to a hang-up or crashing of software and breakdown of the SUT. Due to the required level of the interfering HPEM environment and the relatively low efficiency of backdoor coupling, false information results mainly from front-door coupling.

5.3.7 Transient Upset

It is noted that it typically requires an induced transient signal to have a comparable magnitude to a logic state (i.e., several volts of signal) to affect the logic state of an electronic component. At a frequency of ~1 GHz, an effective coupling height of 0.1m is typical for an unprotected circuit or component. This implies tens to hundreds of V/m of tuned narrowband environment is capable of coupling sufficient magnitude to cause an effect, if the pulse width is assumed to be such that the quality factor Q of the HPEM environment is greater than the victim system Q [13–15]. At the nominal frequency of 1 GHz, approximately 100 cycles or a 100-ns pulse duration has been shown to induce a transient in a circuit at a comparable magnitude to the wanted signal and therefore was sufficient to produce an effect.

The consequences of a transient upset depend on the system design and features of the HPEM waveform such as the repetition rate or burst duration. Some examples of transient upset for a single transient in a digital sequence are shown in in Figures 5.2(a–g).

Figure 5.2(a) shows the base state, a normal digital pulse train sequence. Figure 5.2(b) shows the case where a digital 1 is masked or suppressed by the interfering signal. This could also be the case that an expected 0 results in a 1. Figure 5.2(c) demonstrates a case where the interfering signal introduces a false bit, a 1 in this instance. In practice, the digital circuit response to this false information will depend upon whether the magnitude and duration and even the timing of the false bit are within the expectation of the digital processor. For

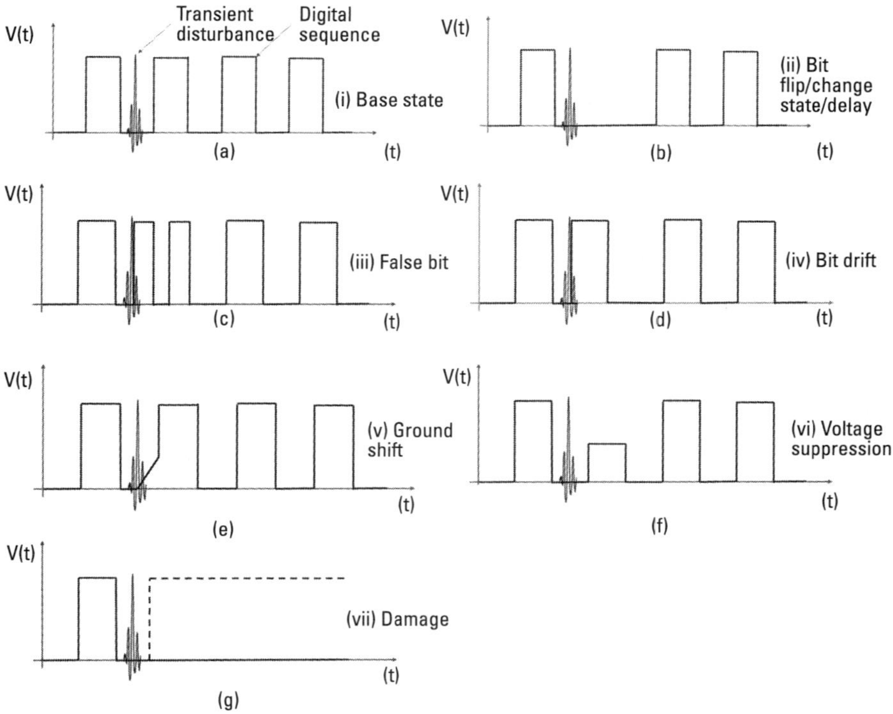

Figure 5.2 (a–g) Examples of the effect response on a digital sequence.

example, many processors employ digital filtering, gating, or other digital signal processing techniques to reject false bits that do not occur at the expected time. Figure 5.2(d) is similar to the case previously though here the magnitude and width of the false bit are comparable to a true bit. Figure 5.2(e) shows a case of a shift in the ground or zero-volt state of the circuit during a bit period. This could be produced by a difference in ground potential across the circuit. Figure 5.2(f) shows a case that, for the processor, would appear to manifest the same as the bit or flip of Figure 5.2(b). Here the magnitude of the bit is suppressed below the expected magnitude. Figure 5.2(g) shows the case where the circuit is damaged. The example here shows that the digital pulse sequence is latched in the high or 1 state, although it is possible for the sequence to be latched in the low or zero state or even in some indeterminate state if, for example, the impedance of the device at either end of the circuit is changed through the damage.

5.3.8 Chaotic Effects

It has been shown by several authors [16–20] that circuits that incorporate feedback to regulate their linear behavior such as phase locked loops (PLLs), automatic gain control (AGC), and even operational amplifiers can be nudged into a chaotic or random state through HPEM exposure. A PLL is a circuit implementation of several components that generate an output signal whose phase is related to the phase of an input signal. PLLs are widely used for synchronization purposes in applications such as demodulation, recovery of clock signals, clock multipliers, and digital signal processing.

Nonlinear systems such as PLLs can exhibit a wide variety of complex behavior including subharmonic, quasiperiodic, and chaotic dynamics. It has been shown that random noise and resonant perturbations can shift the PLL into a chaotic state such that the locking is disrupted.

5.3.9 Damage and Destruction

Damage typically requires higher power than saturation and much higher power than jamming or interference. Some data regarding the damage threshold of RF receivers have been published [21]. These data indicate that power densities of 0.1 mW/m^2 to 10 mW/m^2 and received energy of 20 nJ for a single 100-ns pulse are sufficient to cause in-band damage to the receiver detector. Other studies of permanent damage have shown that pulse energies typically of the order of μJ, corresponding to a field strength of the order of 1 kV/m or less, are sufficient to destroy a receiver [22–25]. The most vulnerable component in a modern receiver seems to be the LNA [26, 27]; once the LNA is damaged, the receiver cannot function, so the effect is permanent until the receiver is repaired or replaced. For an experimental SiGe LNA device, a damage level as low as 20 nJ has been reported [9].

Insulating layers of printed circuit boards or components (e.g., integrated circuits (ICs)) can be destroyed by HPEM waveforms. If the dielectric insulation layer between traces or inside the die of the IC is too thin, induced voltage signals can cause an electrical dielectric breakdown. The resulting component destruction effects range from flashover effects, melted traces, and bond wire destruction. Figure 5.3(a–d) shows some examples of electron microscope images of damage to semiconductor components [private communication with J. Bohl, HPE 201 Short Courses, Switzerland, 2009.]

The identification of which destruction mechanism occurs requires additional detailed examinations of the system (e.g., removing dielectric packaging of the component to expose the semiconductor components and detailed on-chip investigation).

5.3.9.1 Latch-Up

A latch-up is a type of short circuit that can occur in an IC. More specifically, it is the inadvertent creation of a low-impedance path between the power supply rails of a MOSFET circuit, triggering a parasitic structure that disrupts proper

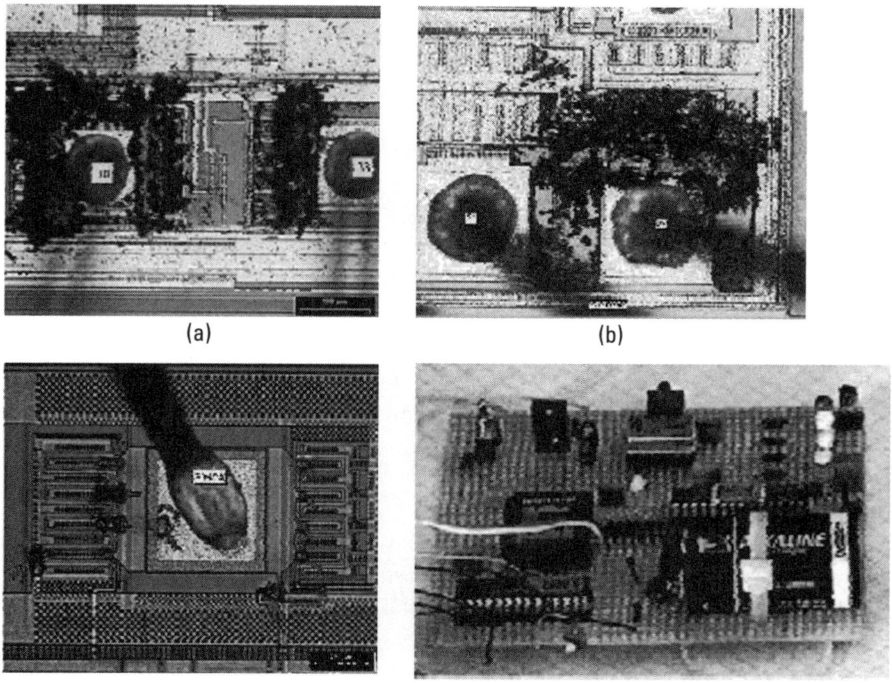

(a) (b) (c) (d)

Figure 5.3 (a–d) Examples of thermal damage to unprotected electronics.

functioning of the part, possibly even leading to its destruction due to overcurrent [28].

The parasitic structure is usually equivalent to a thyristor, a PNPN structure that acts as a PNP and an NPN transistor stacked next to each other. During a latch-up when one of the transistors is conducting, the other one begins conducting too. They both keep each other in saturation for as long as the structure is forward-biased and some current flows through it, which usually means until a power-down. The latch-up can happen at any place where the required parasitic structure exists. A common cause of latch-up is a positive or negative voltage spike on an input or output pin of a digital chip that exceeds the rail voltage by more than a diode drop. Another cause is the supply voltage exceeding the absolute maximum rating, often from a transient spike in the power supply. It leads to a breakdown of an internal junction. The possible sources of a latch-up are:

- Minority carriers injected into the substrate by a transient forward bias on parasitic PN junctions (typically in the input or output circuits);

- Photoelectric generation from ionizing radiation;
- Impact generation from thermal heating.

Generally, a latch-up leads to the burnout of the semiconductor junctions.

5.3.9.2 Punch-Through

Punch-through in a MOSFET is an extreme case of channel length modulation where the depletion layers around the drain and source regions merge into a single depletion region. The field underneath the gate then becomes strongly dependent on the drain-source voltage, as is the drain current. Punch-through causes a rapidly increasing current with increasing drain-source voltage. This effect is undesirable as it increases the output conductance and limits the maximum operating voltage of the device [28].

5.3.9.3 Thermal Damage Mechanisms

The simplest model for damage assumes that the damage mechanism is fast and there is no thermal interaction with the substrate, such that the electrical energy (E) is equal to the thermal energy (Q) [29]. This is often referred to as the fuse-melting model, (5.1) to (5.4).

$$I^2 \cdot R \cdot \Delta t = m \cdot c \cdot T \tag{5.1}$$

where I is the current required to fuse the thin wire (A), R is the electrical resistance of the wire (Ω), Δt is the pulse length, m is the mass of the heated area (Kg), c is the specific thermal capacity of the material (J/Kg.K), T is the melting point of the material (K), κ is the thermal conductivity (Ω^{-1} m^{-1}), A is the area of the heated region (m^2), and ρ is the density of the heated material (kg^{m-3}) where

$$R = \frac{l}{\kappa \cdot A_q} \tag{5.2}$$

and

$$m = \rho \cdot A_q \cdot l \tag{5.3}$$

Therefore,

$$I = \left[\frac{\rho \cdot A_q^2 \cdot c \cdot T \cdot \kappa}{\Delta t} \right]^{\frac{1}{2}} \tag{5.4}$$

This model is very simple and does not include many features of HPEM exposure such as pulse width, pulse repetition frequency, and burst duration, which are very important. The analysis of this model does not correlate well with experience from HPEM effect testing; still, it is a useful initial indicator of some of the parameters in play. More detailed analysis of thermal damage mechanisms as they relate to signal indicators is provided later in this chapter.

5.3.10 Published Device and Circuit-Level Effects Data

Extensive research on HPEM effects on devices by Göransson [30] has shown that, with digital components, there can be large differences in effect level and types between different technologies. Göransson observed small differences between different samples of the same type from the same manufacturer. Differences of up to 16 dB in component effect levels were noted for different manufacturers. Göransson also noted a very strong frequency dependence in the component effects, with the effect threshold level increasing rapidly with increasing frequency. Göransson evaluated two fundamental digital technology types: Transistor-Transistor Logic (TTL) and Complementary Metal Oxide Semiconductor (CMOS). This is illustrated in Figure 5.4.

The chart in Figure 5.5 shows the outcome of a large experimental campaign carried out by Esser and Smailus studying CMOS devices where energy and duty cycle are used to evaluate the effect [31].

In the era of the Cold War, a significant effort was made to understand and quantify the failure and upset levels of components due to HEMP excitation. A

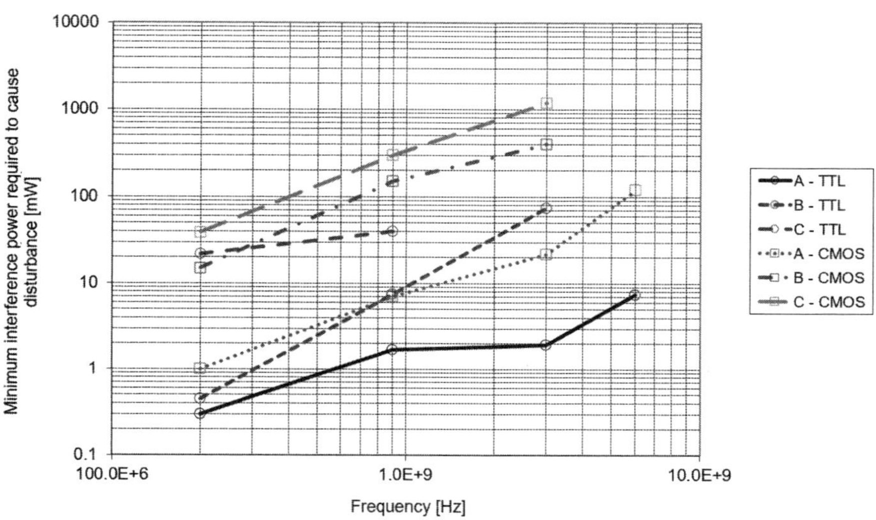

Figure 5.4 Effects on a TTL and CMOS quad 2-input NAND gate as a function of frequency.

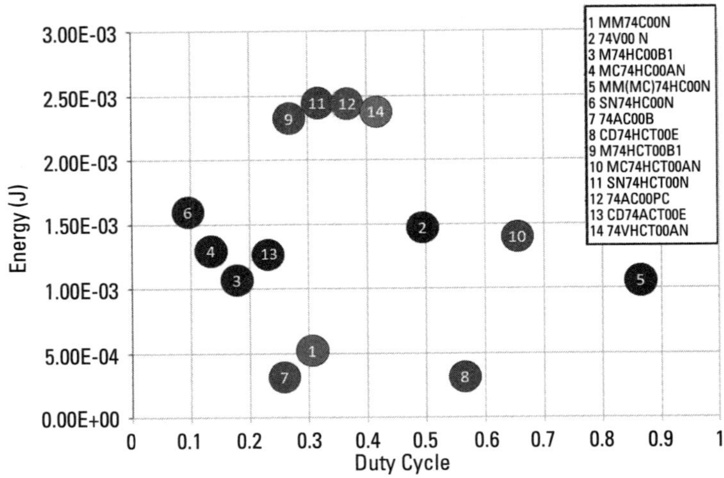

Figure 5.5 Energy density effect on components.

large amount of data relating to component damage due to HEMP exists [32], and the results of extensive testing on communications systems, power systems, and components are available, for example, [33, 34].

Neilsen [21] provided a simple tabulation of damage as a function of energy density (fluence) and power density (intensity), shown in Table 5.1.

Such simplified approximations must be considered with caution. Features of the device or component including parasitic features, the circuit layout, the coupling, and the HPEM signal indicators and other important parameters must be taken into consideration.

5.4 Equipment, System, and Network-Level Effects

It should be clear from the discussion on device-level effects that there are a considerable number of different mechanisms in play and that the degree of ab-

Table 5.1
Simplified Tabulation of Damage to Semiconductors

Type of Damage	Fluence Threshold (J/cm^2)	Intensity Threshold (W/cm^2)
In-band	10^5 to 10^{-2}	10^{-8} to 10^{-6}
Out-of-band		10 to 10^1
Thermal		10^3 to 10^4

straction from what we observe at an equipment or system level is quite acute. Equipment and system-level effects are perhaps easier to comprehend and testing is less abstracted from reality and therefore possesses lower uncertainty. However, at equipment level or higher levels of the system of systems hierarchy it is very difficult to glean exactly which effect mechanism from those listed previously is the actual contributor to the manifestation of the effect.

Chapter 6 looks at classification schemes that have been developed to try to systematically categorize observed effects. In this section, we review some of the published effects test data. The purpose of this section is to show that effect mechanisms are highly complex and that there are important variables in play that must be understood. This section is just a small summary of a vast number of effect studies that can be found in the literature.

The effects of HPEM environments on early personal computer systems was studied by Everett and Everett [35]. They used Hypoband HPEM to test in a mode-tuned chamber and looked for occurrences of digital circuit upset. A simple program to read into and fetch from RAM memory was executed as the device was being continuously illuminated. With each memory fetch, there was a check to see if the recalled number had been affected by the HPEM exposure. A changed number indicated that an upset had occurred. In some cases, the effect was dramatic in exhibiting a wild-running display. The maximum field strength of the test was 200 V/m (RMS), corresponding to an incident average power density of 10.6 mW/cm^2 = 106 W/m^2.

Hypoband HPEM exposure of several unshielded microprocessors and small computers (including the TRS-80 and the ZX81) indicated that upsets for microwave frequencies in the range of 1 GHz to 10 GHz were rare. However, upsets were frequent for frequencies in the range of 250 MHz to 1 GHz. The authors conjectured that the advent of faster microprocessors and small computers operating with clock rates in the gigahertz range, more upsets are expected for frequencies above 1 GHz. For the KIM-1 microprocessor, the equipment with the lowest effect threshold in the test, upsets were observed for field strengths as low as 2 V/m (RMS) and incident average power density of 106 $\mu W/cm^2$.

The Hypoband long-pulse (pulse widths greater than 200 ns) environment is of special interest, as it has been used in the evaluation of effects by several authors. Perhaps the first highly relevant article containing effects thresholds of computer systems was published in 1999 [36]. In this article, standard EMC-type TWT amplifiers were used to illuminate three separate tower computer systems via a standard double-ridge waveguide horn antenna.

The distance between the antenna and the SUT was 1m, and the maximum achievable peak field level was quoted as 100 V/m. A variety of modulation types and antenna polarizations were used. The SUT was exercised, processing data and video images and reading or writing to the hard disk. A variety of upset types or effects were observed including:

- Loss of data;
- Reset (where the computer restarted automatically);
- Disk write error (reported by the operating system);
- Loss of hard disk access (reported by the operating system, power-down required);
- Power-down (computer shuts down).

Another important source of Hypoband long-pulse effect data was provided by Bäckström [37–39]. Bäckström evaluated a variety of systems for effects against Hypoband (HPM) using the MTF facility described earlier and other conventional techniques. The systems tested included motor vehicles, computers, monitors, card readers, missiles, radios, and telecommunications systems. Of these, it is remarked that flat-screen monitors were damaged at 100 V/m at a pulse repetition frequency of 1 kHz and a pulse width of 0.5 ms (50%) duty cycle, and the center frequency appears to be 140 MHz. The card reader was disturbed at 80 V/m using the 1.3 GHz MTF; the upset type was to lock out users.

For the motor vehicle (a 1993 model), it was observed that engine stopping occurred at 500 V/m at 1.3 GHz to 3 GHz with a 5-μs pulse width and 200-Hz pulse repetition frequency. Damage occurred at 15 kV/m (1.3 GHz) and 25 kV/m (2.86 GHz). Damaged devices included engine control units and relays.

This considerable effort over many years of effects testing by Backstrom and his team enabled them to develop some general observations about Hypoband long pulse effects . General observations by Bäckström include:

- System effects are more prominent in the 1 to 3-GHz region than the 5 to 15-GHz region;
- Upset in the 1 to 3-GHz region starts to occur around a few hundred volts per meter (theoretically evaluated at 300 V/m);
- Permanent damage in the 1 to 3-GHz region starts to occur at 15 to 25 kV/m;
- Permanent damage can occur with the system turned off (unpowered);
- Damage to telecommunications receiver components (front-door coupling) at field strength of 2 kV/m.

Bäckström used the 300-V/m (upset) and 15-kV/m (damage) values as a benchmark for system effects to microwave Hypoband-type waveforms. However, it must be assumed that these magnitudes are only valid for the modulation scheme and the particular test configuration used.

Nitsch et al. [40] evaluated the effects of logic devices, micro-controllers, computer motherboards, computer systems, and networks. Hyperband, HEMP E1, and Hypoband waveforms have been generated, although specific details of the generator system are not provided. The Hyperband and HEMP tests have taken place in a TEM waveguide, and E-field strengths of 50 kV/m for EMP and 100 kV/m for the Hyperband waveform were achieved. The Hypoband tests appear to have taken place in a reverberation chamber where the maximum field achievable is 4 kV/m. Three different specifications of computer systems (386.25 MHz, 485.33 MHz, and 486.66 MHz) were tested. However, the hard disk for these systems was removed for the test and an external cable was attached to monitor the direct memory access (DMA) controller and programmable interval timer (PIT) module. This cable is likely to have violated any EM shielding provided by the system enclosure.

The upset criteria are not precisely defined but appear to relate to incorrect program function after interrogation. The upset threshold (Nitsch et al. used the term breakdown threshold) for the three systems is given in Table 5.2.

These data appear to indicate that the more modern systems are more strongly affected to the Hyperband threat. However, it is very unclear whether the computers evaluated for this study were complete systems (i.e., whether the computer motherboard was within an enclosure).

For the computer networks study, only the cable was illuminated with the HPEM waveform with the computer terminal systems effectively screened using absorber walls. This situation is very unlikely to occur in a realistic scenario because the systems connected to the network will be at least partially illuminated. The results of the minimum upset thresholds for various conditions are as shown in Table 5.3.

Again, the actual repetition rate at upset is not given. However, it is stated that the number of lost frames increases linearly with increasing repetition

Table 5.2

Minimum Radiated Effects Threshold of Computers

SUT Type	Modulation	Average Upset Threshold
386.25 MHz	Hyperband (t_r = 100 ps, t_{fwhm} = 2.5 ns) repetition rate not given	17 kV/m
486.33 MHz		13.5 kV/m
486.66 MHz		12 kV/m

Source: [40].

Table 5.3
Minimum Radiated Effects Threshold of Networks

Upset Type	Modulation	Minimum Upset Threshold
Bit errors	Hyperband ($t_r = 100$ ps, $t_{fwhm} = 2.5$ ns) repetition rate not given	200 V/m (10 Base 2)
Lost frames		4 kV/m (10 Base T)
Network upset (denial of service)		6 kV/m (10 Base T)

Source: [40].

frequency. The range of repetition frequencies reported as being achievable from the HPEM test generator was 1 Hz to 200 Hz [41]. It was also observed that shielded network cable types, shielded twisted pairs (STP), offer more protection than unshielded types.

In another study [42], Nitsch et al. evaluated the effects of complete computer systems to HEMP E1 and Hyperband, again with the TEM simulator. Table 5.4 provides a summary of this work.

Nitsch et al. pointed out that the effect threshold of the SUTs is affected by the pulse characteristics of the waveform. Indeed, the effect threshold for the fast rise time, narrow pulse width Hyperband pulse is lower in all cases than the effect threshold for the slower rise time, longer-duration HEMP E1 pulses. This is perhaps counterintuitive because the effective energy in the Hyperband pulse is much less than that of the HEMP E1 pulse. However, it is shown that,

Table 5.4
Minimum Radiated Effects Threshold of Computers to HEMP and UWB

SUT Type	Modulation	Average Upset Threshold
AMD K6 300 MHz	EMP1 ($t_r = 10$ ns, $t_{fwhm} = 400$ ns)	16 kV/m
	EMP2 ($t_r = 1$ ns, $t_{fwhm} = 25$ ns)	7 kV/m
	Hyperband1 ($t_r = 100$ ps, $t_{fwhm} = 2.5$ ns)	3 kV/m
Pentium II MMX 350 MHz	EMP1 ($t_r = 10$ ns, $t_{fwhm} = 400$ ns)	40 kV/m
	EMP2 ($t_r = 1$ ns, $t_{fwhm} = 25$ ns)	12 kV/m
	Hyperband1 ($t_r = 100$ ps, $t_{fwhm} = 2.5$ ns)	7.5 kV/m
Pentium II 400 MHz	EMP1 ($t_r = 10$ ns, $t_{fwhm} = 400$ ns)	9 kV/m
	EMP2 ($t_r = 1$ ns, $t_{fwhm} = 25$ ns)	6 kV/m
	Hyperband1 ($t_r = 100$ ps, $t_{fwhm} = 2.5$ ns)	4 kV/m

Source: [42].

due to the higher-frequency content of the Hyperband pulse, the waveform couples more energy into the SUT geometry (i.e., the Hyperband waveform has a higher coupling efficiency for the specific SUT geometry).

Camp et al. [43] evaluated the Hyperband waveform effects on microcontrollers and different generations of computer motherboards from 8088, 5-MHz technology through to Pentium III, 500-MHz technology. The same TEM simulator used by Nitsch was used for this research. The general trend observed by Camp et al. is a reduction in the effects threshold for the newer technologies, from 21.6 kV/m for the 8085 to 3.2 kV/m for the Pentium III. This trend appears to concur with the prediction that, as technology moves forward, devices become more susceptible. However, it should be noted that these tests were carried out on exposed motherboards that were not integrated within an enclosure. It is well known that containment or shielding is an effective EMC design technique that could perhaps mitigate this fact.

Hoad et al. studied the effects of narrowband or Hypoband waveforms on information communications technology (ICT) systems using a reverberation chamber [44–46]. The Hypoband long-pulse environment used in all cases comprised a 30-μs pulse width a 1-kHz *prf.* The carrier frequency was incremented in logarithmic steps from 400 MHz to 8 GHz. This was achieved using standard EMC-type amplifiers, both solid-state and TWT types.

The systems selected for testing (SUT) comprised standard, commercially available computer systems housed in tower or desktop cases. Several models of computer systems were evaluated from different manufacturers. The manufacturer of a computer under test has been obscured in order to protect the interests of the manufacturer. These systems were selected and tested based primarily on availability. However, the variation of numbers and types enables trends to be developed covering variation in effects with specification/age, technology type, batch, and manufacturer. A brief summary of the computer specifications evaluated is given in Table 5.5.

Table 5.5
Specifications of the Computers Evaluated

Processor Type	Processor Clock Frequency	Case Style	Manufacturer	Number of Units
486	66 MHz	Desktop	C	1
486	100 MHz	Desktop	E	1
Pentium III	667 MHz	Desktop	D	3
Pentium IV	1.4 GHz	Mini Tower	C	1
Pentium IV	1.4 GHz	Mini Tower	I	1
Celeron	2.6 GHz	Mini Tower	D	1

This table comprises eight separate computer systems. An example specification of the manufacturer Brand D computer is: Desktop Case, 667-MHz Intel Pentium III Processor, 64-MB SDRAM, 10-GB HDD, supplied with Windows 98 operating system.

Hoad et al. also evaluated some computer network equipment. Several simple network configurations were developed and implemented. The range of equipment tested included:

- Brand D PIII 667-MHz computers with internal 10/100-MBps Ethernet cards;
- A 10/100-MBps PCI Combo network interface card;
- A dual-speed 10/100-MBps switching hub;
- 2 × 20m Category 5 STP Ethernet cables.

The effects tests were carried out in a reverberation chamber which allow for the SUT to be completely, consistently, and repeatedly evaluated with every EM illumination angle and polarization covered equally during one rotation of the stirrer.

The applied field in a reverberation chamber can be expressed in terms of the ensemble average total field (root of sum of the squares) or a rectangular component. Assuming that the receive antenna is 100% efficient and perfectly matched, it can be shown that the equations for either the total field or its rectangular component can be derived from the following equation:

$$S = \frac{8\pi}{\lambda^2} P_r \tag{5.5}$$

In (5.5), P_r and S are the ensemble average antenna received power and the scalar power density, respectively. Using (5.5) and other relationships, the ensemble averaged total E-field (E_t) is given by (5.6).

$$E_t = \frac{8\pi}{\lambda} \sqrt{15 P_r} \tag{5.6}$$

Equation (5.7) gives the ensemble averaged rectangular component (E_c) of the field:

$$E_c = \frac{8\pi}{\lambda} \sqrt{5 P_r} \tag{5.7}$$

From comparison of (5.6) and (5.7), it can be seen that the ensemble averaged rectangular component and total fields are related by a factor of $\sqrt{3}$. Equation (5.6) was used to derive the curves in Figures 5.6, 5.7, and 5.8.

The same monitor, keyboard, mouse, and cable layout were used for all SUTs tested. In this way, only the differences in effects of the computer main unit were recorded, allowing a fairer intercomparison. In order to further improve repeatability, the processor and hard disk were set working at maximum capacity. Because it is extremely difficult to predict when this will occur in normal use, the computers were exercised using a specialized test program. For the networked SUTs, a network analysis program was used to provide two pieces of raw data on network performance, throughput, and packet error rate (PER).

The types of effects observed were various and complex and are summarized in Table 5.6. For the networked SUT, one other effect specific to the network traffic was recorded. This was the increase of PER to 100% or the decrease in network data throughput to 0%, effectively network failure, a manifestation similar to what cyber specialists term denial of service (DoS). It was found that the indicators changed very swiftly from showing normal traffic to complete network failure. Therefore, no data on gradual network degradation were collected.

In all cases, the magnitude of the Hypoband long pulse environment level was gradually increased to a point where manual intervention was required (i.e., where the SUT must be manually reset). If no effects were observed, the HPEM exposure level was taken to the limit of the amplifier in use. Once an effect was observed at the test frequency, the test frequency was changed to the next test frequency and the test repeated.

It should also be noted that the effect observed at any specific frequency was unpredictable. As the HPEM exposure level is increased, the effect may change from mouse deflection to shut down in an approximately linear manner, but it was equally observed that the first recorded effect could be any of those listed.

For the data displayed next, the effect threshold recorded is representative of the cases when manual intervention was required. It was either necessary to manually restart the test software or to carry out a manual soft or hard reboot of the SUT.

The graph in Figure 5.6 shows the actual data taken for five different specification computers from four different manufacturers.

Each point on the graph represents a frequency where an effect requiring manual intervention was observed. Frequencies where no effect was observed are not marked. Failure to induce an effect only indicates that no effects were observed at the maximum output of the amplifier in use.

It can be seen from Figure 5.6 that the effect threshold increases as the frequency increases in the order of approximately 20 dB/decade. This indicates

Table 5.6
Effects Observed on Standalone and Networked Computer Systems

Effect	During Exposure to EM Stress	After Exposure to EM Stress
No effect	No effect	No effect
Monitor upset	Blanking or interference	Returns to normal function
Mouse pointer deflection	Influence position	Returns to normal function
Program closure	Pop-up menus, program closures, programs moved or deleted	Desktop function may be altered, missing or moved icons
Peripheral crash	Mouse, monitor, or keyboard behaves erratically	Remove plug to reinitialize or manually reboot
Crash with self-restart	The computer stops processing and latches	The computer starts processing again or a soft reset (Ctrl-Alt-Del) is required
Shutdown with self-restart	The computer switches off and attempts to restart without manual intervention	Once EM stress removed, the computer restarts normally; during restart, operating system detects abnormal shutdown, and several files may be affected
Blue screen	An exception error occurs resulting in the customary blue screen error message	The system generally can be rebooted without persistent effect
Crash with manual restart	The computer stops processing and latches	During restart, the operating system detects abnormal shutdown and several files may be affected
Shutdown with manual restart	The computer shuts down or switches off spontaneously	The computer remains nonfunctional; during restart, operating system detects abnormal shutdown and several files may be affected
Peripheral damage	The computer may crash or shut down	Investigation reveals permanent damage to a peripheral component (i.e., monitor, keyboard, mouse)
Functional damage	The computer may crash or shut down	During restart, the computer reports a failure to find the operating system; reinstallation of the operating system cures the fault (expected minimum outage of 2 hours)
Physical damage	The computer may crash or shut down	During restart, the computer either fails to boot or a critical device such as the hard disk malfunctions (expected minimum outage of 1 day)

that the region of highest coupling efficiency is at the lower end of the frequency range tested. As the frequency increases, both the coupling efficiency and rectification efficiency are decreasing, leading to an increase in the effect threshold.

Figure 5.7 shows the effects recorded on a single standalone SUT compared with exactly the same SUT in a networked configuration.

For this test, only the core system components of a hub switch and the necessary interconnecting cable between the hub and the computer were introduced to the original standalone SUT configuration.

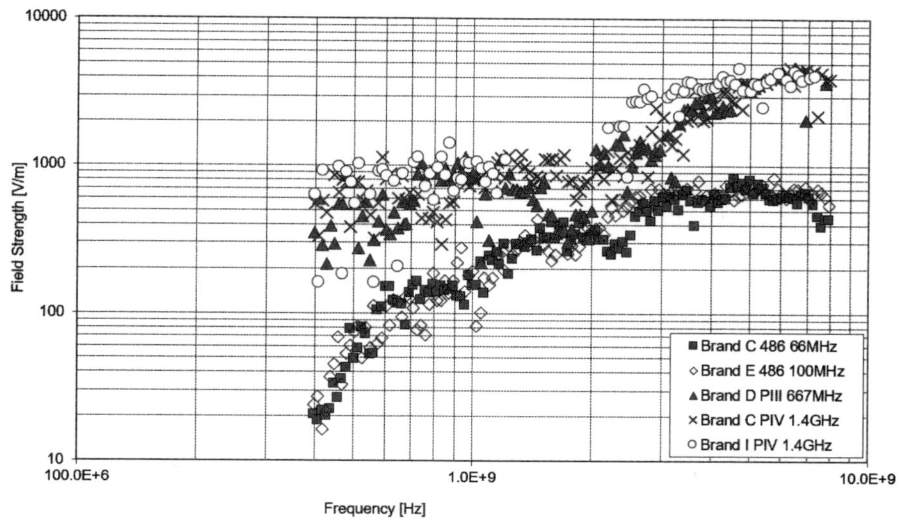

Figure 5.6 Effect threshold of five different specification computers.

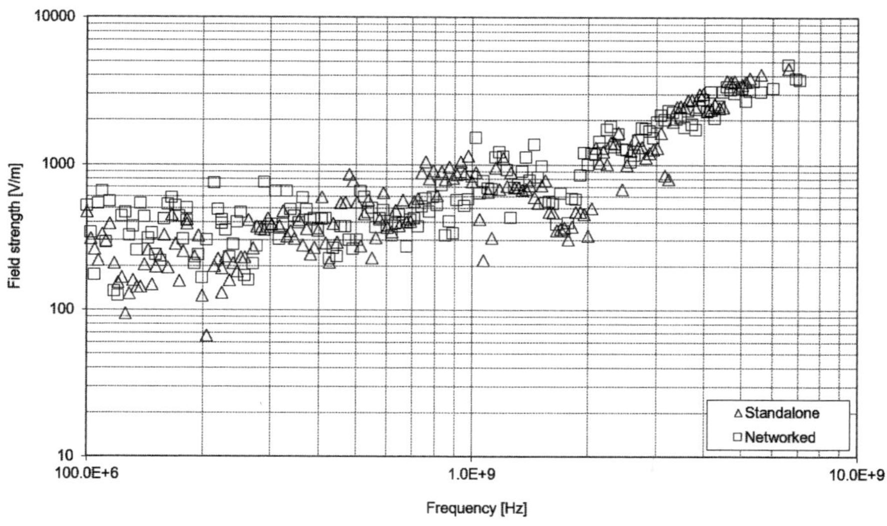

Figure 5.7 Standalone SUT effect threshold compared with exactly the same SUT in a networked configuration.

Interestingly, it can be seen that there are only some minor differences in the effect threshold for the networked and standalone computer configurations, although the difference appears to be greater at the lower-frequency end of the test, probably because of the introduction of the Ethernet cable.

Figure 5.8 compares the effect threshold of the networked computer compared with the effect threshold corresponding to network failure level (i.e., when the network traffic was blocked). Blocking or stopping of the network traffic is similar to a cyber denial-of-service (DoS) effect.

It can be seen from Figure 5.8 that there are strong regions, highlighted here, where effects to the network (DoS) preceded effects to the computer (i.e., the effect threshold to cause network DoS is lower than the effect threshold of the computer).

In these regions, the network outage preceded effects to the computer by a significant margin. The resonant peaks below 700 MHz are likely to be due to coupling to the network cable or network hub switch power supply cable. It was noted that in this region the network took a long time to recover. This perhaps indicates that the effect could be due to thermal protection switching within the hub switch power supply. An increase in effects on the network traffic disruption in the 3–4.5-GHz region probably indicates that either the hub unit or the hub power supply unit was affected rather than the computer.

5.4.1 Summary of Equipment, System, and Network-Level Effects

The data previously described are useful in a general sense in that they improve our general understanding of effects. However, extracting useful scientific meaning which can be used to describe effects in a much wider population of systems or to infer effects thresholds at different levels of the system of systems hierarchy, or to infer something about different HPEM environment types from

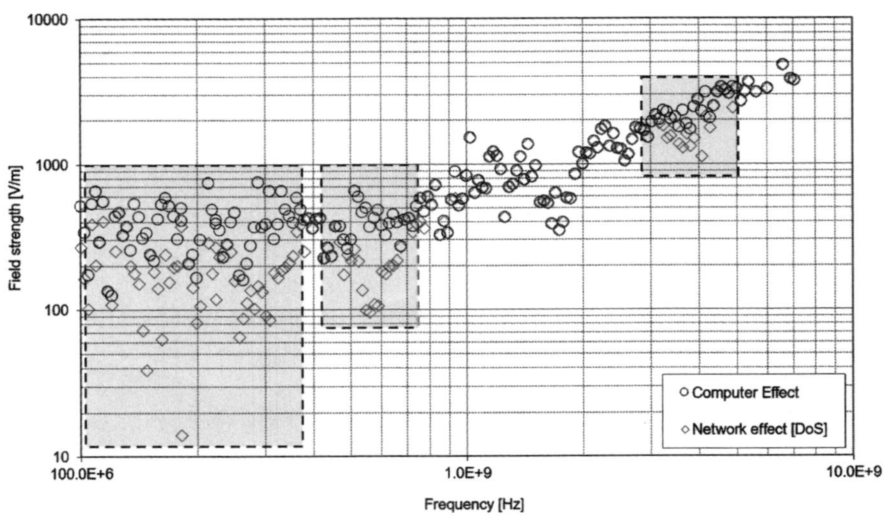

Figure 5.8 Comparison of computer effect with network effect.

the effect test results discussed is very difficult. Comparing and contrasting results from the different data sets are virtually impossible due to factors such as:

- Different test configurations and different ways of exposing the equipment, system, or network were used in each case. A standardized approach to testing was not used.
- There is some variability in the effect results even under highly controlled test conditions for nominally identical test subjects. This was discussed in Chapter 4.
- System-level effects are complex and manifest in many ways. The system function (i.e., how it is used and what it is used for) will have a profound impact on the severity of the effect.
- The experimenters' interpretation of the effect is of intrinsic importance to the understanding of the meaningfulness of the effect, yet no standardized classification scheme was used.
- The magnitude of the electric field is used as the primary descriptor of the effect threshold, yet it is not a metric by itself that can be used to describe the effect threshold of the system because different waveform parameters (signal indicators) require widely different E-field magnitudes for upset of similar systems.

It is clear from this discussion that standardized ways of carrying out effect testing (these are in development, see Chapter 4), a consistent way across the community of classifying effects (see Chapter 6), and a consistent way of analyzing the important features of the HPEM exposure (HPEM signal indicators) are vitally needed.

5.5 HPEM Signal Indicators

The main area of interest in this section focuses on signals that are typical of EM responses of electronic systems. These include E-fields and H-fields at various locations within a system, voltages or currents on cables or in electrical circuits, or perhaps the charge and surface current densities induced on electrically conducting surfaces. All of these responses can exist in either the time domain or in the frequency domain, and measurements in one domain (or perhaps both) are common [47].

It should be noted that other types of signals are also amenable to this type of analysis, although they are not considered here. These include pressure waves, the stock market index, the number of sunspots during the solar cycle, the distribution of frequencies in a cello concert, or any other quantity that can be represented by a time series or by a spectrum. In this section, we will discuss the differences (and similarities) between the transient and frequency-domain signals and the respective signal indicators that can be developed for these two classes of signals.

5.5.1 Transient or Time-Domain Signal Indicators

Time-domain signals are common in HPEM measurements and are perhaps the best understood, because the human experience is essentially a transient one. We are familiar with the concept of a start time and a stop time, with good behavior in between. A transient waveform arising from a physical (causal) system will typically be zero for early times up to a turn-on time, at which point the response grows and develops in time. It may grow monotonically, oscillate about a particular value, or drop back to zero after a specified time. In any case, such a waveform contains much information about the nature of the system that is producing the response, plus information about the excitation function that evoked the response in the first place.

As an example of typical transient waveforms that can be encountered in practice, Figure 5.9 illustrates two different transient responses, both with and without noise added.

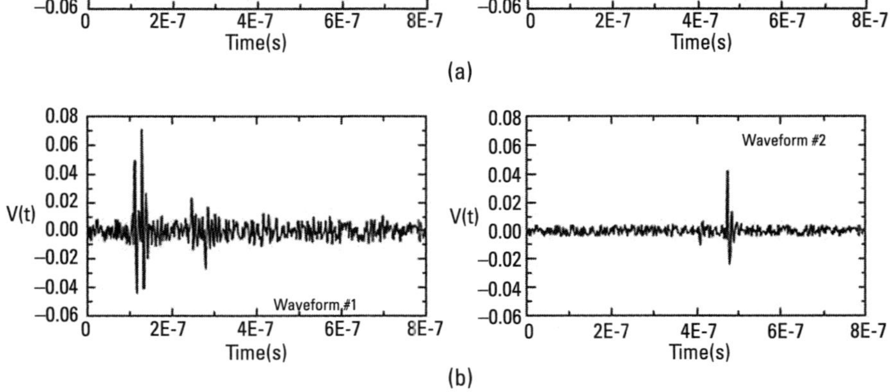

Figure 5.9 (a) Waveforms without noise. (b) Waveforms with noise; example of two transient waveforms encountered in EM field measurements or analysis.

Although these waveforms have resulted from the use of a computer model of a particular EM scattering problem, similar waveforms are encountered in HPEM effects measurement programs, and as such, these waveforms will adequately serve to illustrate the transient response indicators.

Waveform #1 contains several repeating wave packets in the overall transient record, while waveform #2 has just one packet, together with a pre-pulse contribution that occurs slightly before the main waveform. The issue to be examined here is how to distinguish between the two waveforms in Figure 5.11(a) in a simple manner. Clearly, the human eye can do this, but we wish to develop a robust numerical algorithm for performing this task. The answer lies in the definition of various transient signal norms or signal indicators. A type of waveform parameter is the waveform *p-norm*, defined in [2] and shown in (5.8).

$$\|R\|_p = \left\{ \int_{-\infty}^{\infty} |R(t)|^p \, dt \right\}^{1/p} \tag{5.8}$$

where p is an integer: 1, 2, ... Note that these norm quantities generally involve a mathematical operation on the entire waveform through the integration process, as opposed to the simple waveform parameters. Table 5.7 illustrates three commonly used waveform norms for high-power EM response characterizations.

The norms are commonly used in the characterization of HPEM responses.

Several different parameters can be used to distinguish one transient response from another. As an illustration, a simple early-time, monopolar transient waveform $V(t)$ is shown in Figure 5.10.

This figure illustrates the following waveform parameters: the peak value $V_{max}(t)$, the time to the peak value, t_{peak}, the 10% to 90% rise time, the

Table 5.7
Examples of Waveform p Norms

Value of *p*	Waveform Norm Attribute	Physical Quantity
1	$\int_{-\infty}^{\infty} \lvert R(t)\rvert dt$	Rectified value of response
2	$\left\{ \int_{-\infty}^{\infty} \lvert R(t)\rvert^2 dt \right\}^{1/2}$	Square root of energy in response waveform
	$\lvert R(t)\rvert_{max}$	Peak value of response

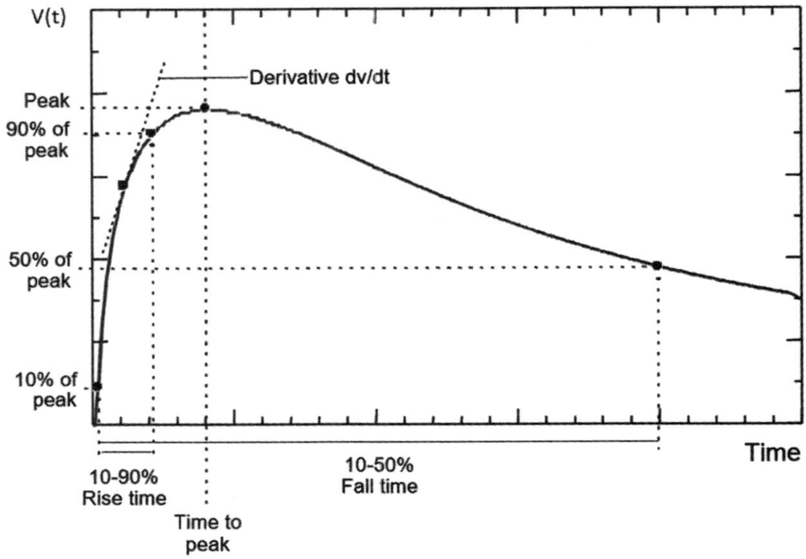

Figure 5.10 Illustration of various parameters used to characterize a transient response waveform $V(t)$.

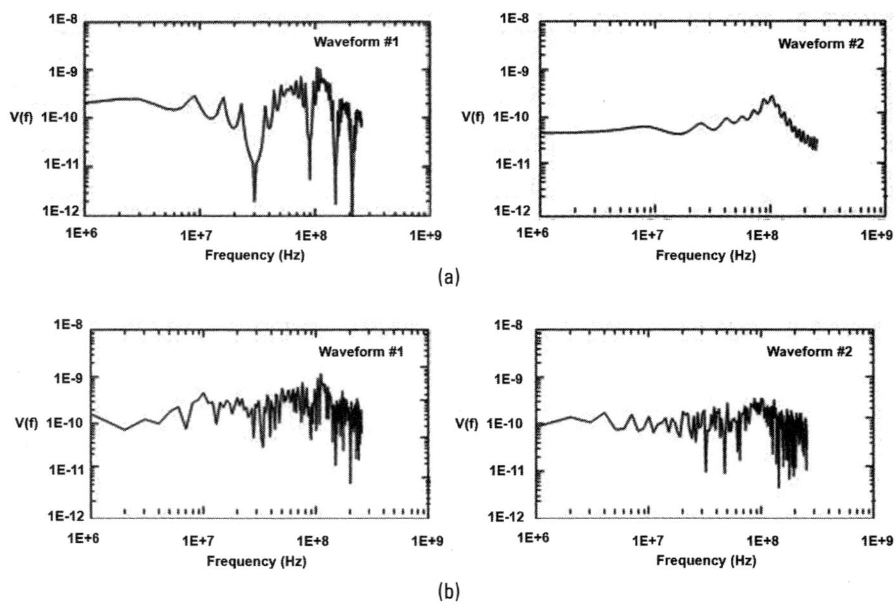

Figure 5.11 (a) Response spectra without noise and (b) response spectra with noise; example of two magnitude functions encountered in EM field measurements or analysis.

maximum rate of rise $(dv/dt)_{max}$, and the energy contained in the waveform. These parameters are discussed further next.

- *The peak value* $V_{max}(t)$: This quantity (often denoted by $V_{\max}(T_0)$ or $E_{\max}(t)$)is the infinity norm of [2] and is perhaps the most widely used waveform indicator. It answers the question of which response is larger. Its application to the more complicated waveform of Figure 5.9(a, b) is straightforward, but as one can imagine, the presence of noise in the waveform may give rise to errors in this quantity.

  From a numerical standpoint, calculating this peak value is a simple task, obtained by simply cycling through all of the points in the waveform, looking for the largest value. Once this value is determined, a more sophisticated analysis can be conducted, if desired, in which the values of the function at the nearest neighbors to the peak, together with the peak value, are fit to a quadratic polynomial to yield a better estimate of the actual peak.

  While this procedure is simple for waveforms having only one peak (as in Figure 5.10), it is more difficult with the waveform having strong oscillations as noted in Figure 5.9, especially if there is noise present. In this case, there can be several peaks present and the peak value indicator may comprise a list of several peak values, sorted from the largest to the smallest.

- *The time of the primary peak value,* t_{peak}: While not a norm in a mathematical sense, another waveform parameter of interest is the time to the waveform peak value. This quantity, t_{peak}, is indicated in Figure 5.10.

  Because this parameter is measured relative to a reference time ($t = 0$), it is necessary to know how to define this reference time for the t_{peak} waveform parameter to be useful. Often, a comparison of two waveform peak-value times is possible (as in the case in Figure 5.9), if the common $t = 0$ reference time is known.

- *The 10% to 90% rise time of the peak value of the waveform:* The time that a waveform takes to rise from 10% of its peak value to 90% of the value is a commonly used parameter to describe a transient response. This quantity is more useful than the time to peak discussed above, in that it does not require a definition of a common time origin for two waveforms.

  As in the case of determining the peak value, a more accurate estimation of this time can be obtained by interpolating the function near the 10% and 90% values with second-order polynomials. While this procedure is relatively simple for the waveform of Figure 5.10, difficulties can arise with the more complicated signals shown in Figure 5.9, where there may be a secondary peak near the primary maximum value. In this case, the two peaks almost merge and it is difficult to determine the true 10% to 90% rise time of the primary peak.

For waveforms with multiple peaks, it is often possible to calculate 10% to 90% rise times for each peak. Thus, these values, together with the peak values, will characterize the waveform. For example, waveform #1 in Figure 5.9(a) could be characterized by three local peaks (and the corresponding rise times), while waveform #2 would have only one peak.

- *The maximum rate of rise* $(dv/dt)_{max}$*:* The local time rate of change (derivative) of the waveform is a parameter that has some importance in determining the flashover of dielectric materials and the firing of surge protection devices. Thus, the maximum value of the derivative $(dv/dt)_{max}$ is a useful parameter to extract from a waveform and tabulate. This quantity is also indicated in Figure 5.10.
- *The 10% to 50% fall time of the waveform:* Another characteristic waveform time that does not depend on the absolute time reference is the time that the waveform takes to develop from its 10% of peak value, rising through the peak, and then falling to 50% of the peak value. This is denoted as the 10% to 50% fall time and is indicated in the figure. Like the rise time, this value is not a true mathematical norm, but it does serve as a useful waveform parameter.

  Like the rise time, the determination of this fall time can be improved for many waveforms by polynomial fitting of the data points and analytically computing this time. Multiple peaks in the waveform also pose problems for determining this response accurately in some cases.
- *The energy norm:* As discussed in [1, 2], there are other types of norms that are also useful in representing waveforms. One is the energy norm, defined in (5.9) as:

$$\epsilon = \left| \int_0^\infty v(t) \right|^2 .dt \tag{5.9}$$

where $v(t)$ represents the transient waveform. This index is a measure of the total energy contained in the waveform and can be used to distinguish between various waveforms.

5.5.2 Frequency-Domain Signal Indicators

A frequency-domain response, consisting of both the magnitude and phase (or equivalently the real and imaginary parts) of the spectrum, contains the same information as the transient response; only it is in a different form. This difference often makes the frequency-domain representation of data useful in ways that are not immediately evident in the transient response.

Using the theoretical development of the Fourier transform [48], a reasonably well-behaved function of time $v_{(t)}$ can be viewed as being composed of a superposition of many different sinusoidal waveforms, each having a distinct amplitude and phase. There are several different forms for this expression, one of which is the exponential form (5.10a) [49]:

$$f(t) = \frac{1}{2\pi} \int_{-\infty}^{\infty} F(\omega) \exp(j\omega t) d\omega \tag{5.10a}$$

The spectral density, $V(\omega)$, is a complex-valued function that represents the amplitude of a sinusoidal waveform component given by $e^{(j\omega t)}$. It may be determined from knowledge of the time function through the inverse relation,

$$F(\omega) = \int_{-\infty}^{\infty} f(t) \exp(-j\omega t) dt \tag{5.10b}$$

As a consequence of these expressions, it is possible to go from the time domain to the frequency domain rather easily, using a simple integration. Most often, this integration is performed using the fast Fourier transform (FFT) technique, as described in [50].

Several different approaches can be used for obtaining the spectral responses $V(\omega)$. Clearly, if the transient response is well known, the numerical evaluation of the spectrum by way of (5.10b) is possible. In other cases, it is possible to measure the spectrum directly, as discussed in [47]. Note that, for real-valued time functions, these spectra have the property that $V(\omega) = V^*(\omega)$, where the * denotes the complex conjugate. Consequently, only the spectrum for positive frequencies is needed. Sometimes, only the magnitude of the complex-valued spectrum is available, and as will be discussed, many of the spectral indicators can be used with this partial spectrum.

Figure 5.11 plots the spectral magnitudes of the two waveforms discussed earlier and illustrates the properties of typical spectra. Notice that these spectra consist of several resonance peaks, generally with one being a dominant peak and the others being slightly less important in determining the response spectrum.

Figure 5.12 illustrates various parameters used to characterize a spectral response.

Focusing our attention on one of the many peaks in the spectrum of Figure 5.11, we note that this peak frequently can be described by a general bell-shaped resonance curve typically found in circuit analysis [51]. This results in the following set of spectral indicators: spectral resonance frequencies, peak spectral magnitudes, resonance Q_s, the DC asymptotic value, and the spectral energy norm. These parameters are discussed next.

- *Resonance frequencies:* Perhaps the most common of all spectral indicators is a simple listing of the most important frequencies (or resonances) contained within the spectrum. This is not a mathematical norm quantity but is very useful in distinguishing between one waveform and another. At times, the resonances of a response are well separated and clearly evident, as in the spectrum for waveform #2 in Figure 5.11. At other times, the resonances can be close together and hidden, making it difficult to see exactly where they are. Resonance data can be expressed as a simple list of frequencies, such as f_0, f_1, f_2, ..., or by the primary frequency f_0, and the frequency deviations $\Delta fi = (f_i - f_0)$.
- *Peak spectral magnitudes:* The peak value of the spectral magnitude at the primary (i.e., most important) resonance frequency f_0, is a useful quantity for comparing spectra. It is an absolute quantity, in that it has units such as A/Hz, or (V/m)/Hz. In cases where there are several resonances

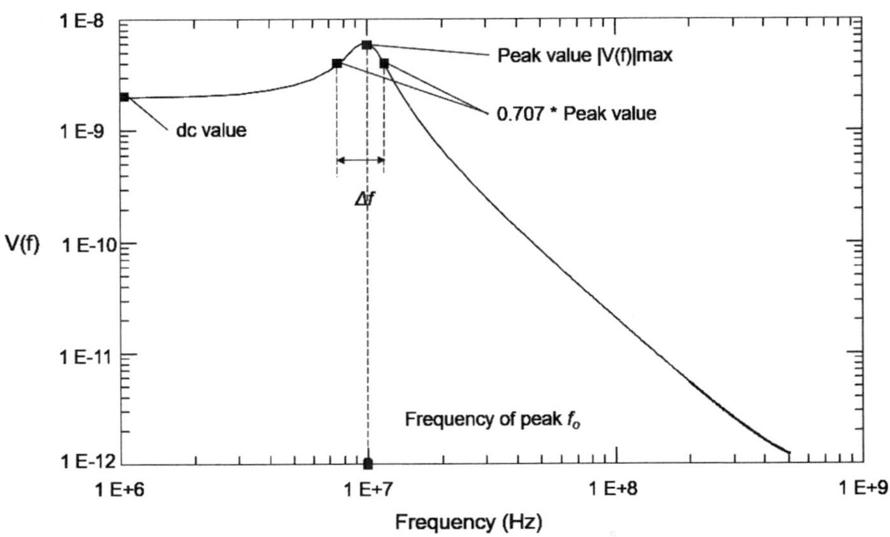

Figure 5.12 Illustration of various parameters used to characterize a spectral response.

present in a spectrum, it is possible to obtain a set of relative spectral magnitude parameters by normalizing the spectrum peak to unity and then taking the relative magnitudes of each of the other spectral peaks. Thus, if $V(f)$ denotes the spectrum, the relative magnitudes of the other peaks are expressed via (5.11) as:

$$R_i = \frac{|V(f_i)|}{|V(f_o)|} \quad for \quad i = 1,2,... \tag{5.11}$$

- *Resonance Q:* The widths of each of the resonances in a spectrum play an important role in the spectral and temporal characterization. Borrowing concepts from circuit theory, we note that very thin (i.e., sharply peaked) resonances correspond to highly oscillatory waveforms in the time domain. This property can be defined in terms of a resonance quality factor Q [51], which is equal to the resonance frequency f_i divided by the bandwidth of the resonance, (5.12).

$$Q_i = \frac{f_i}{\Delta f_i} \tag{5.12}$$

In this definition of the Q, the resonance bandwidth Δf is defined in terms of the half-power points of the local resonance that is equivalent to the half-power spectral amplitude points as illustrated in Figure 5.12.

- *DC asymptotic value:* Another spectrum parameter is its DC value. From (5.11b), it is seen that this is a real-valued quantity and is simply the average value of the transient response, given by (5.13):

$$F(\omega) = \int_{-\infty}^{\infty} f(t) \exp(-j\omega t) dt \tag{5.13}$$

This parameter has been defined as the *one-norm* [2].

- *Spectral energy norm:* While the energy contained in a transient waveform can be determined by the integration given in (5.9), it is also possible to obtain the same energy norm by operating on the spectral magnitude. Known as Parseval's theorem [48], the following relationship, (5.14), between the time-domain and frequency-domain spectra can be developed:

$$\varepsilon = \int_0^{\infty} |v(t)|^2 \, dt = \frac{1}{2\pi} \int_{-\infty}^{\infty} |V(\omega)|^2 \, d\omega \tag{5.14}$$

Thus, the energy contained in the waveform can be estimated from either the transient response or from the spectrum.

5.5.3 Pulsed CW Signals

Pulsed CW signals are usually characterized by the parameters of CW, power, energy, pulse width, duty cycle, pulse repetition frequency (*prf*), and burst duration and the subsequent influence that they have on each other or the spectral density of the waveform. Figure 5.13 shows how a CW signal appears in the time domain.

The time period is simply the inverse of the frequency in hertz. A repetitively pulsed CW signal is shown in Figure 5.14.

Here the CW signal is only present for short periods. The duty cycle is the ratio of the pulse width (τ_p) over the total pulse to pulse period T (τ_p/T) and is also known as the pulse repetition interval or *pri*. The *prf* is simply $1/T$.

This can also be seen as on-off modulation of the original CW signal shown in Figure 5.13.

- *Pulse and pulse width:* The IEC defines a pulse as follows: "transient waveform that usually rises to a peak value and then decays, or a similar waveform that is an envelope of an oscillating waveform" [52]. The envelope in this case is the pulse width τ_p. Typically the pulse width is measured at half the maximum height (often denoted as full-width half-max (FWHM)) of the pulse or envelope. The pulse here is shown as a nice rectangular envelope but other envelope shapes are perfectly feasible

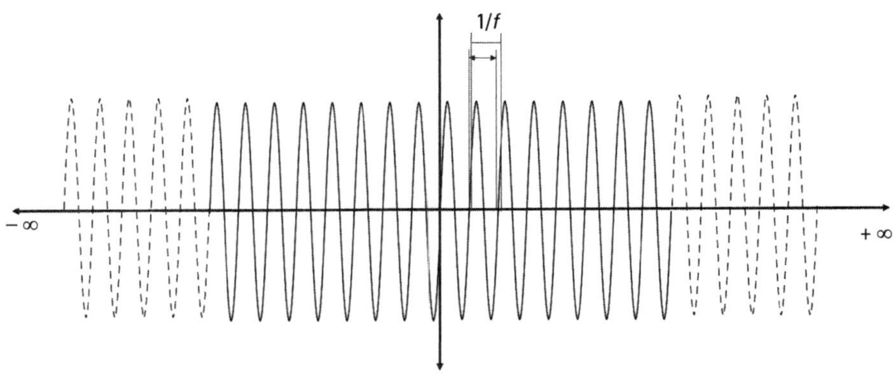

Figure 5.13 A CW signal.

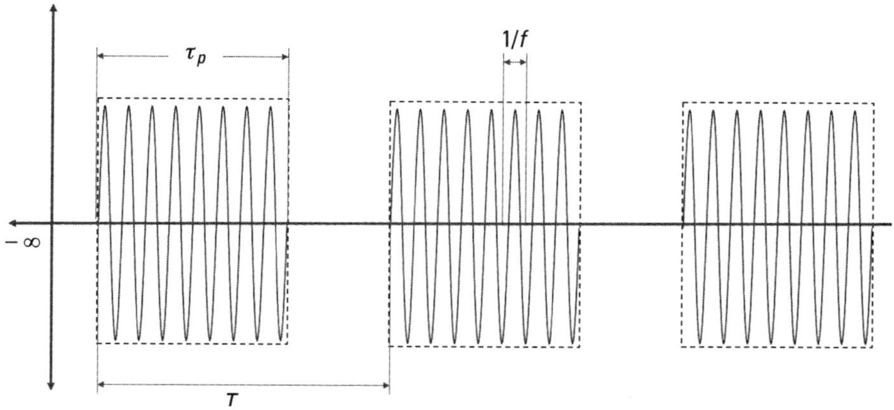

Figure 5.14 A repetitively pulsed CW signal.

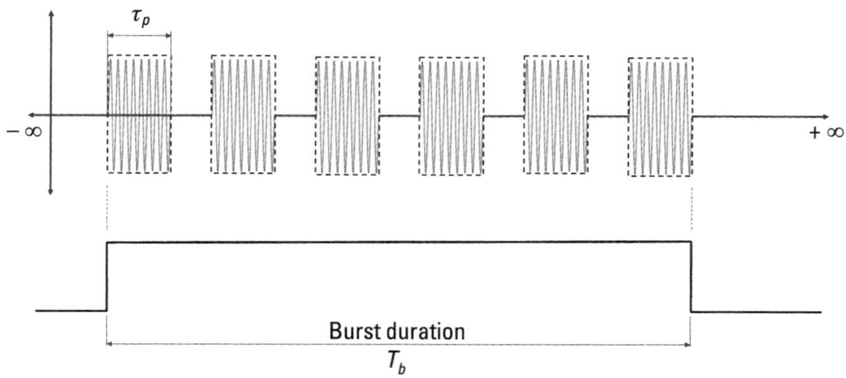

Figure 5.15 Burst duration envelope.

including but not limited to a Gaussian shape (as shown in Chapter 2) a ramp or triangular shape or in the case of a Mesoband signal a double exponential. The shape of the pulse envelope will have a significant impact on the spectral density.

- *Prf:* The *prf* is defined by the IEC as the "number of pulses per unit time, measured in Hz" [52]. The *prf* as shown in Figure 5.14 is the reciprocal of the pulse repetition interval (i.e., the time period between pulses). For Hypoband systems, typically the *prf* is between 1 Hz and 1 kHz; indeed, any HPRF/IEMI source that uses a spark gap switch in the output stage is limited to 1 to 2 kHz. Some Hyperband solid-state sources are capable of *prfs* of over 10 MHz.

- *Duty cycle:* The duty cycle is simply the ratio of on-time or pulse width to the time interval between pulses or pulse repetition interval. Duty cycle is typically expressed as a percentage. A duty cycle of 100% implies that the pulse width and the interval are of equal duration. Duties typically lie in the range of less than 1% for Hyperband systems, increasing up to 50% for Hypoband long-pulse systems.
- *Burst duration:* A burst is defined by the IEC as the "a sequence of a limited number of distinct pulses or oscillations of limited duration" [52]. A burst of pulses with a rectangular burst envelope is shown in Figure 5.15. It can be seen that a burst has a limited duration and a limited number of pulses per burst. The burst envelope is shown as a nice rectangular envelope, but other envelope shapes are perfectly feasible including but not limited to Gaussian, a double exponential, a ramp, or a triangular shape. The shape of the burst envelope has an impact on the spectral density.
- *Power and energy:* The peak power in an HPRF/IEMI waveform is directly related to the magnitude that is typically expressed in volts per meter or watts per meter squared. The concept of average power is also sometimes used, which is equal to the peak power divided by the pulse width or, for continuously pulse-modulated waveforms, the peak power divided by the duty cycle. The concept of average power can be misleading because the shape of the pulse envelope will have a bearing on the true average power.

Many measuring instruments such as spectrum analyzers display average or root mean square (RMS) values for the magnitude of a measured waveform. The RMS function is strictly a shape factor for pure sinusoidal signals.

The power in the waveform relates to the rate of energy delivery. Energy (Joule) is simply power (watts) multiplied by the time period (seconds), or $J = W.s$. Power is therefore energy per second.

For a given time-domain waveform with magnitude v over time t, the energy in the waveform can be found by (5.15). This is the N5 norm often referred to as the root action integral. Energy is sometimes referred to as fluence and power is sometimes referred to as intensity. To visualize this in a simple way, imagine the light beam from a spotlight; the energy is the heating effect of the light and the power is the brightness of the light.

$$\varepsilon = \int_0^\infty |v(t)|^2 \ dt \tag{5.15}$$

We may look at the power and energy contents of the three waveforms, CW (Figure 5.13), pulsed CW (Figure 5.14), and a burst of pulsed CW (Figure 5.15) that only lasts for a finite duration. First, we observe that there can be no true CW signal in practice that lasts from minus infinity (*t*) to plus infinity (*t*). In Table 5.8, we summarize the relevant results for the three waveforms.

For a burst duration limited series of repetitive pulses, the peak power in the waveform is unaffected by the burst duration. However, the energy over the burst period is affected by the burst duration, because the waveform is time-limited. These distinctions become important when we consider some effects such as charging or heating and probability of intercept, which is related to *prf*, and component damage or burnout, which strongly correlates with energy.

5.5.4 Use of Response Indicators

The corollary of signal indicators that are important parameters which describe the HPEM exposure waveform, are response indicators that can relate to signal indicators but deal with the system response to the exposure. In effects tests, experiments, or campaigns involving a large amount of acquired data, one or more of the above signal indicators, and response indicators may be calculated and summarized in a final test report. It is often thought that by just calculating and tabulating these quantities, the system becomes more understandable and has been suitably characterized by the test. Unfortunately, this is not true, because the derived norms must themselves be compared with something else, either with other derived quantities for the same system or with similar quantities from another system having better known properties or attributes.

Therefore, it is useful to illustrate several different ways to compare and contrast EM responses of systems using the derived norm or signal indicator quantities. A few of the ways such indicators may be used are suggested here.

5.5.4.1 Tabulation Indicators

Perhaps the simplest use of norm and signal indicator data is in listing the parameters in tabular form. For example, considering a set of m waveform or spectral responses from which n response indicators are derived for each response, an $n \times m$ array of parameters results as listed in Table 5.9.

This process can be thought of as taking a much larger set of values, say, 2,048 data points in a transient waveform, and condensing the essence of the data into a much smaller set of numbers (n in the above example). There is no guarantee that the norm or signal indicators used are really adequate in describing the data, and the analyst should strive to ensure that the selected indicators are truly appropriate for describing his or her data or system.

Table 5.8
Environment, Power, and Energy Relationships for the Three Waveforms

Parameter	CW Signal (Figure 5.13)	Pulsed CW Signal (Figure 5.14)	Pulsed CW as a Burst (Figure 5.15)
Duration	Infinitely long in time	Infinitely long in time	Finite duration in time
Analytical representation	$v(t) = \sin(w_o t)$ $w_0 = 2\pi f_0$	$v(t) = \begin{bmatrix} \sin(w_o t) \ \textit{for}\ t = nT \ \textit{to}\ (nT + \tau_p) \\ 0 \quad \textit{for} \quad t = (nT + \tau_p) \quad \textit{to}\ (n+1)T \\ \textit{for} \quad n = 1,2,3........\infty \end{bmatrix}$	$v(t) = \begin{bmatrix} \sin(w_o t) \ \textit{for}\ t = nT \ \textit{to}\ (nT + \tau_p) \\ 0 \quad \textit{for} \quad t = (nT + \tau_p) \quad \textit{to}\ (n+1)T \\ \textit{for} \quad n = 1,2,3........N \\ N = (\tau_b - \tau_p + T)/T \end{bmatrix}$
Power	$v^2(t) = \sin^2(w_0 t)$	$P = \begin{bmatrix} \sin^2(w_o t) \ \textit{for}\ t = nT \ \textit{to}\ (nT + \tau_p) \\ 0 \quad \textit{for} \quad t = (nT + \tau_p) \quad \textit{to}\ (n+1)T \\ \textit{for} \quad n = 1,2,3........\infty \end{bmatrix}$	$P = \begin{bmatrix} \sin^2(w_o t) \ \text{for}\ t = nT \ \text{to}\ (nT + \tau_p) \\ 0 \quad \text{for} \quad t = (nT + \tau_p) \quad \text{to}\ (n+1)T \\ \text{for} \quad n = 1,2,3........N \\ N = (\tau_h - \tau_n + T)/T \end{bmatrix}$
Energy	$\varepsilon = \int_0^\infty \lvert v(t)\rvert^2 \, dt$	$\varepsilon = \begin{bmatrix} \int_0^\infty \lvert v(t)\rvert^2 dt \ \textit{for}\ t = nT \ \textit{to}\ (nT + \tau_p) \\ 0 \quad \textit{for} \quad t = (nT + \tau_p) \quad \textit{to}\ (n+1)T \\ \textit{for} \quad n = 1,2,3........\infty \end{bmatrix}$	$\varepsilon = \begin{bmatrix} \int_0^\infty \lvert v(t)\rvert^2 dt \ \textit{for}\ t = nT \ \textit{to}\ (nT + \tau_p) \\ 0 \quad \textit{for} \quad t = (nT + \tau_p) \quad \textit{to}\ (n+1)T \\ \textit{for} \quad n = 1,2,3........N \\ N = (\tau_b - \tau_p + T)/T \end{bmatrix}$

5.5.4.2 Multiparameter Scatter Plots

In many cases, it is difficult to see important trends in the response indicator data of Table 5.9 due to the large amount of data and the inability of the human eye to easily comprehend lists of numbers. A graphical representation is often easier.

A useful approach for understanding the response data is to try to determine if any of the response indicators are highly correlated with another indicator. One way to do this is to develop a *x-y* scatter plot consisting of *n* data points as shown in Figure 5.16(a, b), with one response parameter along the *x*-axis and another along the *y*-axis.

Figure 5.16(a) illustrates the case when two time-domain signal indicators or frequency domain spectral signal indicators are highly correlated, and a linear regression can be performed to draw a straight line through the data points. In other, more complicated cases, the correlations can be high but a nonlinear relationship between the parameters can be noted. Figure 5.16(b) illustrates two parameters that are not highly correlated.

5.5.4.3 Statistical Distributions

Another way of using response indicators is by developing statistical distributions for the parameters. Once these distributions are determined, they can be used as an alternate method for comparing system responses.

As an example, consider calculating a particular response indicator for different responses of a system. This might represent the peak value of a transient response of a cable induced by *m* different incident fields exciting the line. Such an example was considered in [13]. These responses can be sorted into various response bins and then plotted as a density function, as shown in Figure 5.17.

This curve, which illustrates the number of occurrences of an imaginary response indicator for a particular value of the indicator, is the probability density function (PDF) and is a standard technique used for statistical analysis [15].

Table 5.9
Listing of *m* Responses and Their *n* Norms or Response Indicators

Waveform	Indicator 1	Indicator 2	Indicator 3	...	Indicator *n*
Response 1	X_{11}	X_{12}	X_{13}	...	X_{1n}
Response 2	X_{21}	X_{22}	X_{23}	...	X_{2n}
...	...	...	...	...	...
Response *m*	X_{m1}	X_{m2}	X_{m3}	...	X_{mn}

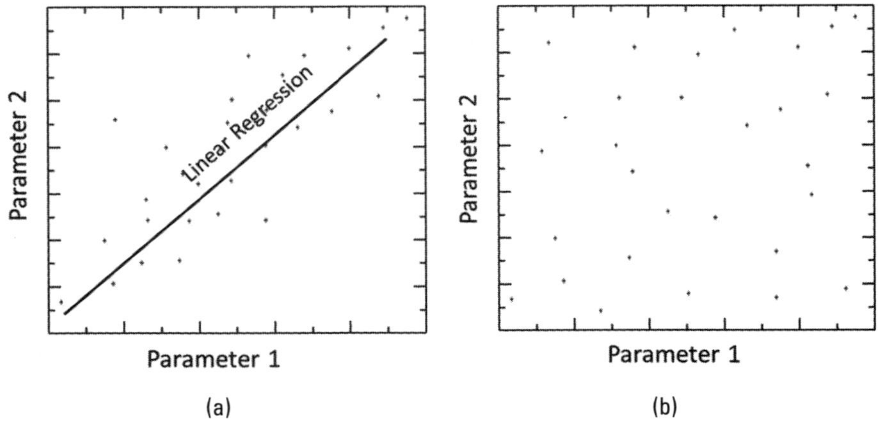

Figure 5.16 Example of scatter plots for two response indicators showing (a) good correlation and (b) poor correlation between the two parameters.

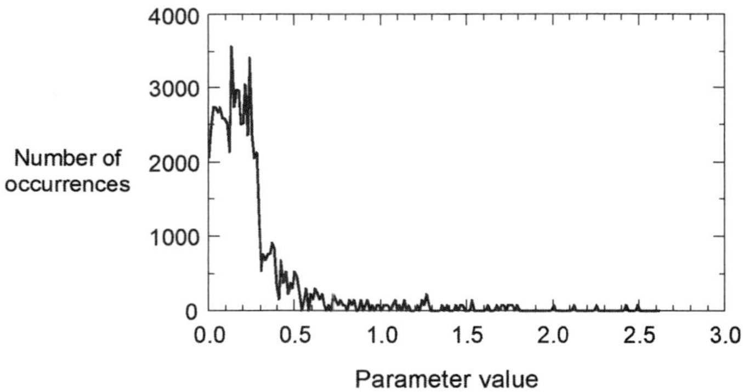

Figure 5.17 Plot of the PDF (probability density function) of a response indicator.

The cumulative probability density (CPD) curve shown in Figure 5.18 is particularly useful in summarizing the behavior of the response indicator in this particular problem. From this curve, we can describe the results as follows. The absolute worst-case of the response indicator is on the order of 3, although the probability of obtaining this response is very small. Perhaps a better choice of a worst-case response might be found for the 95% level, for which 95% of the responses will be lower than the specified value of about 0.87. Other levels of peak responses are the 90%, 50%, 10%, and 5% levels, which have responses of 0.5, 0.29, 0.05, and 0.02, respectively.

While the complete CPD curve of Figure 5.18 can be used to represent the statistics of the response parameter in a practical problem, in many instances, it may be adequate to use only a small number of CPD values, say, the values

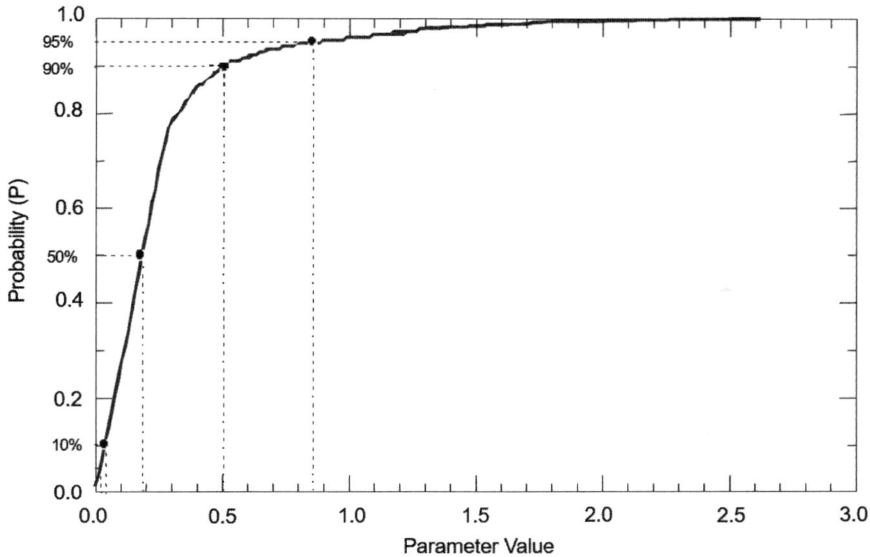

Figure 5.18 Plot of the CPD function for the response indicator shown in Figure 5.17.

of the distribution for the probabilities of 5%, 10%, 50%, 90%, and 95%. In this way, the large database of *m* waveforms (or spectra), each consisting of possibly several thousand individual data points, has been collapsed down 5 numbers, which characterize the trends in the data.

A simple example of the calculation and display of the response indicators using the statistical approach was described in [13], where the HEMP response of overhead and buried lines has been examined. Figure 5.19 illustrates a sample computer program output screen that was developed for that study for the purpose of displaying this condensed summary response data.

It is important to keep in mind that these statistical results must ultimately be compared with another quantity in order to say anything meaningful about the operation of the system. For example, if a system is tested and a particular response indicator is calculated, it is useful to compare this quantity with the same norm quantity determined from another system having a known behavior. For example, the second system may be known not to fail in a certain EM environment, and if the measured norm is less than the known norm for the hard system, one can infer that the SUT is also hard.

To describe this concept, suppose that in addition to measuring the CPD of a response parameter as shown in Figure 5.18, similar cumulative probability distributions for the failure rate of the system are measured or determined by analysis. This results in a failure CPD for the system based on the response parameter that may be compared with the CPD of the parameter (i.e., the stress experienced by the system).

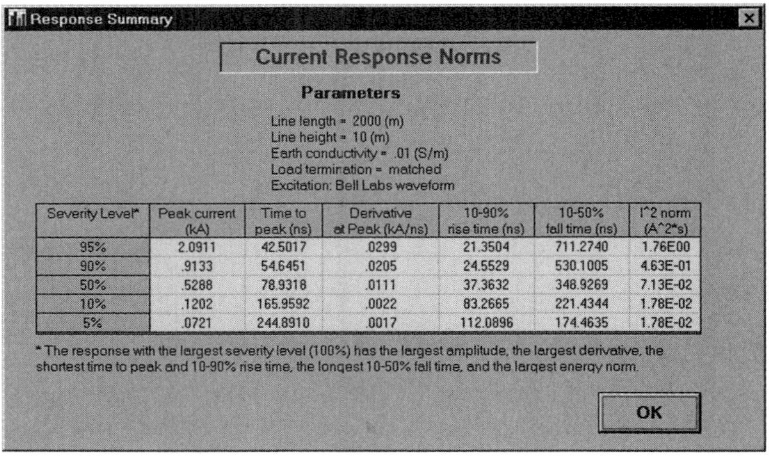

Response Summary

Current Response Norms

Parameters

Line length = 2000 (m)
Line height = 10 (m)
Earth conductivity = .01 (S/m)
Load termination = matched
Excitation: Bell Labs waveform

Severity Level*	Peak current (kA)	Time to peak (ns)	Derivative at Peak (kA/ns)	10-90% rise time (ns)	10-50% fall time (ns)	I^2 norm (A^2*s)
95%	2.0911	42.5017	.0299	21.3504	711.2740	1.76E00
90%	.9133	54.6451	.0205	24.5529	530.1005	4.63E-01
50%	.5288	78.9318	.0111	37.3632	348.9269	7.13E-02
10%	.1202	165.9592	.0022	83.2665	221.4344	1.78E-02
5%	.0721	244.8910	.0017	112.0896	174.4635	1.78E-02

*The response with the largest severity level (100%) has the largest amplitude, the largest derivative, the shortest time to peak and 10-90% rise time, the longest 10-50% fall time, and the largest energy norm.

OK

Figure 5.19 Illustration of the results screen for the norms program.

An illustration of this concept is shown in Figure 5.20. The center curve represents the CPD for the measured norm or indicator response, while the upper and lower curves represent the failure CPDs for the system, based on the same observable parameter. The lower curve denotes a hard (or survivable) system, in that for any chosen value of the parameter value (the stress), there are

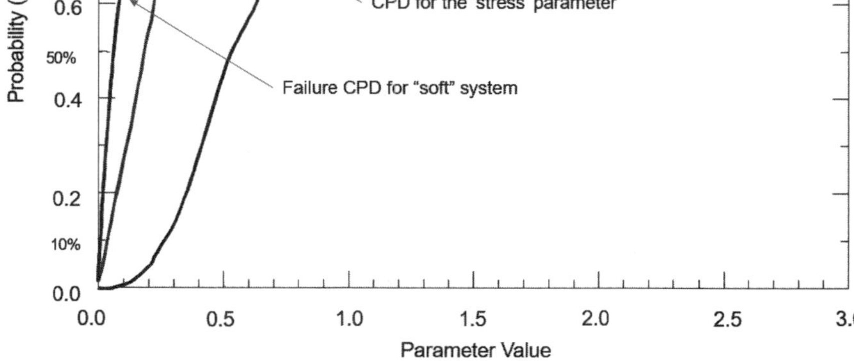

Figure 5.20 Illustration of the measured CPD for a parameter response in a system, together with two CPDs for the failure of the system, based on the same measured parameters.

statistically fewer systems experiencing failure than there are systems experiencing the stress. On the contrary, the upper curve represents a system that is soft and would be prone to error.

Instead of using the CPDs for this stress-strain comparison, a comparison of the PDFs of the stress and failure levels are frequently used. While this provides exactly the same information as the CPD comparisons, some analysts prefer this approach to that illustrated here. This approach was discussed and illustrated in [47].

5.5.4.4 Neural Nets

As a final example of the use of response indicators, consider the problem of determining if a system is hard against the effects of an external EM excitation using neural networks [15]. Assume that for the m measured EM responses discussed above are taken on a known hard system and the n response indicators are defined for each measurement. Using these n parameters as inputs into the neural network shown in Figure 5.21, the system can be trained using the m data sets.

By training, it is meant that the internal connection weights $w_{i,j}$, and the node firing thresholds, t_j, are dynamically adjusted, based on a knowledge of the system status (hard or not hard) for each of the training sets.

Once the training is complete, the neural network can be used as a classifier for newly acquired response indicators for another similar, but unknown, system.

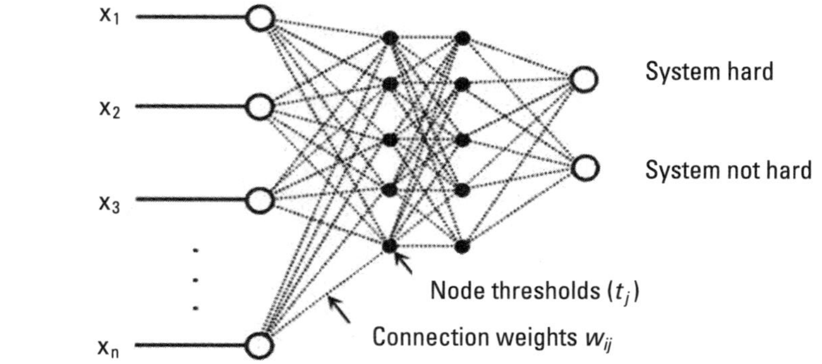

Figure 5.21 Example of a simple neural network based on response indicators for determining the EM hardness of a system.

5.6 Impact of Signal Indicators on HPEM Effect Mechanisms

This section briefly discusses the impact of the extended set of the signal indicators discussed previously on some of the HPEM effect mechanisms also discussed previously.

5.6.1 Effects of Pulse Repetition: Charging and Heating

There are many examples of effects where the pulse width and repetition (duty) of a repetitively pulsed waveform have been identified as the effect driver in analog and digital circuits. Consider a simple circuit in which a nonlinear junction (e.g., diode) is operating in series to a capacitor. The parasitic charging effect for a single pulse is illustrated in Figure 5.22(a) and that for a burst of pulses is illustrated in Figure 5.22(b).

The horizontal dashed line in Figure 5.22(a) near the *x*-axis represents the quiescent state of the system (i.e., the temperature or charge in the system is

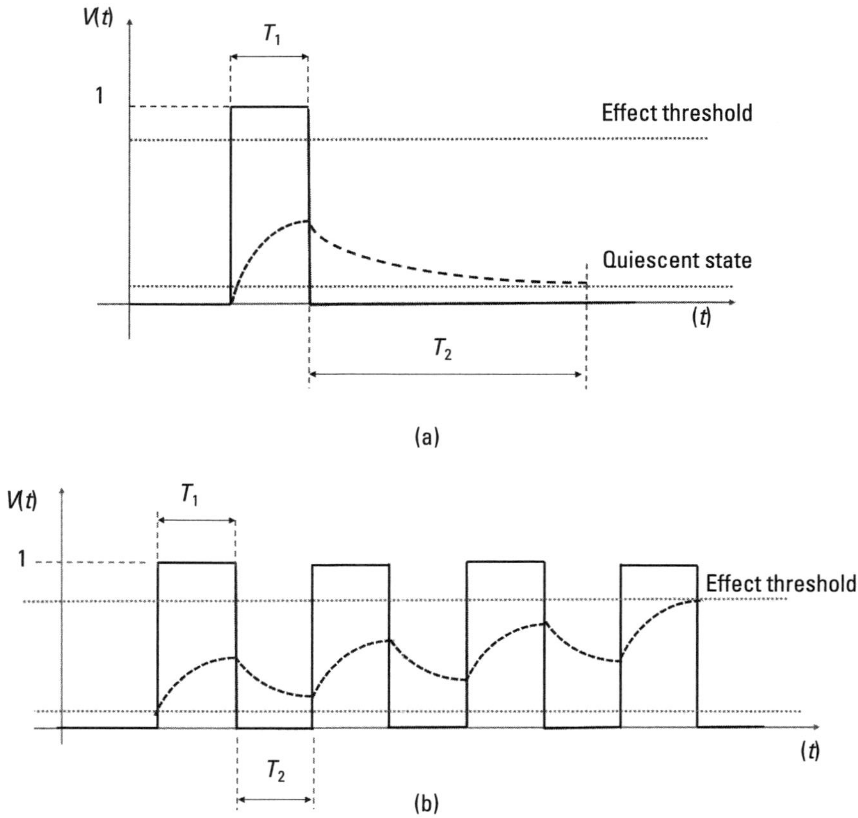

Figure 5.22 Parasitic charging: (a) single pulse and (b) repetitive pulses.

nonzero). The HPEM pulse in Figure 5.22(a) causes the charge or temperature to rise as shown by the dashed curve. Once the pulse is removed, the energy dissipates and the charge or temperature returns to the quiescent state. For the circuit, the voltage across the capacitor after time period τ_1 is given by (5.16):

$$\tau_1 = \left(\frac{R_1 \cdot R_2}{R_1 + R_2}\right) \cdot C \tag{5.16}$$

and the voltage across the capacitor V_c in the charging phase at some instant of time t is given by:

$$V_c(t) = \varepsilon(t) \bullet \frac{R_2}{R_1 + R_2} \bullet \left(1 - e^{-\frac{t}{\tau_1}}\right) \tag{5.17}$$

The voltage across the capacitor V_c in the discharging phase is given by:

$$V_c(t) = V_c(\tau_1) \bullet \varepsilon(t - \tau_1) \bullet e^{-\frac{(t-\tau_1)}{\tau_2}} \tag{5.18}$$

If the pulse width is small compared to τ_1, then the exponential time function can be approximated by a linear ramp that is proportional to the pulse area: $A{\cdot}\tau$. Clearly, this effect mechanism is highly dependent on pulse width, *prf*, energy, and burst duration. This effect mechanism has been used to illustrate the power to failure or damage of LNAs [23–27]. Often in experimental effect test campaigns, the burst duration aspect is not considered.

5.6.2 Effects of Pulse Repetition: Thermal Damage

Early work in understanding semiconductor device damage specifically from HPEM exposure and including important parameters relating to the signal indicators was published by Wunsch and Bell in 1968 [53]. The authors tested a large number of devices over pulse lengths ranging from ~0.1 to 20 μs. Wunsch and Bell produced a simple model for semiconductor damage that assumes that the breakdown is due to a temperature rise at the device junction,

$$P_f = \frac{C}{\sqrt{t_f}} \tag{5.19}$$

where P_f is the power required to destroy the component, t_f is the pulse length, and C is a constant for the component. The constant C is summarized in (5.20).

$$C = A \cdot \sqrt{\pi \cdot \rho \cdot C_p \cdot \kappa} \cdot \Delta T \tag{5.20}$$

where A is the area of the heated region, ρ is the density of the heated material, C_p is the specific heat of the material, κ is the thermal conductivity, and ΔT is the temperature increase required to induce failure. Again, the Wunsch and Bell model only holds for a narrow range of pulse widths.

Empirically, experimenters have observed that, for a typical circuit, the peak power of the HPEM environment required to induce an effect decreases as the pulse width increases, as indicated by the solid curve of Figure 5.23.

Increasing pulse width obviously correlates with increasing power (P). Broadly speaking, the right side of Figure 5.23 is referred to as the constant power regime and the left side of the figure is referred to as the constant energy regime. For very long pulses, the delivered energy creates a continuous (as long as the pulse exists) temperature increase in the conductors and the surrounding dielectric mass. A steady state exists and the temperature (destruction temperature) only depends on the delivered pulse amplitude (power). For very short pulses, the delivered energy is stored during the pulse and is transferred to temperature. This is sometimes referred to as the adiabatic region where adiabatic refers to a condition in which heat does not enter or leave the system concerned. Here the destruction level is proportional to the delivered energy, proportional

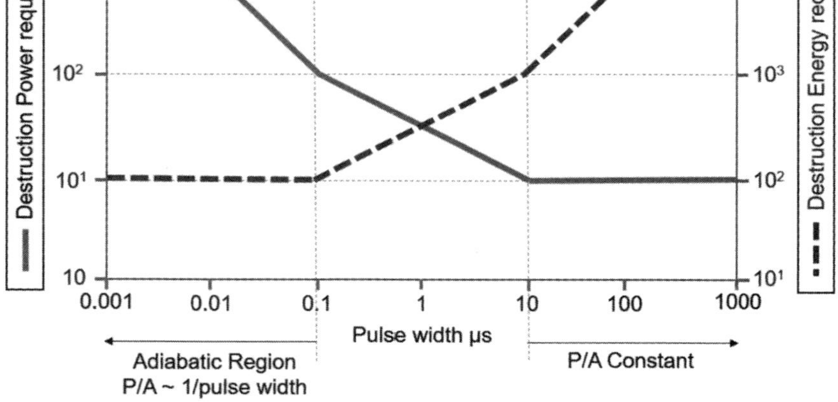

Figure 5.23 Empirically observed effect of pulse width on effect.

to the product of pulse width, *prf,* and burst duration. Between the two extreme parts of the curve exists the Wunsch-Bell region. Here the destruction energy is proportional to the square root of the pulse length.

Tasca [54] proposed a model that considers the energy required to induce failure, (5.21):

$$E_f = \left[\frac{4}{3}\pi r^3 \rho C_p + 4\pi r^2 \sqrt{\rho C_p \kappa t_f} + \frac{8}{3}\pi r \kappa t_f\right] \cdot \Delta T \tag{5.21}$$

where E_f is the failure energy and all the other parameters are the same as those discussed above.

This is a 3-D model because in this model the volume where failure occurs is defined by a sphere with a radius *r.*

Hoijer et al. [23] expanded this further to include parameters relevant to the HPEM waveform injection, (5.22).

$$E_f = P_f t_f = \frac{\frac{a}{k}\left(W_f - W_0\right)}{\left(1 - |\tilde{A}|^2\right)\left(1 - e^{-at_f}\right)} \tag{5.22}$$

where P_f is the power required for failure, *a*/*k* is a constant with an empirically fitted value of 3.7×10^5, W_f is the thermal energy require for failure, W_0 is the quiescent energy level, and Γ is the reflection coefficient at the input of the device.

Hoijer showed good correlation of this model with effects tests conducted on LNA semiconductor devices, which are often used in receiver front ends.

More recently, a very comprehensive analysis of power-to-failure of LNAs for Hypoband environments was performed by Zhou et al. [26]. Zhou et al. showed that pulse thermal resistance, thermal capacitance, and breakdown temperature are important parameters in understanding thermal damage behavior of semiconductors.

5.6.3 Effects of Pulse Repetition: Probability of Intercept

The probability of intercept, also referred to as the probability of failure, is an effect mechanism in digital systems. Digital devices exchange information bits over a bus. If an HPEM signal of sufficient magnitude and duration couples onto the bus, then a disturbance to the process or bit errors can occur. Excluding magnitude and pulse width for a moment, it can be shown that the likelihood or probability of bit errors is dependent upon the ratio of the repetition rate of the interfering pulse compared to the repetition rate of the bits (i.e., the data rate).

The probability of a bit error P can therefore be expressed [55, 56] by the following equation:

$$P = 1 - \left(1 - P_e(z)\right)^{f \cdot \frac{N}{R}} \tag{5.23}$$

where P_e is the probability of an error, f is the pulse disturbance repetition rate, N is the number of bits in the data packet, R is the data rate, effectively 1 over the data period, and z is the error function, given by:

$$z = \frac{V_s}{V_d}\sqrt{\frac{T_s}{T_d}} \tag{5.24}$$

where V_s is the amplitude of the data signal, T_s is the bit duration, V_d is the amplitude of the pulse disturbance, and T_d is the pulse disturbance duration.

This is illustrated in Figure 5.24(a) for a single disturbing pulse and Figure 5.24(b) for a disturbing pulse rate that is higher than the data rate.

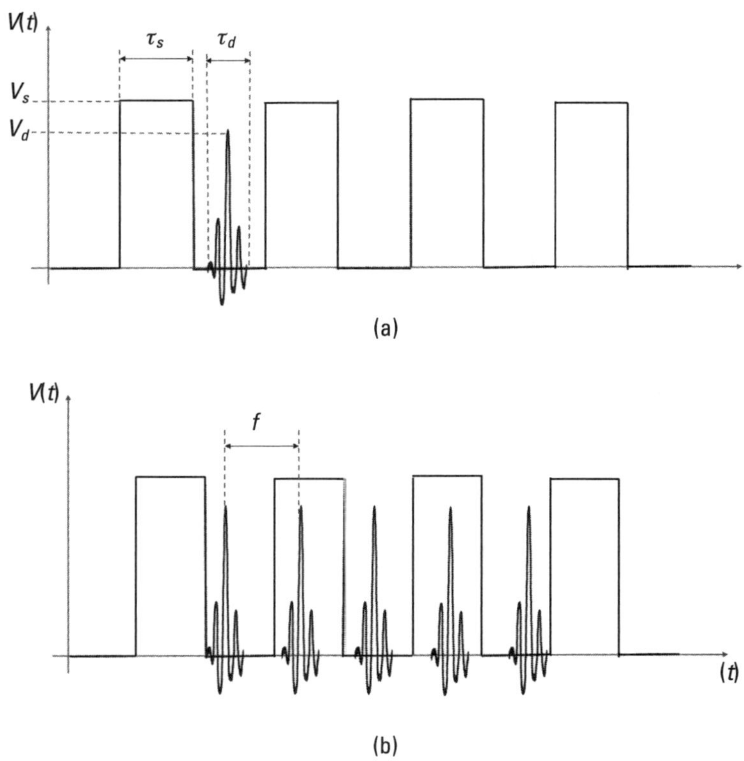

Figure 5.24 (a) Single disturbance pulse and (b) repetitive disturbance pulses.

Modern digital circuits have error correction algorithms that can, to some extent, cope with the loss of bits and even packets of data as long as the data packet is re-sent. However, the burst duration will also have a bearing on the ability of the error correction. Figure 5.25 shows the disruption effect (similar to denial of service or jamming) on a Wi-Fi link for three different amplitudes of the electric field *U*. A bit error rate (BER) of 1 relates to the complete disruption of the communication channel.

This chart shows the importance of the f/R ratio from (5.23). Essentially, when the disturbing signal rate f exceeds the data rate R, a much lower amplitude can cause disturbance of the data communications channel.

5.6.4 Effects of Spectral Density

A typical simplified continuous repetitive pulsed waveform and the envelope of its frequency spectrum are shown in Figure 5.26.

It is seen that there is a significant DC content in the time-domain waveform of Figure 5.26. The DC content is the area under the time-domain curve. The breakpoints f_1 and f_2 in the frequency spectrum are inversely proportional to the pulse width and pulse rise time, respectively. Cleary pulsed CW waveforms (e.g., Hypoband waveforms) have a different spectral content to pure CW waveforms and impulsive or burst modulated will also have further differences in their spectral characteristic.

From Figure 5.26, the first break point f_1 in the spectral envelope for a trapezoidal pulse occurs at a frequency that is inversely proportional to the pulse width. This figure also shows that, for a fixed *prf*, the maximum amplitude of the spectral envelope is proportional to the pulse width.

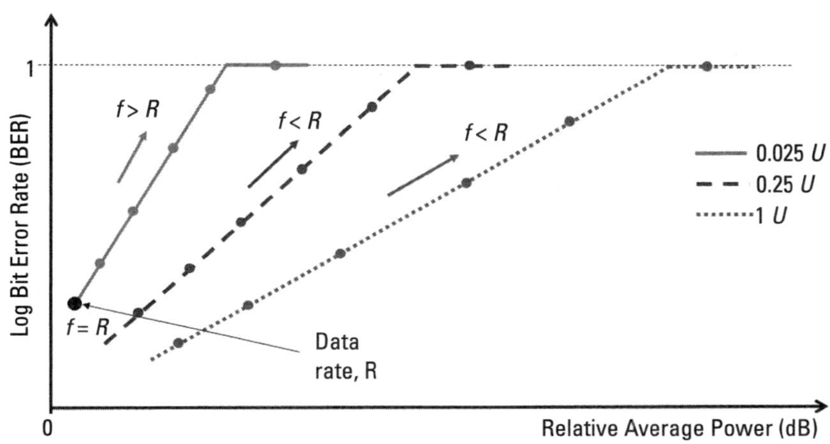

Figure 5.25 BER of a Wi-Fi communications channel in the presence of a repetitive pulse disturbance.

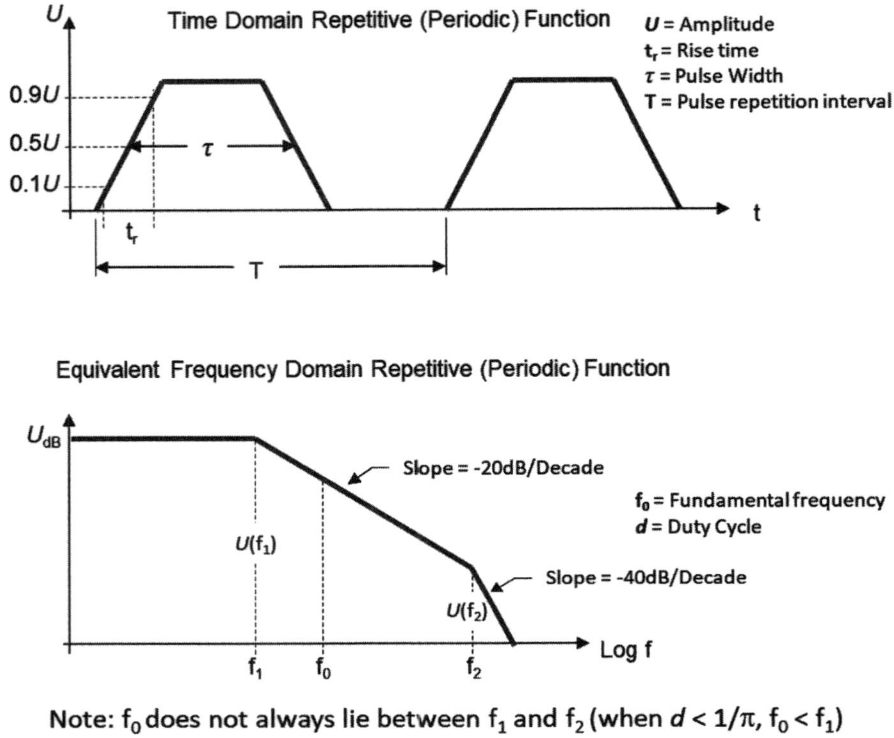

Figure 5.26 Envelope of repetitive pulse spectrum in the frequency domain.

Using these principles, the example in Figure 5.27 shows that every tenfold increase in the pulse width results in a decade reduction in the first corner frequency and a 20-dB increase in the associated spectral amplitude.

It is observed that, with only 20 cycles of a sine wave, we are already approaching the spectrum of a truly CW signal, which has a single frequency spectrum [57].

Figure 5.27 shows that the (one-sided) frequency spectrum of a CW signal is a single spectral line at the CW frequency. If the CW signal is gated, or pulsed, then the signal can be represented in the frequency domain by discrete harmonics. Although Figure 5.27 shows that pulsing the waveform has the effect of broadening the frequency spectrum slightly compared to the truly narrowband continuous case, it is apparent that the waveform is still narrowband relative to the carrier frequency, as long as we have a sufficient number (20 or more) of sinusoidal oscillations.

The harmonic content and hence spectral density are therefore strongly affected by the time-domain features of the waveform [58]. Consider the time-

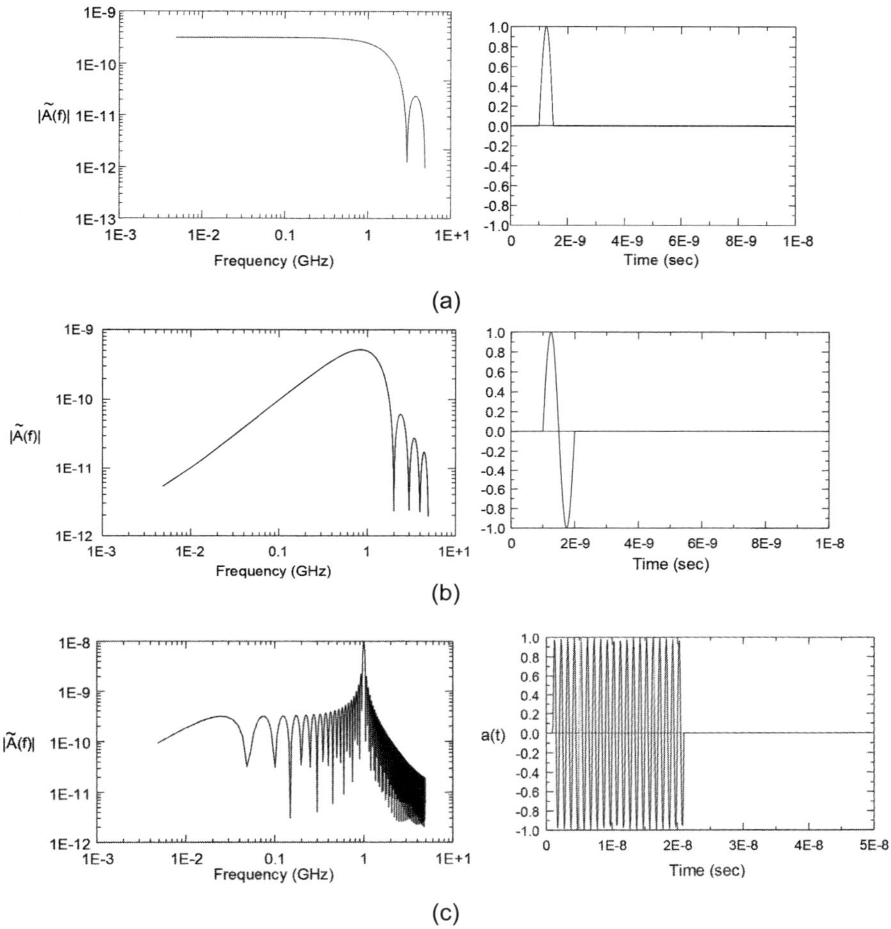

Figure 5.27 (a) Half-cycle sine wave and its magnitude spectrum; (b) one-cycle sine wave and its magnitude spectrum; and (c) 20-cycle sine wave and its magnitude spectrum.

domain expression of a repetitively pulsed waveform (5.25) and its continuous Fourier transform equivalent (5.26):

$$f(t) = \sum_{n=-\infty}^{+\infty} F_n e^{jn\omega_0 t} \tag{5.25}$$

$$F_n = \frac{1}{T}\int_{-T/2}^{T/2} f(t) e^{-jn\omega_0 t} \tag{5.26}$$

The continuous Fourier transform of a repetitive pulsed waveform is presented in Figure 5.28.

If the pulse shape is kept the same (constant V and τ) but the duty cycle is decreased by increasing the pulse spacing T, then the following effects occur:

- The spectral amplitude decreases ($V\tau / T$);
- The spacing between harmonics decreases ($1/T$);
- The number of harmonics up to the first zero crossing increases (T/τ);
- The frequency of the first zero crossing remains unchanged ($1/\tau$);
- Apart from amplitude scaling, the general shape of the spectrum ($\sin x/x$) remains unchanged.

In the simple example illustrated in Figure 5.29, the spectral density and harmonic content of a Mesoband waveform is modeled for a range of pulse repetition frequencies with a constant pulse width using the discrete Fourier transform of (5.26).

The discussion above demonstrates that the harmonic content and spectral density change significantly with change in duty cycle (*prf* and pulse width ratio).

5.6.4.1 Significance with Respect to the Effect Bandwidth Theory

The waveform norms discussed earlier, describe those features of a waveform that have an influence on effects; however, these norms were developed from the consideration of a single pulse such as a HEMP E1 pulse. Many of the

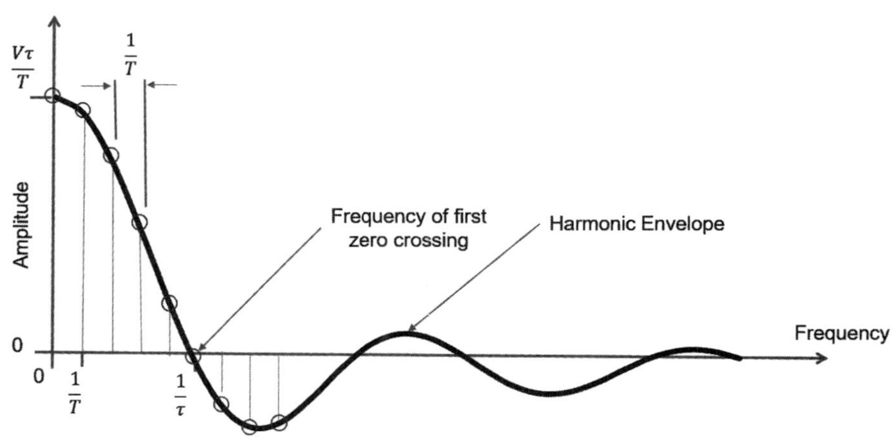

Figure 5.28 Fourier Transform of a repetitively pulsed waveform.

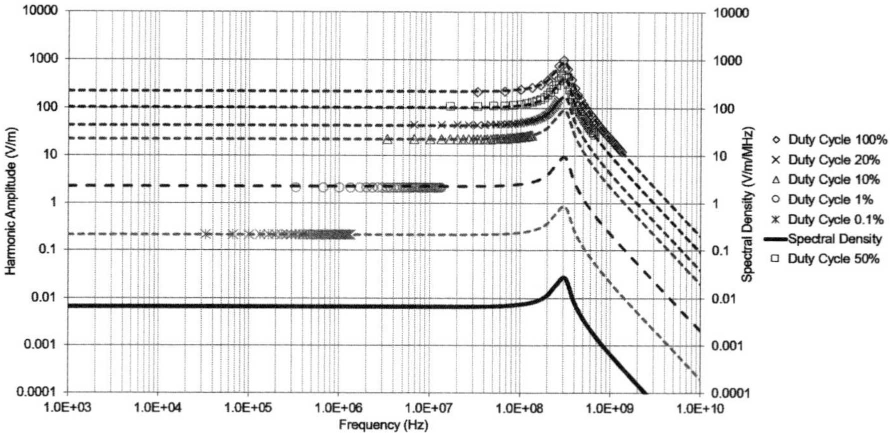

Figure 5.29 DFT of a Mesoband waveform with varying duty.

HPEM environments produced by IEMI and HPRF DE sources are a series of pulses at some *prf* usually for a finite time or burst duration.

The pulse width, pulse envelope shape, *prf*, and burst duration of a given IEMI or HPRF DE source are constrained by the physical design of the source, for example:

- The pulse width and envelope for Hypoband sources can be limited by complicated physics inside of the microwave tube, which cause pulse shortening;
- The *prf* can be constrained by the physics of the output switch, if, for instance, it is a spark gap;
- The burst duration can be limited through thermal constraints or just the limited capacity or energy density of the prime power source, which may be a capacitive energy store or battery, for example.

These physical constraints limit the parameter space, but specific combinations of each of these parameters can have a huge influence on the observed effect. This means that any conclusions drawn from experiments that try to evaluate effects by varying one parameter or a combination of parameters should be considered with caution because these parameters are highly interdependent on each other and result in very different spectral content.

Because repetitively pulsed waveforms are not strictly addressed by the waveform norms discussed previously, it is important to understand how other signal indicators such as, the pulse width, *prf*, or the burst duration affect other important parameters related to effects such as the power and energy delivered and the spectral density.

A concept that can be used to understand the importance of the spectral density is hereby termed the effect bandwidth. The effect bandwidth is different from the canonical coupling efficiency model discussed in Chapter 3, but there are parallels. Figure 5.30 illustrates the concept.

In this figure, which is heavily correlated with coupling, the total system effect is shown as a function of the effect of items (a), (b), and (c) within the system, each of which has its own frequency response range. For example, an automotive system can be considered to comprise of many electronic components or devices important to its function. One such device could be the primary battery voltage regulator—an analog device, perhaps item (a) in the figure. Another such device could be the vehicle sensors that are connected over long cables, perhaps item (b) in the figure. The digital engine management unit could perhaps be item (c) in the figure. In this theory, it is assumed that essentially any electronic item will have a passband containing those frequencies (including RF) that the item expects to see, and an out-of-band region where a significantly greater level of interference is required for the system to respond. This concept seems logical. Most modern electronic systems are a hybrid of analog devices used, for example, in the power supply and digital devices used, for example, as the microprocessor. Analog devices such as voltage regulators and power diodes and transistors do not require a large bandwidth to operate whereas digital devices must facilitate very fast switching speeds, so their bandwidth is correspondingly much wider.

Important parameters of the HPEM waveform such as pulse width, *prf*, and burst duration that have a consequential impact on energy and spectral

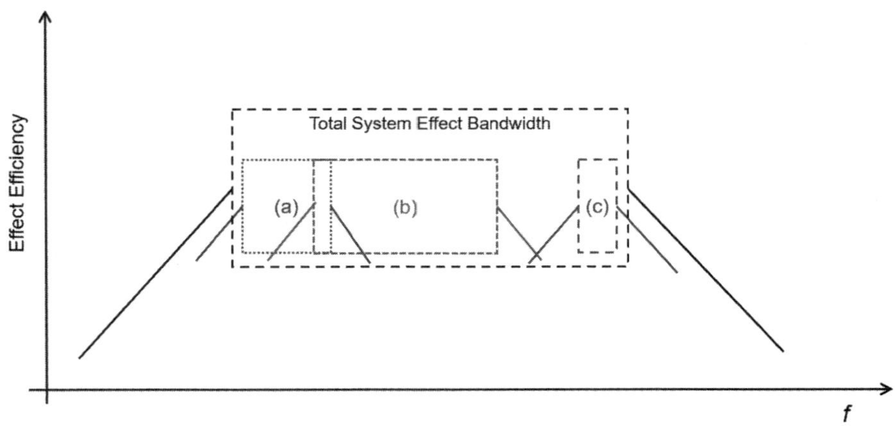

Figure 5.30 Effect bandwidth concept.

density of the waveform change considerably and hence can have a large bearing on the effect threshold.

Extending this further, it is inevitable that more energy is required to disturb a device out-of-band than within the devices passband. Further still, the maximum efficiency for an effect will occur when the disturbing waveform bandwidth corresponds with the devices effect bandwidth. This was discussed in Chapter 3, and illustrated in Figure 3.4. The effect bandwidth concept can also be described in terms of the quality factor of the disturbing signal (source) Q_s in comparison with the quality factor of the disturbed circuit or the target Q_t. In what follows, we discuss the relationship between Q_s and Q_t.

From a time-domain perspective, if the pulse width narrows but the pulse amplitude remains the same, then the energy contained in each pulse will decrease. Providing the *prf* does not change, the average power of a sequence of such pulses will decrease in proportion to the energy of the individual pulses. From a frequency-domain perspective, the effect of reducing the pulse width while preserving its amplitude will be to increase the bandwidth occupied by the major part of its frequency spectrum and reduce the amplitude of the harmonics to satisfy the lower-power requirement.

If it is assumed that an effect depends on the energy of the individual pulses, it follows that an effect can only be guaranteed by maintaining the average power of the pulse sequence. For a constant *prf*, this can only be achieved by increasing the pulse amplitude to compensate for the reduction in duty cycle caused by the pulse width reduction.

To this end, it is possible to explain the impact of this theory by describing two regimes: a short-pulse regime where the disturbing pulse bandwidth is wider than the effect bandwidth (Figure 5.31(a)), and a wide regime where the disturbing pulse bandwidth is narrower than the effect bandwidth (Figure 5.31(b)).

The quality factor is given by the effect bandwidth divided by the center frequency f_o. In Figure 5.31(a), we observe that the target Q is larger than source Q or $Q_t > Q_s$. In Figure 5.31(b), we observe that the source Q is larger than the target Q or $Q_s > Q_t$. For efficient coupling into the target, it is desirable to have the condition of $Q_s > Q_t$. This points out that for efficient coupling of the disturbing signal to the target system, we will need:

- The disturbing signal to be a high-power moderate bandwith (i.e., mesoband) or damped sinusoidal waveform;
- The center frequency of the damped sinusoidal waveform to match a primary vulnerable frequency of the target system;
- The Q of the disturbing signal to be larger than the Q of the target system.

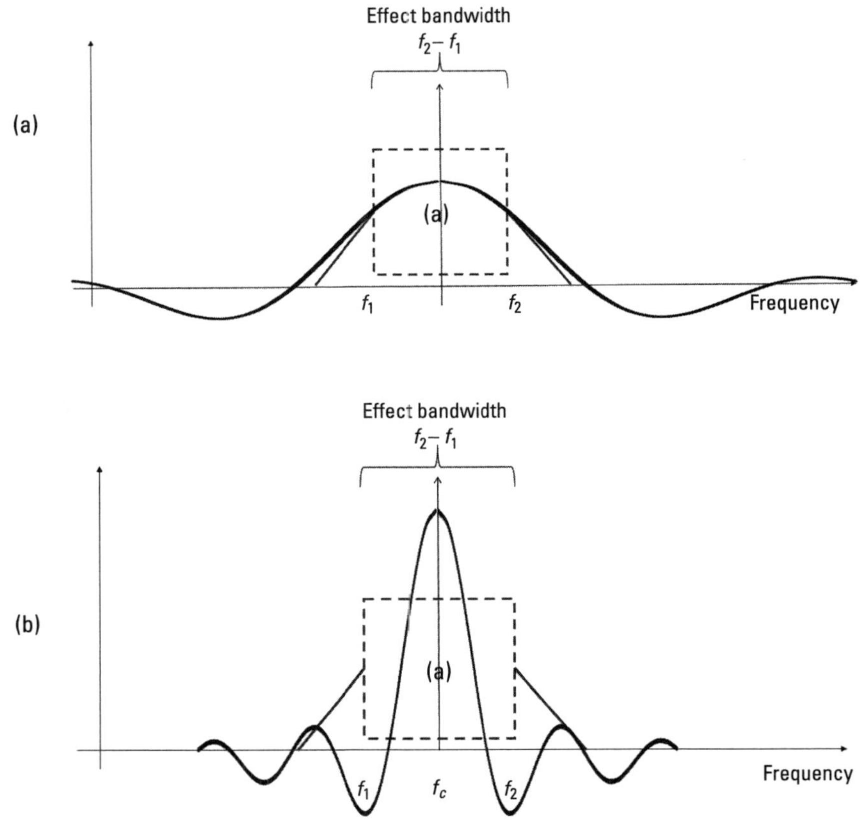

Figure 5.31 (a) Effect bandwidth for component (a) when the disturbing waveform bandwidth is wider than the effect bandwidth. (b) Effect bandwidth for component (a) when the disturbing waveform bandwidth is narrower than the effect bandwidth.

This is also the reason why many damped sinusoidal or Mesoband sources and antennas have been built by many nations. Given these observations about the damped sinusoidal waveform, it is prudent to review its characteristics. In Figure 5.32(a), we show a damped sinusoidal waveform in the time domain. We show its magnitude spectrum in Figure 5.32(b).

The analytical expression for the time-domain waveform shown in Figure 5.32(a) is given by (5.27):

$$E(t) = E_o e^{-\alpha t} \sin(\omega_o t) u(t) \tag{5.27}$$

The analytical expression for the frequency-domain waveform shown in Figure 5.32(b) is given by (5.28):

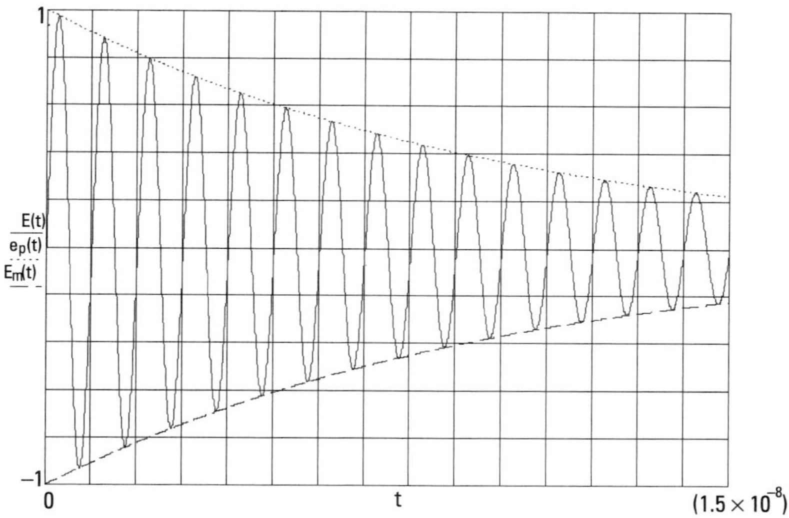

(a)

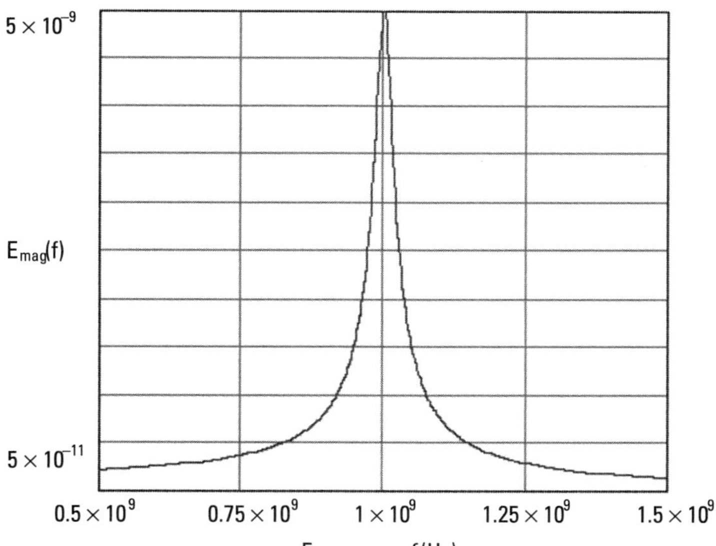

(b)

Figure 5.32 Damped sinusoidal waveform in (a) the time domain and (b) the frequency domain.

$$\tilde{E}(f) = \frac{\omega_o E_o}{\left(\alpha^2 + \omega^2 - \omega^2\right) + 2j\alpha\omega} \tag{5.28}$$

In practice, the effect bandwidth is very difficult to resolve as the in-band or passband response of a device will not only be a function of specific device parameters but also be composed of parasitic capacitances and inductances that are a function of the location or situation of the component.

5.7 Summary

This chapter has summarized effects mechanisms and introducted the concept of response norms and signal indicators. A response norm is a scalar parameter defined from a waveform, spectrum, or other function that obeys certain mathematical requirements and serves to classify the response. Signal indicators are similar parameters to norms in that they also are single, scalar values that are derived from a more complicated signal. Although they are not true norms in a mathematical sense, nevertheless, they are also useful in describing and classifying signals.

Table 5.10 provides a tabulation of the norms N1–N5, which are typically used for waveform characterization, together with an indication as to why the norm is of particular interest.

In this chapter, some of the most frequently used norms and parameters for transient and spectral responses in the electromagnetics area are reviewed, and various ways of using these parameters are suggested. It should be remembered that simply calculating a norm or a response indicator is not the final goal. The calculated parameters must be used in some sort of comparative way to infer a fact about the EM behavior of the system being tested.

Table 5.10
Waveform Norms Used for HPEM Waveforms

p-Norm	Norm Quantity	Name	Use
$\|R\|_\infty$	$N_1 = \lvert R(t) \rvert_{max}$	Peak (absolute) value	Circuit upset
NA	$N_2 = \lvert \partial R(t) / \partial t \rvert_{max}$	Peak (absolute) value of derivative	Out-of-band highpass filter
NA	$N_3 = \left\lvert \int_0^t R(x)dx \right\rvert_{max}$	Peak (absolute) impulse	Out-of-band lowpass filter
$\|R\|_1$	$N_4 = \int_0^\infty \lvert R(x) \rvert dx$	Rectified total impulse	Equipment discharge
$\|R\|_2$	$N_5 = \left\{ \int_0^\infty \lvert R(x) \rvert^2 dx \right\}^{1/2}$	Root of action integral	Component burnout

We have reviewed effects data and effects mechanisms from front-end effects such as interference or jamming through to damage and destruction. Damage mechanisms are of two types, thermal and nonthermal. Thermal effects result in the destruction of semiconductor junctions. Thermal damage mechanisms have been reviewed.

One of the important aspects of effects on electronics lies in the wide variation of the incident parameters of the disturbing environment such as pulse amplitude, rate of rise, pulse width, *prf*, and burst duration. The impact on effects of these incident parameters has been examined.

For electronics under Hypoband illumination, rectification is usually the principal mode by which the microwave energy is coupled into the systems. It is the cause of computer malfunction under radar illumination, of erroneous output from an EKG when a physician's paging system is activated, and of a stereo system receiving a citizen's band (CB) signal. It is also because of rectification that warnings are posted for heart pacemaker interference from microwave ovens and airport radar transmitters. Microwave energy coupling via rectification generally occurs by the signal entering a victim amplifier and digital circuit through interconnecting signal and/or power cables and perhaps being enhanced by parasitic resonances. A typical response is then characterized by the microwave signal being propagated to a nonlinear device, such as the detector of an AM radio or the bipolar junction transistor input to a digital gate. The resulting nonlinear response produces a video pulse or a wideband signal that propagates throughout the electronic system, perhaps as a legitimate signal upsetting the normal data transmission and storage. In some cases, overstresses may damage system components. The resulting effects may be temporary (i.e., ceasing as soon as the source is removed), or they may be permanent.

This chapter has provided a scientific or physical understanding of different effects mechanisms and some theories about how HPEM interacts and causes effects. However, an effect is only real if it sufficiently disturbs the function of the device or circuit to a degree that a human can observe or experience the effect as discussed in the preceding paragraph. The next chapter discusses the more subjective human aspect and functional aspects of effects manifestation and how a classification scheme can be applied to a subjective human experience of effects.

References

[1] International Electrotechnical Comission (IEC), *Electropedia: The World's Online Electrotechnical Vocabulary*, http://www.electropedia.org/. Accessed September 2019.

[2] Baum, C. E., "Norms and Eigenvector Norms," *AFWL Mathematics Notes*, Note 63, November 1979.

[3] Horowitz, P., and W. Hill, *The Art of Electronics*, 3rd ed., Cambridge, U.K.: Cambridge University Press, 2015.

[4] Lui, P. L., "Passive Intermodulation Interference in Communication Systems," *IEEE Electronics & Communication Engineering Journal*, Vol. 2, No. 3, June 1990, pp. 109–118.

[5] Schleher, D. C., *Electronic Warfare in the Information Age*, Norwood, MA: Artech House, 1999.

[6] Poisel, R., *Modern Communications Jamming Principles and Techniques*, 2nd ed., Norwood, MA: Artech House, 2011.

[7] Xu, W., et al., "The Feasibility of Launching and Detecting Jamming Attacks in Wireless Networks," *Proceedings of the 6th ACM International Symposium on Mobile Ad Hoc Networking and Computing*, New York, 2005, pp. 46–57.

[8] Pelechrinis, K., M. Iliofotou, and S. V. Krishnamurthy, "Denial of Service Attacks in Wireless Networks: The Case of Jammers," *IEEE Communications Surveys & Tutorials*, Vol. 13, No. 2, 2011, pp. 245–257.

[9] Nilsson, T., R. Malmqvist, and M. Bäckström, "Investigation of HPM Susceptibility Levels on Low Noise Amplifiers," *Proceedings of EMC Europe 2006*, Barcelona, Spain, September 4–8, 2006.

[10] Månsson, D., et al., "Susceptibility of Civilian GPS Receivers to Electromagnetic Radiation," *IEEE Transactions on Electromagnetic Compatibility*, Vol. 50, No. 2, May 2008, pp. 434–437.

[11] Stenumgaard, P., L. Pääjärvi, and K. Fors, "Radiated Emission from Personal Computers–A Safety Risk for GPS Receivers?" *Proceedings of EMC Europe Workshop 2007*, Paris, France, June 14–15, 2007.

[12] Razavi, B., *RF Microelectronics*, 2nd ed., Boston, MA: Pearson Education, 2012.

[13] Tesche, F. M., "Voltage and Current Surge Characteristics for Buried and Above-Ground Cables Excited by a NEMP," DPA NEMP Laboratory, Spiez, Switzerland, October 1, 1997.

[14] Tesche, F. M., "Methodology and Models for Estimating HPM Responses Conducted into a Protective Enclosure," *Interaction Notes*, Note 518, August 1996.

[15] Muller, B., and J. Reinhardt, *Neural Networks*, New York: Springer-Verlag, 1990.

[16] de Moraes, R. M., and S. M. Anlage, "Effects of UHF Stimulus and Negative Feedback on Nonlinear Circuits," *IEEE Transactions on Circuits and Systems I: Regular Papers*, Vol. 51, No. 4, 2004.

[17] Kohlberg, I., "A Stochastic Process and Chaos Interpretation of HPE and HPM Effects on Electronic Systems," *Asia-Pacific International Symposium on Electromagnetic Compatibility*, 2010.

[18] Jha, A. G., A. P. Das, and A. Kumar, "Effects of Electromagnetic Interference on Non-Autonomous Chaotic Circuits," *4th International Conference on Computers and Devices for Communication (CODEC)*, 2009.

[19] Kandangath, A., et al., "Inducing Chaos in Electronic Circuits by Resonant Perturbations," *IEEE Transactions on Circuits and Systems I: Regular Papers*, Vol. 54, No. 5, 2007.

[20] Booker, S. M., et al., "Designing Input Signals to Disrupt Commercial Systems in Band—A Nonlinear Dynamics Approach," *IEEE Transactions on Circuits and System*, Vol. 49, No. 5, May 2002, pp. 639–645.

[21] Nielsen, P. E., *Effects of Directed Energy Weapons*, Washington, D.C.: National Defense University Press, 1994.

[22] Rudolph, M., et al., "Analysis of the Survivability of GAN Low-Noise Amplifiers," *IEEE Transactions on Microwave Theory and Techniques*, Vol. 55, No. 1, January 2007, pp. 37–43.

[23] Hoijer, M., et al., "Pulse Length and Power Dependency of the Failure Threshold of a Low Noise Amplifier," *EUROEM 2008*, Lausanne, Switzerland, July 2008.

[24] Colangeli, S., et al., "GAN-Based Robust Low-Noise Amplifiers," *IEEE Transactions on Electron Devices*, Vol. 60, No. 10, October 2013, pp. 3238–3248.

[25] Nilsson, T., and R. Jonsson, "Investigation of HPM Front-Door Protection Devices and Component Susceptibility," FOI Technical Report, FOI-R--1771--SE, Swedish Defence Research Agency FOI, Sensor Technology, Linköping, Sweden, November 2005.

[26] Zhou, L., et al., "Experiments and Comparisons of Power to Failure for SiGe-Based Low-Noise Amplifiers Under High-Power Microwave Pulses," *IEEE Transactions on Electromagnetic Compatibility*, Vol. 60, No. 5, 2018, pp. 1427–1435.

[27] Baek, J. -E., Y. -M. Cho, and K. -C. Ko, "Analysis of Design Parameters Reducing the Damage Rate of Low-Noise Amplifiers Affected by High-Power Electromagnetic Pulses," *IEEE Transactions on Plasma Science*, Vol. 46, No. 3, 2018, pp. 524–529.

[28] Taylor, C. D., and D. V. Giri, *High-Power Microwave Systems and Effects*, United Kingdom: Taylor and Francis, 1994.

[29] Dwyer, V. M., A. J. Franklin, and D. S. Campbell, "Thermal Failure in Semiconductor Devices," *Solid-State Electronics*, Vol. 33, No. 5, 1990, pp. 553–560.

[30] Göransson, G., "HPM Effects on Electronic Components and the Importance of This Knowledge in Evaluation of System Susceptibility," *Proceedings of the IEEE EMC Symposium*, Seattle, WA, 1999.

[31] Esser, N., and B. Smailus, "Measuring the Upset of CMOS AND TTL Due to HPM-Signals," *Digest of Technical Papers, 14th IEEE International Pulsed Power Conference (PPC-2003) (IEEE Cat. No.03CH37472)*, Vol. 1, 2003.

[32] Bell Laboratories, *EMP Engineering and Design Principles*, Technical Publications Department, 1975.

[33] Roe, J. M., and V. G. Puglielli, "Using the Integrated Circuit Electromagnetic Susceptibility Handbook to Assess the Susceptibility of Electronic Systems," *Proceedings of the Symposium and Technical Exhibition on EMC*, Rotterdam, Holland, 1979.

[34] Miller, D. B., et al., "The Effects of Steep-Front, Short-Duration Impulses on Power Distribution Components," *Digest of the IEEE/PES Summer Meeting*, Long Beach, CA, July 10–14, 1989.

[35] Everett, III, W. W., and W. W. Everett, Jr., "Microprocessor Susceptibility to RF Signals-Experimental Results," *Proc. of the 1984 Southeast Conference*, April 1984, pp. 512–516.

[36] LoVetri, J., A. T. M. Wilbers, and A. P. M. Zwamborn, "Microwave Interaction with a Personal Computer: Experiment and Modelling," *Proceedings of the 1999 Zurich EMC Symposium*, 1999.

[37] Bäckström, M., et al., "Susceptibility of Electronic Systems to High-Power Microwaves: Summary of Test Experience," *IEEE Transactions on Electromagnetic Compatibility*, Vol. 46, No. 3, August 2004.

[38] Bäckström, M., B. Nordstrom, and K. G. Lovstrand, "Is HPM a Threat Against Civil Society?" *Proceeding of the International Union of Radio Science (URSI) General Assembly*, Maastricht, Netherlands, 2002.

[39] Bäckström, M., "HPM Testing of a Car: A Representative Example of the Susceptibility of Civil Systems," *Workshop W4, Proceedings of the 13th International Zurich Symposium and Technical Exhibition on EMC*, February 1999.

[40] Nitsch, D., et al., "Susceptibility of Some Electronic Equipment to HPEM Threats," *IEEE Transactions on Electromagnetic Compatibility*, Vol. 46, No. 3, August 2004, pp. 380–389.

[41] Mojert, C., et al., "UWB and EMP Susceptibility of Microprocessors and Networks," *EMC Zurich Symposium*, February 2001.

[42] Nitsch, D., "The Effects of HEMP on Complex Computer Systems," *Proceedings of the 17th International Zurich Symposium and Technical Exhibition on EMC*, February 2005.

[43] Camp, M., H. Garbe, and D. Nitsch, "Influence of the Technology on the Destruction Effects of Semiconductors by Impact of EMP and UWB Pulses," *Proceedings of the IEEE International Conference on EMC*, Minneapolis, MN, August 2002.

[44] Hoad, R., et al., "Trends in EM Susceptibility of IT Equipment," *IEEE Transactions on Electromagnetic Compatibility*, Vol. 46, No. 3, August 2004.

[45] Hoad, R., et al., "An Investigation into the Radiated Susceptibility of IT Networks," *Conference Proceedings of EMC Europe*, Eindhoven, The Netherlands, September 2004.

[46] Hoad, R., A. Lambourne, and A. Wraight, "HPEM and HEMP Susceptibility Assessments of Computer Equipment," *EMC Zurich in Singapore*, Singapore, Asia, February 2006.

[47] Tesche, F. M., "CW Test Manual," NEMP Laboratory, Spiez, Switzerland, December 7, 1994.

[48] Papoulis, A., *The Fourier Integral and Its Applications*, New York: McGraw Hill, 1962.

[49] Hildebrand, F. B., *Advanced Calculus for Applications*, Englewood Cliffs, NJ: Prentice-Hall, 1963.

[50] Bingham, E. O., *The Fast Fourier Transform*, Englewood Cliffs, NJ: Prentice Hall, 1974.

[51] Chang, D. C., *Analysis of Linear Systems*, Reading, MA: Addison Wesley, 1962.

[52] IEC 61000-4-36, "Electromagnetic Compatibility (EMC) – Part 4-36: Testing and Measurement Techniques – IEMI Immunity Test Methods for Equipment and Systems," 2019.

[53] Wunsch, D. C., and R. R. Bell, "Determination of Threshold Failure Levels of Semiconductor Diodes and Transistors Due to Pulse Voltages," *IEEE Transactions on Nuclear Science*, Vol. 15, No. 6, February 1968, pp. 244–259.

[54] Tasca, D. M., "Pulse Power Failure Modes in Semiconductors," *IEEE Transactions in Nuclear Science*, Vol. 17, 1970, pp. 364–372.

[55] Parfenov, Y. V., et al., "The Probabilistic Analysis of Immunity of a Data Transmission Channel to the Influence of Periodically Repeating Voltage Pulses," *Asia-Pacific EMC Week and Technical Exhibition*, Singapore, May 19–23, 2008.

[56] Chepelev, V. M., Y. V. Parfenov, and Y. -Z. Xie, "One of Ways to Choose UWB Pulse Repetition Rate for Assessment of the Electronic Devices Immunity," *7th IEEE International Symposium on Microwave, Antenna, Propagation, and EMC Technologies (MAPE)*, 2017, pp. 240–242.

[57] Sabath, F., E. L. Mokole, and S. N. Samaddar, "Definition and Classification of Ultra-Wideband Signals and Devices," *URSI Radio Science Bulletin*, Vol. 2005, No. 313, June 2005, pp. 12–26.

[58] Balmer, L., *Signals and Systems: An Introduction*, 2nd ed., Upper Saddle River, NJ: Prentice Hall, 1997.

6

Classification and Implications of HPEM Effects

6.1 Introduction

The last two decades have witnessed an increased interest in HPEM, particularly the generation of high-power electromagnetic fields and their effects on electronics. As components for Hypoband (i.e., HPM), mesoband, sub-hyperband, and hyperband technologies have achieved notable progress, high-power systems difficult or impossible to build 10 years ago are now being used for an increasingly wide variety of applications. With the advent of HPEM, sources capable of producing peak output powers in the gigawatt range, there has been interest in using HPEM generators in military defense applications to disrupt or destroy offensive electronic systems. In numerous publications, it has been reported that malicious actors such as criminals and terrorists have the possibility to interrupt and/or damage sensitive electronics by generating IEMI [1–3]. Natural HPEM environments such as lightning have also long been known to cause interference and damaging effects.

Electronic components and subsystems (e.g., microprocessor boards) are essential parts of modern civilian and military systems such as aircraft, communications, IT infrastructure, traffic management, and safety systems. Since these electronic components now control safety or security-critical functions, concern has grown over the vulnerability or failure of electronic systems from a variety of different externally impacting disturbances. Therefore, the effects of safety and mission-critical systems is of vital interest since an upset or failure in these systems could cause major accidents or economic disasters. The increase of nonmetallic materials like carbon-fiber composites, which are now extensively used in automotive and aircraft applications to house electronic systems, as well as the decrease of signal levels and increase in clock speeds in modern semiconductor-based electronic devices are likely to result in a decreased effects threshold level of electronic systems from HPEM exposure. Consequently, the investigation of the HPEM effect on electronic systems as well as their protection and hardening against HPEM environments is of great interest.

Due to the design and the functionality of electronics, the employed HPEM test environment and the setup of effects tests, the observed effects differ significantly from each other. In addition, manufacturers of electronic systems are reluctant to have the effects data of their systems published and discussed in public. Therefore, a scientific discussion is needed that provides a categorization or classification scheme for HPEM effects that performs two key objectives: (1) the scheme must summarize the essential information without giving away too much detail on the system, and (2) the scheme must enable a comparison of different manifestations of HPEM effects in different systems.

As we showed in the previous chapter, effects caused by an HPEM environment can be characterized in a variety of ways. For example, observed effects can be described by attributes of the physical mechanism such as:

- Interference: noise background that masks the wanted signal;
- Bit flip/false information: inversion of a data bit generating false data;
- Transient upset: extending the data bit period, perhaps indefinitely;
- Flashover: dielectric breakdown of insulating structures that may be as small as a few microns thick, by high voltages;
- Wire melting: exceeding some thermal limit of a conductive structure;
- Bond wire destruction: fusing or vaporizing semiconductor bond wires through a combination of high voltage and thermal effect.

Alternatively, one can classify EM effects by the duration of the effect and the need for human intervention (i.e., power cycling: power off/power on) to recover normal operation. Such a scheme could range from no effect through temporary effect to permanent effect (e.g., destruction).

A third approach for the classification of EM effects is by the implication on the main (or critical) function of the system [4, 5]. The categories can range from no effect through interference and degradation to loss of main function or mission kill.

This chapter will discuss advantages and disadvantages of each of these effect classification techniques regarding system-level assessment. A categorization scheme that is most useful to classify system-level effects regarding the comparison and analysis will be presented.

6.2 Classification of EM Effects

Effects tests demonstrate that HPEM environments are capable of causing interference, malfunctions, upsets, and destruction of electronic equipment, even if they comply with standard EMC specifications or military-hardening requirements.

Over the last few years, a large variety of electronic and electrical devices, equipment, and systems has been scrutinized to collect information on the physical mechanism caused through HPEM exposure, the critical parameters of the HPEM disturbance, and possible protection measures [6–11]. In order to develop a knowledge base that is of general value and independent of specific systems, diverse systems and categories of systems (e.g., automotive, computer, IT systems, motherboards) were tested. The wide variety of systems that have undergone HPEM effect testing has resulted in a large database covering the various manifestations of HPEM effects. However, comparison and analysis of the obtained raw data remain difficult and are still a great challenge. The effect data are still relatively sparse compared to other data sets. A way to analyze the effect data and identify tendencies or principles is to extract the essential information by use of a classification scheme.

6.2.1 Effect Classification by Mechanism

Initially, HPEM effect investigations were performed to discover the physical mechanism of interference and destruction as well as the effective EM coupling. Consequently, the characterization used to analyze effects data focused on the physical mechanism, causing the observed effect. Such a mechanism-based classification scheme is shown in Table 6.1.

The first category U has an academic quality and is of minor interest for practical work. No effect data is useful in HPEM analysis, from a hardening and protection perspective, as it can be used to infer the immunity level of the equipment or system to the HPEM environment. All other categories can be subdivided into interference and destructive effects. The shown two-stage classification follows the practical classification process. During HPEM effects testing, an observed effect can be easily mapped to interference (e.g., the system recovers) or destruction (e.g., the system does not recover). The identification of which the destruction mechanism occurred often needs additional detailed examinations of the system (e.g., opening of ICs and detailed on-chip investigation). As the destructive effects are of a permanent nature, usually the

Table 6.1
Classification of EPEM Effects by Mechanism

Category		Effect	Description
	U	Unknown	Unable to determine due to effects on another component or not observed
	N	No effect	No effect occurs
Interference	I.1	Noise	Raised noise level on system signal and power lines, which results in flashing of displays or reduced data rates
	I.2	Bit flip	HPEM exposure causes corruption of bits in a data stream
	I.3	Failure	Malfunction of the system/component due to HPEM exposure
	I.4	Breakdown	Hang-up or crashing of software
Destruction	D.1	Latch-up	HPEM exposure causes latch-up in semiconductor components
	D.2	Flashover	On-chip flashover/flashover in components
	D.3	On-chip wire melting	Wires on chip are melted by coupling of HPEM energy
	D.4	Bond wire destruction/wire melting on PCB	Wires on PCB and/or bond wires in semiconductor devices are melted by coupling of HPEM energy

time consumption of such additional detailed examination is not a problem. In contrast, the temporary nature of interference-type effects requires a real-time measurement of signals and additional measurements that enable observation, storage, and analysis of internal stages of the system under investigation.

The classification by effect mechanisms enables the analysis of the impact of different HPEM environment parameters on a particular mechanism. For example, Camp and Garbe [11] used the quantities' breakdown failure rate (BFR) and breakdown threshold (BT) to describe the impact of signal attributes of fast transient pulses (e.g., peak electric field) on the breakdown (category I.4) of personal computer systems.

Table 6.1 presents an example of a classification of HPEM effects by mechanism. As the categories are defined in regard to observed effects on the systems under test, some explanation and discussion are necessary.

Usually, a bit flip results in corruption of a data stream, which can be identified and/or corrected by appropriate data coding or bit error correction. Since most of the observed bit flips impact on the transfer or storage of user data, this effect is of a temporary nature and results in a reduction of the data transfer rate or, for example, in a temporary degradation of video images on display screens. It should be mentioned that a bit flip has the potential to cause a latch-up (hang-up) of software.

The categories of failure and breakdown, as described in Table 6.1, are part of the transition from interference to destruction. The focus of the categorization, shown in Table 6.1, is on the physical mechanism. Therefore, every effect on software is classified as interference and only hardware-related effects are classified as destruction. From the viewpoint of a user, a hang-up or crashing of software might be as bad as the physical destruction of a hardware component. For example, if a software operating system fails, it could take many hours of support from an IT expert to resolve perhaps by reinstalling the operating software.

Generally, a latch-up can be resolved by power cycling of the affected component or system. Latch-up is classified as a destructive effect due to its destructive nature since not every system or application (e.g., avionics) permits power cycling.

The main drawback of the classification by effect mechanism is the fact that the classification categories contain sensitive (and classified if conneccted to a military security system) information on the system under investigation. While talking about real-life military used electronic equipment or avionics, no one would admit a destruction of hardware or a damage of software. As the affected component is useless in any case, the specific realization of destruction is of no interest from the operational point of view. In addition, the description of any kind of interference effect (I.1–I.4) does not contain sufficient information to assess the effect in regard to the operational efficiency or functional/mission criticality. For example, a bit flip that occurs only during the exposure to the Hyperband environment can be detected and corrected by a capable channel coding scheme. Even if the coding scheme is not able to correct the bit flip, the system will be back to full operation after the Hyperband exposure is over.

6.2.2 Effect Classification by Duration

Nitsch and Sabath [12] presented a classification scheme that tried to overcome the discussed disadvantages of the classification by effect mechanism and suggested a classification scheme that assesses the observed effects from the user or operator point of view.

The duration of an HPEM effect provides the user with information on how long the desired function will be disturbed. The duration of an effect as a function of the HPEM environment affords an estimation of status of an electronic system. The initial version of the scheme, as well as the version improved by Nitsch and Sabath, used a mixture of the effect duration, the need for human intervention, and the destruction of components as differentiators. From a systematic point of view, it is desired to use a single differentiator only. Consequently, a revision results in the classification scheme shown in Table 6.2.

Table 6.2
Classification of EM Effects by Duration

Category	Duration	Description
U	Unknown	No effect occurs or the duration of an effect has not been observed (e.g., observer was unable to determine the duration due to effects on another component)
E	During exposure only	Observed effect is present only during exposure to the HPEM environment; system functionality is completely available after HPEM exposure is over
T	Some (follow-up) time after exposure	Effect is present sometime after HPEM exposure is over, but the system recovers without human intervention; follow-up time is shorter or equal to the typical reaction/operation cycle of the system
H	Resistant until human intervention	Effect is present until human intervention (e.g., reset, restart of function); due to the effect, the system is not able to recover to normal operation within an acceptable period (e.g., typical reaction/operation cycle of the system); no replacement of hardware or reload of software is necessary
P	Permanent or until replacement of hardware and software	Effect is permanent; the usual interactions of an operator or user do not recover normal operation; effect has damaged hardware to the point that is must be replaced or software to the point that it must be reloaded

The two main advantages of this classification scheme are that: (1) effects are characterized independently from the particular system and the main function, and (2) the criteria are objective rather than subjective. Only the decision between categories T and H does not support aspect (2) without restrictions. At this point, the need for human intervention requires some explanation. In most cases, a hang-up in a software or program (e.g., in the system software) can only be solved by a manually initiated reboot of the computer or a restart of the software. The situation becomes more complicated if the system software of an IT system (e.g., computer network or server) runs through an automatic reboot, but the status of normal operation requires a manual start of application software (or data stream). Some test engineers tend to classify this situation with category T as the system itself recovers without human intervention. As the main function needs the manual start of software, the situation can be categorized as H. In the practical case, the decision depends on whether the test focuses on the main application (this will lead to H) or the basic system (T).

The fact that the classification by duration does not provide information on the operational impact of the HPEM effect is disadvantageous when the assessment is particularly focused on the main or critical function or mission of the system. For example, a faulty video display, which is present for some follow-up time after the HPEM exposure is over (category T), can be critical in one case and of no interest in another case. Similarly, a destroyed maintenance data panel (category P) does not influence a running engine and has therefore no operational, functional, or mission impact.

6.2.3 Effect Classification by Criticality

If HPEM effects are analyzed and assessed regarding the operational impact and the functionality of the system, the operational condition (e.g., critical periods of time, critical functions, minimum performance) must be taken into consideration. However, such an analysis is working on a higher level of abstraction, as no details on the physical mechanism are needed.

Nitsch and Sabath introduced a classification of effects by its criticality for the main function or mission in [12, 13], shown in Table 6.3. This classification scheme provides the essential information on the functionality isolated from its duration.

Compared with previously discussed classification schemes, the classification by criticality of Table 6.3 requires analysis of the observed effect and its impact on the function of the system regarding a particular application. Therefore, this classification scheme depends on the application and its operational conditions. Due to this fact, test engineers require the support of system function or mission specialists to be able to accurately map an observed effect on to a criticality level.

In addition, the classification scheme contains aspects that are of interest for system function or mission specialists. Particularly, the effect description provides a relation between observed effects and the hardening status of the system under consideration. For example, if no effect *N* has been observed, the system can be assessed as immune to the specific HPEM environment. If observed effects belong to the level interference *I* or degradation *II*, the system is susceptible to the specific HPEM environment. Finally, a system that shows a level *III* effect (mission kill) must be assessed as vulnerable.

Table 6.3
Classification of EM Effects by Criticality

Level	Effect	Description
U	Unknown	Unable to determine due to effects on another component or not observed
N	No effect	No effect occurs or the system can fulfill its mission without disturbances
I	Interference	The appearing disturbance does not influence the main mission
II	Degradation	The appearing disturbance reduces the efficiency and capability of the system
III	Loss of main function (mission kill)	The appearing disturbance prevents that the system from fulfilling its main function or mission

From the aspect of the operational efficiency or operational restrictions, which are caused by the HPEM environment, sometimes criticality and the duration of the status (effect) are desired parts of the information. As the classification by criticality (Table 6.3) and the classification by duration (Table 6.2) present the information as a function of one isolated criterion, both classification schemes can be combined. Combinations with a practical relevance are listed in Table 6.4.

At this point, it should be noticed that the combined classification scheme of Table 6.4 carries two of four parts of information that are needed to determine the impact of the observed effect on complex systems. The missing parts of the information are the operational value (functional or mission criticality) and the requirements that are articulated by the system authority. For example, an effect of category *H.II* of a subsystem results in a classification of *E.I* if it is acceptable that the subsystem is running on a lower performance level for some time. In another case, any effect *E.II* or higher results in a criticality level of *III* if an undisturbed function or mission is required (i.e., the system must work through HPEM exposure).

6.3 Conclusions

This chapter discussed the need for a scientific method for categorization of HPEM effects, which provides the essential information and enables a comparison of different manifestations of HPEM effects. The classification of effects by physical mechanism, duration, and criticality were described. The advantages and disadvantages of the presented classification schemes regarding a system-level assessment on HPEM effects were also discussed.

Finally, it was shown that a combination of the classification by duration and the classification by criticality is most useful to assess the impact of HPEM effects. The combined classification scheme allows a determination of the classification of the effect at the system level in combination with the operational value of the system function or mission.

Table 6.4
Combination of Duration and Criticality

		Criticality Level				
		U	**N**	**I**	**II**	**III**
Duration Category	**U**	U	N	—	—	—
	E	—	—	E.I	E.II	E.III
	T	—	—	T.I	T.II	T.III
	H	—	—	H.I	H.II	H.III
	P	—	—	P.I	P.II	P.III

References

[1] Wik, M. W., R. L. Gardner, and W. A. Radasky, "Electromagnetic Terrorism and Adverse Effects of High-Power Electromagnetic Environments," *Supplement to Proceedings of the 13th International Zürich Symposium on EMC*, Zurich, Switzerland, 1999, pp. 181–185.

[2] Radasky, W. A., "Intentional Electromagnetic Interference (EMI) – Test Data and Implications," *Proceedings of the 14th International Zurich Symposium on EMC*, Zürich Switzerland, 2001.

[3] Ianoz, M., and H. Wipf, "Modelling and Simulation Methods to Assess EM Terrorism Effects," *Proceedings of the 13th International Zurich Symposium on EMC*, Zürich, Switzerland, 1999.

[4] Hoad, R., et al., "Trends in EM Susceptibility of IT Equipment," *IEEE Transactions on Electromagnetic Capability*, Vol. 46, No. 3, August 2004, pp. 390–395.

[5] Giri, D., and F. Tesche, "Classification of Intentional Electromagnetic Environments (IEME)," *IEEE Transactions on Electromagnetic Capability*, Vol. 46, No. 3, August 2004, pp. 322–328.

[6] LoVetri, J., and A. Wilburs, "Microwave Disturbance of a Personal Computer: Experimental and FDTD Simulations," *13th International Zurich Symposium and Technical Exhibition on EMC*, February 1999, pp 203–206.

[7] Backstrom, M., "HPM Testing of a Car: A Representative Example of the Susceptibility of Civil Systems," *Workshop W4, 13th International Zurich Symposium and Technical Exhibition on EMC*, February 1999, pp. 189–190.

[8] Nitsch, D., et al., "Susceptibility of Some Electronic Equipment to HPEM Threats," *IEEE Transactions on Electromagnetic Capability*, Vol. 46, No. 3, August 2004, pp. 380–389.

[9] Backström, M., and K. G. Lovstrand, "Susceptibility of Electronic Systems to High-Power Microwaves: Summary of Test Experience," *IEEE Transactions on Electromagnetic Capability*, Vol. 46, No. 3, August 2004, pp. 396–403.

[10] Mojert, C., et al., "UWB and EMP Susceptibility of Modern Computer Networks," *Proc. EMC Zürich 2001*, Zurich, Switzerland, February 2001.

[11] Camp, M., and H. Garbe, "Susceptibility of Personal Computer Systems to Fast Transient Electromagnetic Pulses," *IEEE Transactions on Electromagnetic Capability*, Vol. 48, No. 4, November 2006, pp. 829–833.

[12] Nitsch, D., and F. Sabath, "Electromagnetic Effects on Systems and Components," *AMEREM 2006*, July 2006.

[13] Sabath, F., "Classification of Electromagnetic Effects at System Level," *Ultra-Wideband, Short Pulse Electromagnetics 9*, New York: Springer Science + Business Media, 2010, pp. 325–334.

7

HPEM Protection Concepts and Methods

7.1 Introduction

Over the past 5 decades, significant progress has been made in understanding and mitigating the effects of HPEM environments, including lightning. Lightning protection design largely evolved for the protection of high value or highly symbolic structures such as churches and initially involved simply shunting hazardous lightning transient currents to ground through appropriate use and positioning of lightning conductors. Today, there is a wide range of protection technologies available including transient and surge protection devices.

Lightning protection of equipment and installations continues to be a topic of interest and recent developments have seen the introduction of standards that take a risk-based approach to installation lightning protection implementation.

For HEMP E1, most of the early work focused on the mitigation of HEMP effects on military electrical systems and equipment. Starting from early documents on the characteristics of HEMP [1, 2] mainly for the military and running through International Electrotechnical Commission (IEC) work, mainly for the civilian community, standards and guidelines for HEMP protection have been developed [3–5] and continue to be improved upon.

With the emergence of other HPEM environments produced by HPRF DE and IEMI sources such as hyperband, mesoband, and hypoband environments [6], whose operating frequency spectra extend well beyond several gigahertz [7], new protection techniques have been identified and have started to be developed.

The concept of HPEM resilience, which is informed by HPEM detection, is an emerging technique and is discussed herein.

7.2 Shielding Topology Protection Concept

The overall shielding topology protection concept for an installation is shown in Figure 7.1.

Different elements of this design approach mitigate either radiated or conducted HPEM environments. The overall principle is to enclose all of the potentially vulnerable electronics within a closed metal shield and to either:

- For shielded cables and conducting penetrations: to decouple conducted disturbances at the shield by making very good quality electrical bonds of the cable shield to the outside of the enclosure shield;
- For unshielded conducting penetrations: to use good bonding and a nonlinear protective device to shunt the coupled current to Earth via the enclosure shield.

The shielding topology concept is tried and tested and has been shown to be verifiable and capable of continued or progressive assurance. As long as the shielding performance is maintained and new penetrations into the shielded enclosure are managed so that they do not violate the shield boundary, then

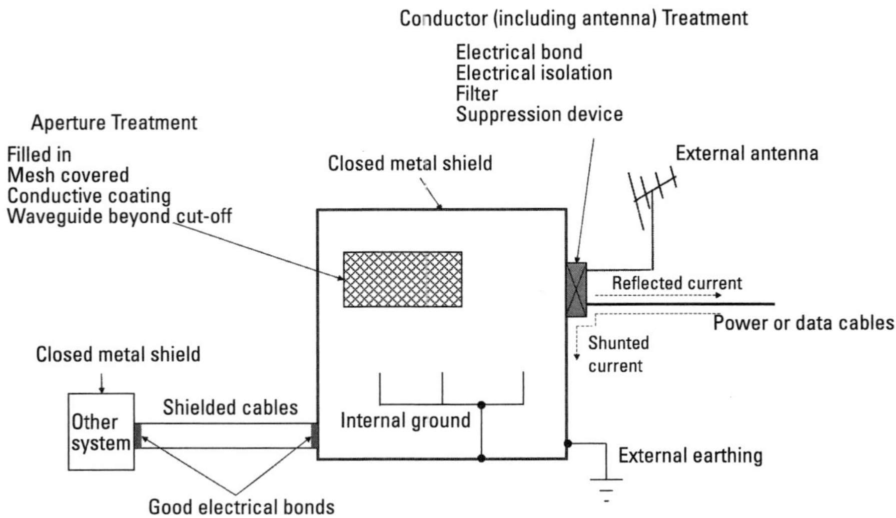

Figure 7.1 Shielding topology protection concept.

this protection concept can provide better than 80 dB of isolation for sensitive electronics inside of the shield from external HPEM environments.

The essence of protection against HPEM environments is to maintain shielding integrity and not allow interfering currents to flow into the protected systems [8]. The first step is to understand the shielding topology and identify surfaces where unwanted signals can penetrate. A way of formalizing or visualizing the complex interaction problem is via shielding topology modeling.

7.2.1 Shielding Topology Modeling

One of the ways of analyzing HPEM interaction mechanisms to systems is by using EM shielding topology modeling. EM shielding topology modeling, which is both qualitative and quantitative, provides rough estimates or bounds to the system response.

EM topology modeling requires the description of the shielding enclosure(s) surrounding potentially vulnerable devices, circuits, and equipment for the purpose of providing EM protection. It involves a description of the size, shape, and other properties of the enclosing shields, the locations and properties of imperfections (both deliberate and unintentional) in the shield, and a description of the signal propagation paths in and through the shields. This idea was defined by Baum and was later formalized by Tesche and others in the literature [9–11]. To understand HPEM effects on a complex, shielded system, we can think of the system as having several layers of conducting surfaces that shield the interior. We can then use topological concepts in estimating the system response. The process is broadly as follows:

- The system is examined for the principal shields or EM barriers;
- Imperfections (openings) in these barriers are noted and categorized;
- An EM signal flow diagram is constructed;
- Models are developed for the most important aspects of the EM signal paths;
- An estimation of the equipment responses to the EM excitation is determined using a variety of methods.

There are also other uses of EM shielding topology in that it can:

- Assist in the design of new systems with EM hardening or protection requirements;
- Provide guidance for EM hardness verification testing;
- Aid in the determination of hardness critical items;

- Provide a starting point for hardness surveillance and maintenance (HM/HS) programs;
- Help in configuration control of a system.

The first step in model development is to determine the topological diagram. This is a description of the principal shielding surfaces in the system and their interrelations to each other. A very simple EM shielding topology model diagram, in this example, an aircraft, is shown in Figure 7.2.

The exact shape or geometry of the shielded enclosure is unimportant in drawing the topological diagram. The shape is important in estimating the external interaction. Shapes can be morphed by stretching, compressing, and twisting, while cutting and joining are prohibited. In the HPEM analysis, the modeling of EM field interaction with a complex system or facility has been accomplished by the introduction of the concept of EM topology.

To develop a computational model for the EM responses, the concept of the interaction sequence diagram is introduced. To determine this diagram, the topological diagram of the protected item of interest is analyzed for all penetrations and the paths that the energy can take in propagating from the outside EM field source to the internal components and they are catalogued. Realizing that the most important paths are usually the hard-wired or conductive paths, a considerable amount of work has been conducted on developing ways of computing the behavior of current and voltage on interconnected wire networks.

By using transmission-line concepts, which result from a simplification of a complete solution of Maxwell's equations, a computationally simple model

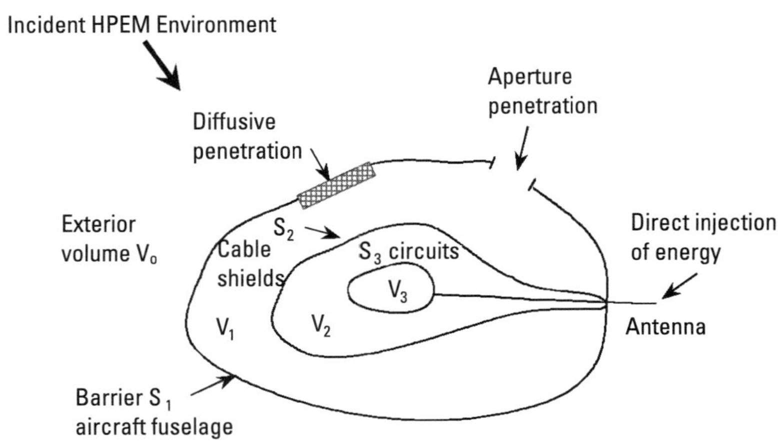

Figure 7.2 Simple EM topology model for an aircraft showing three ways of EM energy penetration into a system.

that predicts the propagation of EM energy into the system has been developed. This model is based on the Baum-Liu-Tesche (BLT) equation, which is a matrix equation describing the behavior of the voltage and/or current at all of the junctions (or interconnections) of the conductors in the network. This equation is similar to a node analysis performed in conventional circuit theory, except that in the BLT equation, the effects of EM propagation along the individual conductors are considered.

This methodology permits any system to be viewed as a collection of conducting surfaces (or barriers) that attenuate the incident EM field environment and effectively shield the interior regions and sensitive electronics from the external EM field. In most systems, the conductive portions of the EM barrier are not perfect, as there will be conductors passing through the surfaces. External HPEM environments can penetrate into the system via currents flowing on these conductors. For a building installation, for example, these conductive penetrations arise from the telephone, communications, and power conductors and other nonelectrical conductors. In addition to these penetrations, there are several apertures (holes) in the system enclosure (e.g., the entrances or the windows of a building), and EM fields may penetrate into the interior at these points. Finally, EM field diffusion through the building walls and roof is a possibility. The above are three fundamental ways by which external EM energy can penetrate into the system to be protected.

7.2.1.1 Interaction Sequence Diagram (ISD)

To develop a way for estimating the HPEM responses inside the system, the interaction sequence diagram is used. ISD is a representation of the principal paths that the EM energy takes as it penetrates into the system. To determine this diagram, the topological representation of the system (Figure 7.2) is examined for all penetrations, and the propagation paths from the outside HPEM field source to the internal components are identified.

Because the most important paths are the conductive paths, many models are available to predict the behavior of current and voltage on such paths and ultimately, at internal equipment or components that may be affected. ISD is essentially an EM coupling path wiring diagram for the system. It shows how externally produced HPEM environments can:

- Couple to external parts of the system;
- Penetrate through the shielding barrier;
- Propagate to equipment and components inside the system;
- Ultimately deposit energy at the equipment or component level.

Ultimately, the signals at the component or equipment located in a subvolume of interest are estimated. This is done by developing signal transfer functions for the coupling, penetration and propagation phenomena that occur on the ISD. Analysis can be performed, involving more electrical details of the ISD signal paths.

7.2.2 Shielding Mechanisms

EM topology modeling and practical measurements have shown us that real shields are not perfect, and the external HPEM environment can enter into a system by one or more of the following mechanisms, using the acronym CAD:

- Conductive penetrations, formed by wires, cables, or other conductors;
- Aperture penetrations through holes in the shield;
- Diffusion through the barrier material.

These points pertaining to the CAD mechanisms will subsequently discussed, in reverse order.

7.2.2.1 Shielding Barrier Materials and Diffusion (D)

EM field diffusion through the shield material is a possibility, if the shield material is not highly conducting or not thick enough at the frequencies or wavelengths relevant to the HPEM environment of interest. A shield is defined as a layer, or surface, within a system that impedes the passage of EM energy from the outside to the inside. Shielding forms the basis of EM topology control, as illustrated in Figure 7.2. As noted in the illustration, a shield is often thought of as a highly conducting surface surrounding the protected volume. The behavior of a physical shield can be compromised in several different ways. EM fields can diffuse through the shield material to provide an internal response. Their contribution to the internal response is generally much less than that of the conductive or radiated aperture penetration.

The shielding topology concept can be visualized as a loop or current flow over the surface of the shield boundary. Even a simple loop is a shield. In reality, shielding can be visualized as many loops in different directions, which effectively become a spherical shield. For the spherical or global shield concept, there is a back EMF inducing a current that creates a scattered field opposing and reducing the incident field.

We can define the magnetic field shielding factor or shielding effectiveness by (7.1) as:

$$S_H = -20\log_{10}\left(\frac{B_{inside}}{B_{incident}}\right) \quad (dB) \tag{7.1}$$

Similarly, we can define a shielding factor or shielding effectiveness for the electric field by (7.2) as:

$$S_E = -20\log_{10}\left(\frac{E_{inside}}{E_{incident}}\right) \quad (dB) \tag{7.2}$$

The shielding factor or shielding effectiveness generally describes the attenuation of the E-fields, the H-fields, or even the power density in the fields due to the topological barrier. It can involve static (DC), time harmonic, or transient EM fields. For physical shields, the shielding material is described by the following electrical parameters: electrical conductivity of the metal σ and magnetic permeability $\mu = \mu_o \mu_r$.

In addition, the following physical parameters are also required to define the shielding: frequency f (or $\omega = 2\pi f$), material thickness Δ, and shape of shield.

Table 7.1 presents a list of the relative permeability and conductivity of common shield materials.

Maxwell's curl equations decouple at DC and the shielding becomes two separate problems of electrostatic and magnetostatic shielding. Electrostatic shielding is easy to accomplish by surrounding the region with a good electrical conductor as seen in Figure 7.3(a). Here electrical charges flow freely and cancel the internal fields. The internal electrostatic field is determined by field penetration through apertures or shield imperfections.

Thus, for electrostatic fields, it is not the thickness of the shield that is important, but the optical coverage or number of holes that determines the shielding, as shown in Figure 7.3(b).

Inside the shield, we note that the shielding will be impaired in a certain volume around the hole. Near the hole, gap in shielding, or aperture shown in Figure 7.3(b), the electrostatic field looks much like that of an electric dipole charge. Such a representation of the internal electrostatic field permits the development of an analytical representation of the electrostatic field penetrating through the shield. At a large distance r from a circular hole with a diameter d, the electrostatic field magnitude is given as

$$|E| \approx \frac{E_o}{6\pi}\left(\frac{d}{r}\right)^3 \tag{7.3}$$

Table 7.1
Constitutive Parameters of Common Material

Material	Relative Conductivity* (σ_r)	Relative Permeability (μ_r)
Silver	1.05	1
Copper-annealed	1	1
Gold	0.7	1
Chromium	0.664	1
Aluminum (soft)	0.61	1
Aluminum (tempered)	0.4	1
Zinc	0.32	1
Beryllium	0.28	1
Brass	0.26	1
Cadmium	0.23	1
Nickel	0.2	100
Bronze	0.18	1
Platinum	0.18	1
Tin	0.15	1
Steel (SAE 1045)	0.1	1,000
Lead	0.08	1
Monel	0.04	1
Conetic (1 kHz)	0.03	25,000
Mu-metal (1 kHz)	0.03	20,000
Stainless steel (430 Hz)	0.02	500

*Note: The listed conductivity is relative to copper, for which $\sigma = 5.827 \times 10^7$ S/m.

where E_o is the strength of the unperturbed, or incident, normal electrostatic field component. From examination of (7.3), it is possible to infer that the attenuation is more than 40 dB, only 2-hole diameters away.

At DC, only ferromagnetic materials permit shielding of the magneto-static field, and even then, this shielding is far from perfect. The shielding of magneto-static field is achieved by the high relative permeability (μ_r) of the shield attracting the magnetic field and causing a slight reduction of the interior magneto-static field. The magneto-static field distribution in the vicinity of a thin, permeable shell is shown in Figure 7.4.

One can derive the internal magneto-static field via (7.4) as:

$$H^{in} = \frac{H^{inc}}{1 + \left(\frac{2(\mu_r - 1)^2}{9\mu_r}\right)\left(1 - \frac{a^3}{b^3}\right)} \tag{7.4}$$

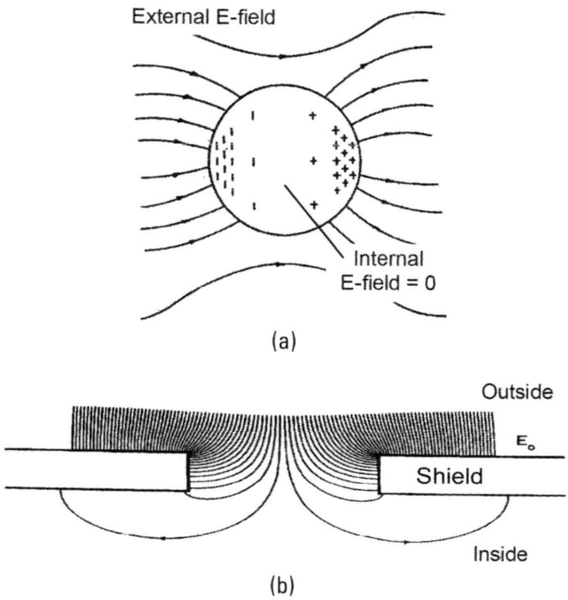

Figure 7.3 Electrostatic shielding. (a) Static shield and (b) compromised shield.

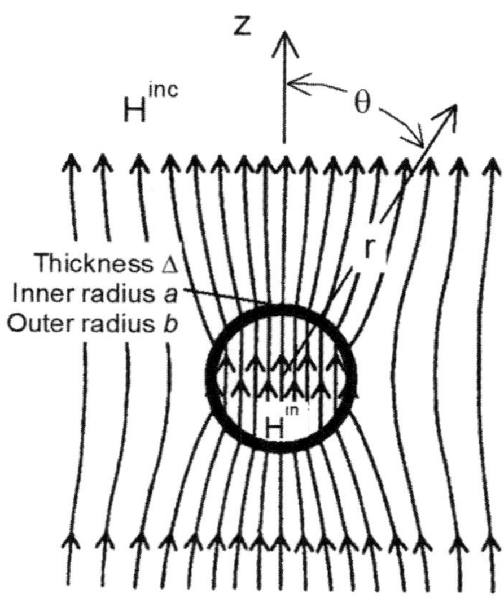

Figure 7.4 Magneto-static field around a spherical shell.

It is seen that the interior field becomes the incident field if $\mu_r = 1$. The shielding effectiveness for the case of $\mu_r = 400$ for two different shield thicknesses is shown in Figure 7.5.

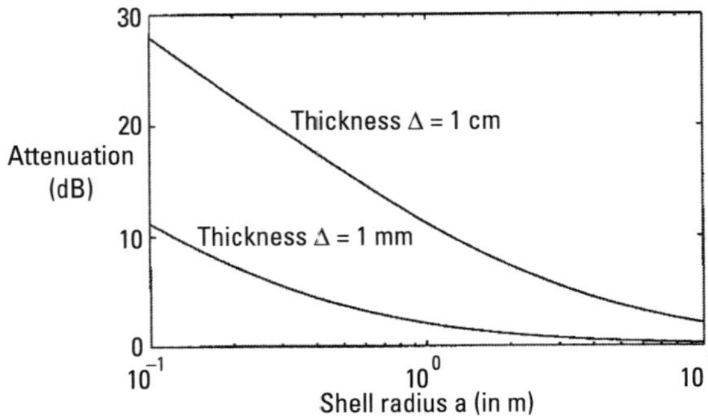

Figure 7.5 Magneto-static field attenuation for a spherical shell with $\mu r = 400$.

We now turn our attention to shielding of time varying plane-wave fields by planar shields. One way to explain dynamic shielding is by considering an infinite slab of conducting material, shown in Figure 7.6. The slab material is assumed to have a thickness Δ, an electrical conductivity σ, and a relative permeability μ_r.

The shielding factor or effectiveness for an incident plane wave field is denoted as S and can be defined [12] by (7.5) as:

$$S = A + R + B(\text{in dB}) \tag{7.5}$$

The term A is the absorption loss in the shield material (or attenuation), R is the reflection loss from both front and back surfaces, and B is a correction factor to account for internal reflections in the shield (required only if $A < 15$ dB). Note that each of these terms are in decibels.

For the incident plane-wave field, the shielding factor S applies to both the electric and magnetic fields.

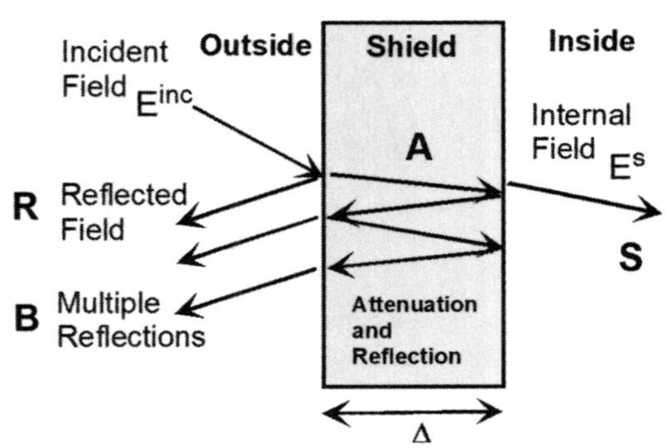

Figure 7.6 Dynamic shielding by an infinite slab of lossy material.

Analyzing this problem as a transmission line [12], (7.5) can be expanded to (7.6):

$$A = 3.34\Delta\sqrt{\mu_r \sigma_r f}\ (dB) \quad R = 168 - 20\log\sqrt{\mu_r f / \sigma_r}\ (dB)$$
$$B = 20\log\left[1 - \eta \times 10^{-A/10} \times \left(\cos 0.23A - j\sin 0.23A\right)\right]$$
with
$$\eta = 4\frac{\left(1-m^2\right)^2 - 2m^2 - j2\sqrt{2m}\left(1-m^2\right)}{\left[1+\left(1+\sqrt{2m}\right)^2\right]^2} \text{ and } m = 0.545\Delta\sqrt{\sigma_r f / \mu_r} \tag{7.6}$$

As an example, the plot of the total EM shielding factor $S_{dB} = R_{dB} + B_{dB} + A_{dB}$, for a plane wave normally incident on an infinite plate of different materials ($\Delta = 0.01$ mm $= 0.0004$ inch), is shown in Figure 7.7.

Several different nonplanar shielding enclosures can be envisioned, as illustrated in Figure 7.8 with Δ as the shield thickness.

For these shields, we can summarize the H-field shielding effectiveness via (7.7) as follows:

$$S_H = \cosh\left(\sqrt{j\omega\tau_d}\right) + F\ \sinh\left(\sqrt{j\omega\tau_d}\right) \tag{7.7}$$

with $\tau_d = \mu\sigma\Delta^2$, where for

A spherical shield:

$$F_s = \frac{1}{3}\left[\frac{(1+j)a}{\mu_r\delta} + \frac{2\mu_r\delta}{(1+j)a}\right] \tag{7.8}$$

A cylindrical shield with H parallel:

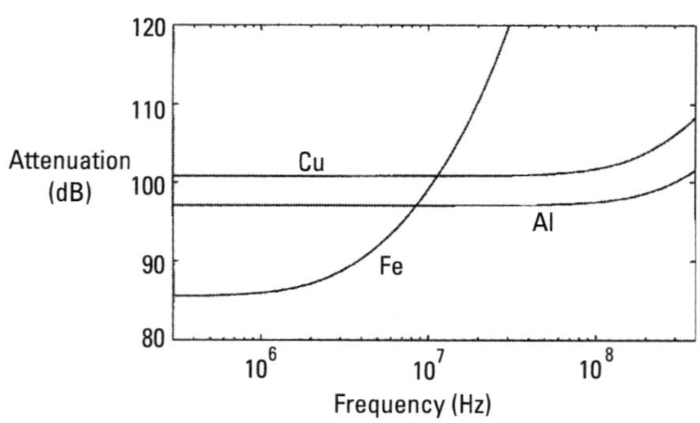

Figure 7.7 Shielding factor for three different materials.

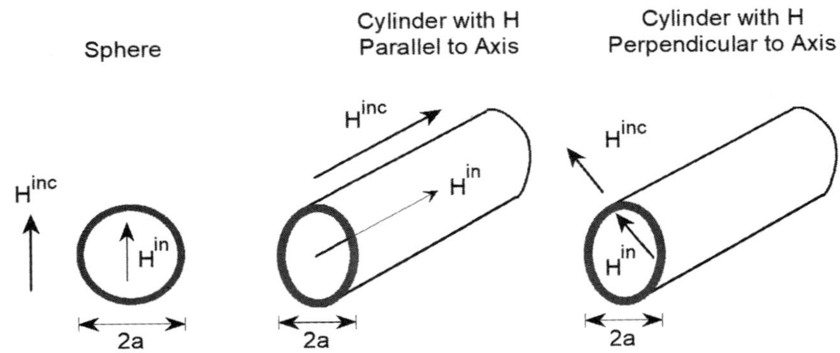

Figure 7.8 Cylindrical and spherical shields of wall thickness Δ

$$F_{c_{\parallel}} = \frac{1}{2}\left[\frac{(1+j)a}{\mu_r \delta}\right] \tag{7.9}$$

A cylindrical shield with H perpendicular:

$$F_{c_{\perp}} = \frac{1}{2}\left[\frac{(1+j)a}{\mu_r \delta} + \frac{\mu_r \delta}{(1+j)a}\right] \tag{7.10}$$

Due to the duality between electrical and magnetic fields, it is possible to infer the electrical shielding efficiency S_E from the magnetic shielding efficiency S_H by making the following change of variables, (7.11).

$$\begin{aligned} &S_H \Rightarrow S_E \\ &\mu_r \Rightarrow \frac{\sigma}{j\omega\varepsilon_o} \\ &\sigma \Rightarrow j\omega\varepsilon_o\mu_r \end{aligned} \tag{7.11}$$

7.2.2.2 Treatment of Apertures (A)

Apertures (holes, slots, and gaps) in the shielding (e.g., windows, connector clearance holes, screens, mating surfaces, and other openings in the shield boundary) will allow EM fields to penetrate into the interior.

Apertures may allow field penetration as shown in Figures 7.9(a, b) [2]. The routing of internal cables away from apertures is one technique that mitigates this issue.

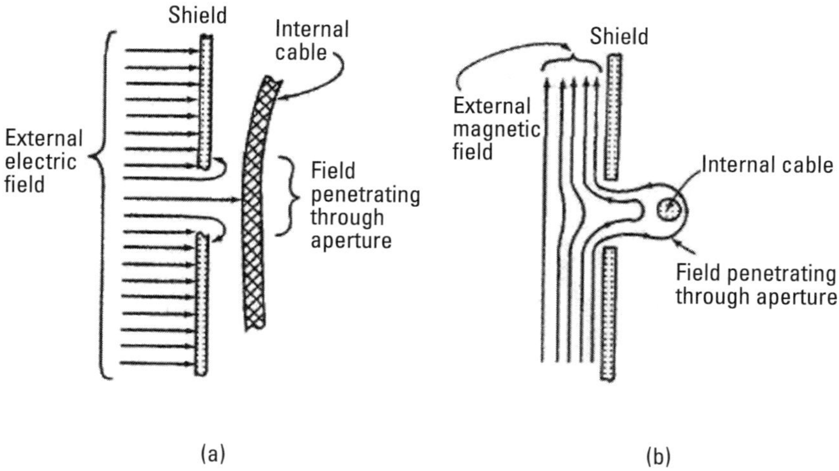

Figure 7.9 Field coupling to interior conductors via an aperture: (a) electric field coupling and (b) magnetic field coupling.

Other techniques include breaking a large aperture into several smaller apertures or extending the depth of the aperture so that it becomes a waveguide beyond cutoff. This is shown in Figures 7.10(a–c). The concept of waveguide beyond cutoff is subsequently discussed.

Joints and Seams

Large shields necessarily have joints and seams. In the joints and seams, it is important that there is very good electrical conductivity through joining surfaces [13, 14]. The mating surfaces should be frequently joined or bonded to minimize the risk of forming a gap or slot aperture. Care should be taken with mating surfaces at the beginning, because, for shields exposed to the environment, corrosion is inevitable and will degrade the shielding performance. Corrosion occurs at metal surfaces due to the presence of electrolytes and/or moisture. MIL-STD-889B and MIL-STD-1250A [13] presented certain materials for industrial applications and test methods. In [13], several materials were tested and reported for their efficacy in providing corrosion-resistant gaskets or fingerstock, which are forms of mating filler material. Corrosion has the effect of increasing the contact resistance, but tin plating seems to be effective in fighting the adverse effects of corrosion.

Waveguide Beyond Cutoff

In Figure 7.10, an aperture is shown as being treated by a series of waveguides beyond cutoff. It is noted that a rectangular waveguide acts like a highpass filter.

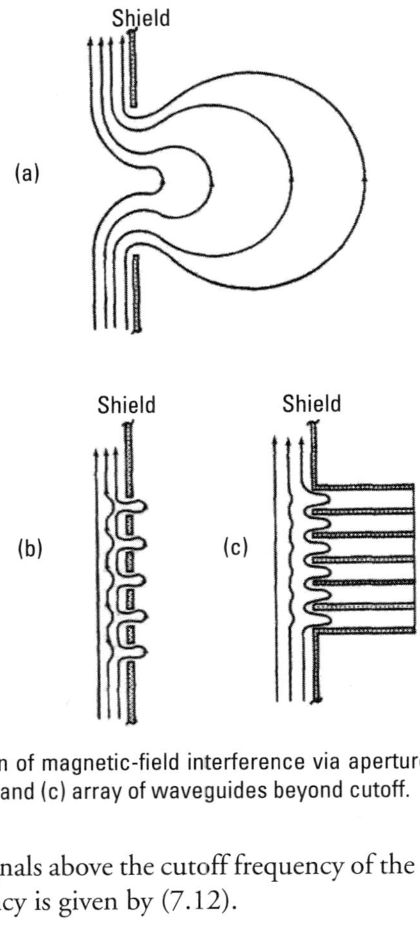

Figure 7.10 Mitigation of magnetic-field interference via apertures: (a) single aperture, (b) many small apertures, and (c) array of waveguides beyond cutoff.

It can propagate signals above the cutoff frequency of the dominant TE_{10} mode. This cutoff frequency is given by (7.12).

$$\textit{cut off frequency } f_c\ (TE_{10}\ \mathrm{mod}\, e) = [c/(2a)] \tag{7.12}$$

where c = speed of light in vacuum = 3×10^8 m/s, and a = long dimension of the waveguide.

Basically, the concept is to make the longer dimension of the waveguide small enough so that the waveguides act like highpass filters and attenuate all frequencies below the cutoff frequency. This is very efficient way to protect apertures like air vents and hatches.

7.2.2.3 Dealing with Conductive Penetrations (C)

In any real system, the conductive portions of the system are not perfect, as there will be conductors passing through the surfaces, and external EM energy can penetrate into the system via currents flowing on the conductors. For an automobile, for instance, such conductive penetrations can be in the form of the radio antenna and headlight cables. Buildings and installations will have many cable-type penetrations providing electrical power and data.

From the techniques described above, it is clear that the fundamental principle applied is to prevent potentially harmful HPEM coupled environments from penetrating the shield. For a shielded conductor, the outside shield of the conductor must be joined to the outside of the shield with a good-quality, low-impedance circumferential bond. Some examples of how to deal with penetrating conductors are shown in Figures 7.11 and 7.12 [9–11]. However, when an unshielded conductor must penetrate the shield, as is almost invariably required, then a nonlinear element is needed.

7.3 Conducted Protection Via Nonlinear Elements

One class of protection device that is commonly used for electrical installations is the nonlinear circuit element. Operating either as a voltage clamp or as a short circuit (crowbar), this concept enjoys widespread use in both civilian and military systems. This section discusses several issues pertaining to the use of nonlinear elements for the protection of circuits, such as filters, clamping diodes, metal oxide varistors (MOVs), and gas discharge tubes (GDTs).

There are several concepts for protecting a load (or circuit) from the effect HPEM. These are based on the parameters of the HPEM environment (the signal indicators) and on the function of the circuit and include the following: amplitude reduction, frequency discrimination and isolation, and circumvention.

As an example of the primary protection approach, if the normal signals on the circuit are pulsed waveforms, as in a computer circuit, the protection could be based on sensing and then eliminating HPEM signals that are much

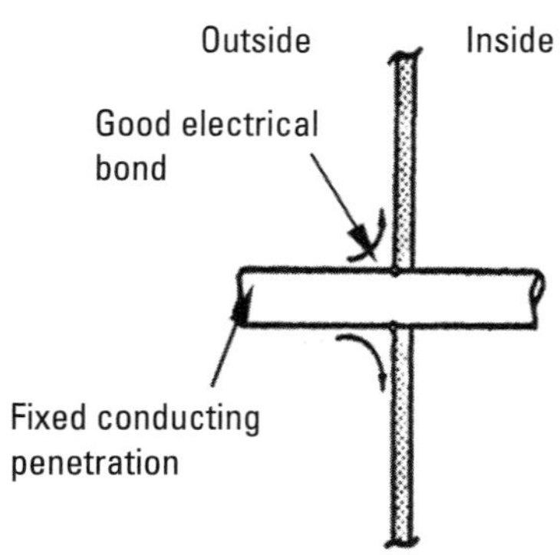

Figure 7.11 Fixed or shielded conductor penetration.

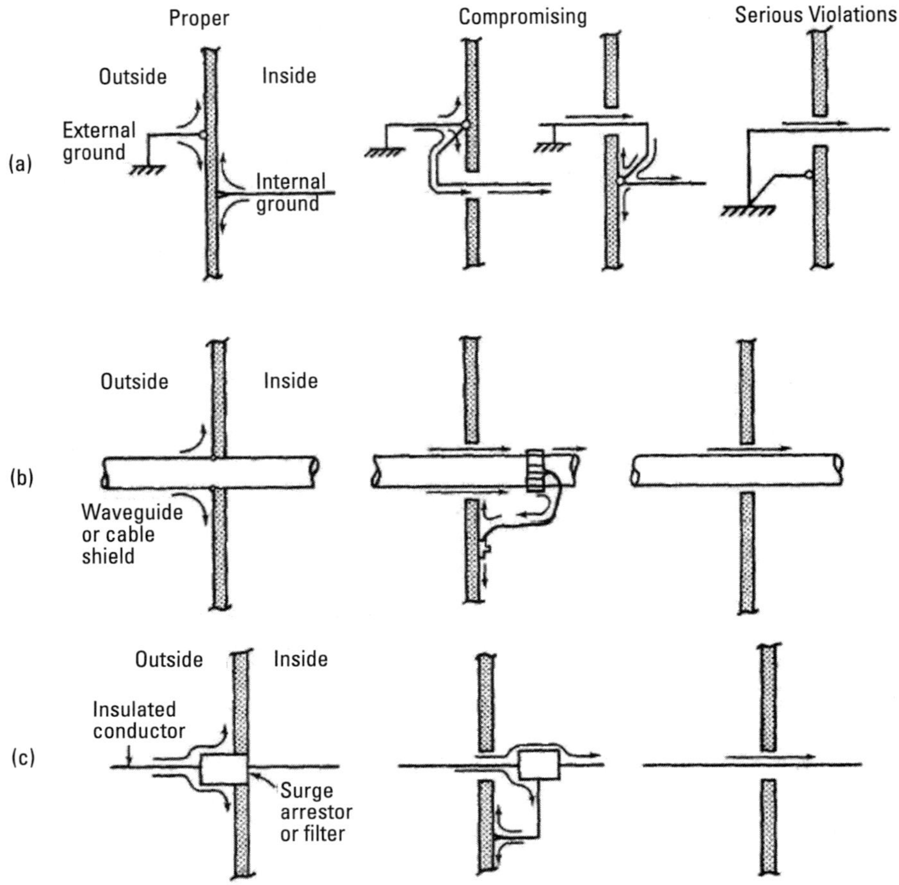

Figure 7.12 Shielding integrity near penetrating conductors: (a) grounding conductors, (b) groundable conductors, and (c) insulated conductors.

larger than the normal operational signals. This discrimination on the basis of signal amplitude forms the design foundation of nonlinear surge protection devices.

However, if the normal operational signals are narrowband, it may be possible to filter out the offending HPEM signal and retain only the desired signal. This forms the basis of the secondary protection approach.

The last approach involves developing a smart circuit that is able to sense the presence of the HPEM environment and suitably alter its functioning to protect itself. This technique is useful if there is a cue or pre-cursor event to the HPEM environment, but in many practical situations this is not the case; therefore, this issue is not discussed further.

7.3.1 Protection Using the Amplitude Reduction Approach

Amplitude reduction and filtering protection elements can be visualized with the circuit of Figure 7.13.

Normally, the circuit is designed with a signal excitation source (the Thevenin source is shown in the figure), which consists of an open-circuit voltage generator and a source impedance. This is connected to a load impedance, Z_L, by a transmission line or other signal path.

When noise or other forms of HPEM are injected onto the circuit (shown in the figure as a lightning or electrostatic discharge), there must be suitable protection devices inserted in the line to protect the load. In Figure 7.13, this protection is represented most generally by series and shunt elements across the line, Z_s and Y_s, respectively.

7.3.1.1 Nonlinear Elements

To discriminate large-amplitude signals from the lower-amplitude normal signals in a system, nonlinear devices are typically used. Following the guide of the protection circuit configuration of Figure 7.13, perhaps the simplest and most common protection element is the electrical fuse, which is used in the position of the series element Z_s, with Y_s open circuited. This fuse will pass the normal operating signals, but when the HPEM overvoltage exceeds a certain amount (depending on the rating of the fuse), the current through the fuse will cause it to fail (in an open-circuit mode), thereby isolating the load impedance and eliminating any possible electrical overstress of this element.

Other types of nonlinear elements are also found for surge protection. For example, one can insert a gas-tube arrestor, an MOV, or other nonlinear elements like Zener diodes for the element Y_s, with Z_s short-circuited. Unlike the fuse, which opens circuits in the event of a large surge, these latter nonlinear elements normally have a very high impedance, but when the voltage across

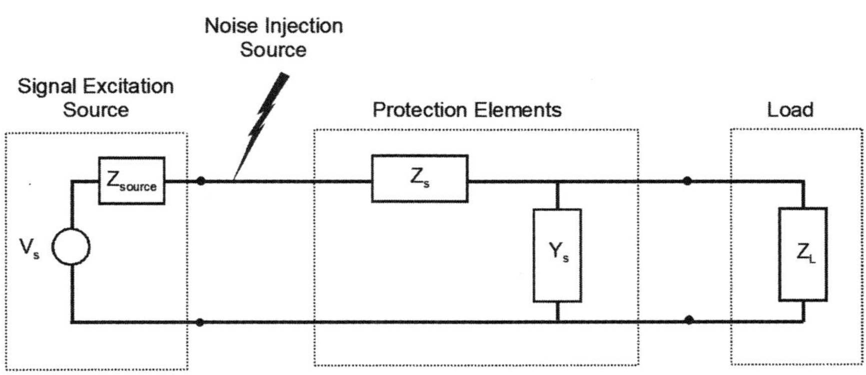

Figure 7.13 General circuit for the conducted protection of load equipment.

them exceeds a particular value, they change state and offer a low-impedance shunt across the line. In this manner, an overvoltage propagating on the line towards the load will see a near short-circuit and be reflected back to the source.

Generally, these nonlinear devices fall into two categories: crowbars and clamps. The crowbar is a shunting device like the gas tube that changes state from a good insulator (high impedance) to nearly zero impedance. A clamp is also a shunting device, which, when activated by the large-amplitude surge, maintains a nearly constant clamping voltage across its terminals, regardless of the amount of current flowing through it.

Table 7.2, adapted from [15], summarizes the properties of several of the most popular nonlinear protection devices. This section is not meant to be a treatise on the properties of such devices, and as such, no further information is provided here about their electrical characteristics. However, there are many references on this subject, including [16–22], and the interested reader is referred to one or more of these reports.

7.3.1.2 Protection by Filtering and Isolation

The other form of circuit protection involves the use of filters, which limit the passage of certain frequency components of signals through the protection device. These filters can pass high frequencies, low frequencies, and frequencies only in a specified band or can serve as a bandstop device. The details of the design of these filters and the choice of the correct parameters are described in

Table 7.2
Summary of Nonlinear Protection Components

Family	Advantages and Disadvantages
Clamps	
MOVs	Fast response (<0.5 ns); large energy absorption; can safely conduct large currents (1 kA for 20 ps); inexpensive; large parasitic capacitance (1 to 10 nF)
Avalanche diodes	Fast response (<0.1 ns); selection of precisely determined clamping voltages (between about 6.8V and 200V); small maximum allowable current (<100A for 100 pF); large parasitic capacitance (1 to 3 nF)
Switching and rectifier silicon diodes	Small clamping voltage (0.7V to 2V); inexpensive; small parasitic capacitance
Crowbars	
Spark gaps	Can safely conduct large currents (5 kA for 50 pF); low voltage in arc mode; very small parasitic capacitance (<2 pF); requires large voltage (~100V) to conduct; can be slow to conduct; possible "follow current" (sustained short-circuit)
Silicon-controlled rectifier (SCR) and Triacs	Small voltage (0.7V to 2V) across conducting switch; can tolerate sustained large currents; slow to turn on or turn off (2 ps); possible sustained conduction

Source: [15].

many standard texts on electrical engineering, for example, [23]. In addition to these types of simple filters, various filter sections can be cascaded together to provide sharper cutoff characteristics.

As mentioned in [15], another protection concept is isolation. This amounts to isolating the load from the source using electrical or mechanical methods. In this manner, the HPEM is prevented from affecting the load.

Table 7.3 summarizes some of the important aspects of filter and isolation components. The subject of the design and application is a large area in its own right and could be discussed to a much larger extent than done here. However, we concentrate on the design of the nonlinear surge protection elements, and filters will not be pursued further.

7.3.1.3 Multistage Protection Using Nonlinear Elements

As described by Büchler [24], it is a common practice to use two nonlinear protection devices together to provide protection for sensitive electronics. This is done because nonlinear elements that can clamp a voltage across a load at relatively low levels of tens of volts can themselves be damaged by a sufficiently large overvoltage excitation. Consequently, the protection device itself may need protection. As an example, such a protection circuit, Figure 7.14 shows two nonlinear devices connected through a coupling impedance Z_s.

The first protection element (on the left of the figure) is commonly referred to as coarse protection, because it is a device chosen to fire at a relatively high voltage (several hundreds of volts). Examples of these types of devices are the crowbar elements of Table 7.2, like the gas tube arrestor. Similarly, the second nonlinear element to the right of the figure is called the fine protection and it usually is a clamping device that limits the voltage across the load to its nominal clamping voltage. MOV devices are an example.

With an overvoltage applied to the composite protection circuit of Figure 7.14, the voltage across the load increases to the point where the low-level

Table 7.3
Summary of Filter and Isolation Components

Family	Advantages and Disadvantages
Filters and Isolation Devices	
Opto-isolators	Large isolation voltages (5 kV), good common-mode rejection, easy to use to receive data, difficult to use to transmit data, fast devices (<1-ps switching time) are expensive
Isolation transformers	Large isolation voltages (5 kV), good common-mode rejection, no attenuation of differential-mode overvoltage
Common-mode filters	Can be effective in attenuating short-duration, common-mode overvoltage, many problems when used alone as SPD

Source: [15].

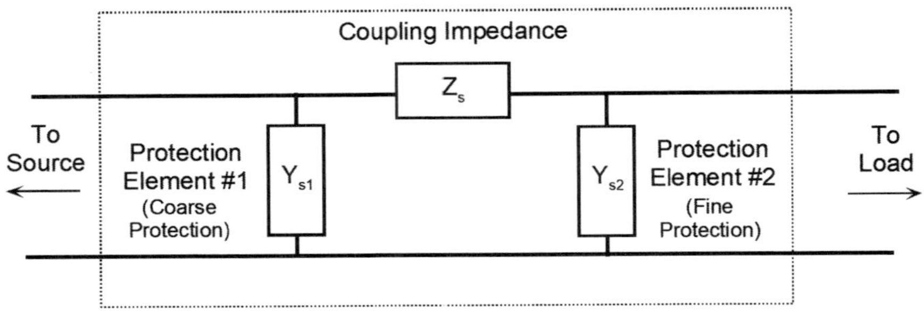

Figure 7.14 A two-stage conducted protection device.

device (element #2) conducts, thereby clamping the voltage at the level of the diode or MOV. As the overvoltage continues to grow, the current though the coupling impedance causes an increasing voltage to be applied across the high-level device (element #1). Eventually, if the overvoltage is large enough, this coarse element fires, and as it is a crowbar type of element, it effectively shunts all of the current from the overvoltage excitation, thereby protecting both the load and protection element #2.

Several problems with this circuit can arise. First, if the value of the coupling impedance Z_s is not chosen properly, the voltage across the coarse protection element #1 may not reach the specified firing level for the device, and most of the HPEM current will flow through element #2, possibly causing damage to this second element. Consequently, the rather complex interaction between the coarse protection firing level and the fine protection clamping level, together with the value of the coupling impedance, must be understood. Second, the presence of parasitic inductances in this circuit can be important in determining its effectiveness for fast-rise pulses. Büchler [24] provided an example of a simple two-stage protection circuit using a UC230 spark gap, having a firing voltage of 230V, together with a 9.1-V clamping diode of the 1.5-kW type. This circuit is shown in Figure 7.15 [24].

In its conducting state, the protective diode shown in Figure 7.15 is represented by its junction resistance, $R_d \approx 40$ mΩ, a junction capacitance $C_d \approx$ 10 nF, and its parasitic lead inductance $L_d \approx 14$ nH. The appropriate value of the coupling resistance R_s must be large enough so that with a slowly increasing input voltage (rising just up to the firing voltage of the gas tube) the diode protection element will not be disabled.

For slow waveforms in the range of 1 to 10 ms in duration, which is typical of a lightning surge, the designed values of the coupling resistance R_s are on the order of ohms. Table 7.4 shows the required minimum coupling resistance R_s for different protection devices for these waveforms.

For faster waveforms, the coupling inductance L_s becomes important, as the protection comes principally from the voltage splitting between the parasitic

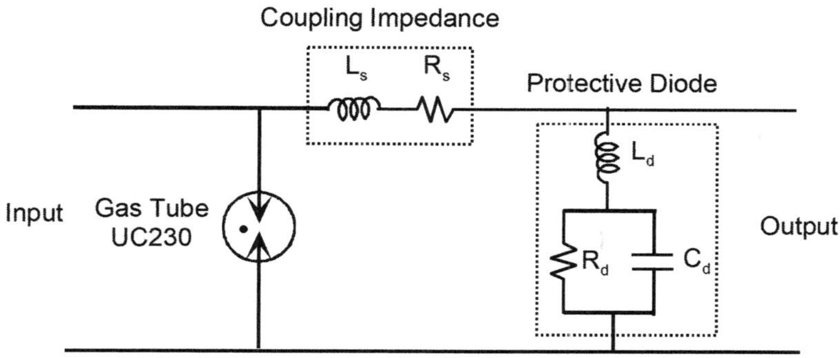

Figure 7.15 Equivalent circuit of a protection circuit with a 9.1-V protective diode in a conductive condition. (From: [24].)

Table 7.4
Minimum Coupling Resistances R_S for Circuits Using Various Primary and Secondary Protection Elements

Protection Diode	Diode Clamping Voltage (V)	Peak Power Capacity (Ω)	R_S (minimum) for 90-V Gas Tube Arrestor (Ω)	R_S (minimum) for 230-V Gas Tube Arrestor (Ω)
P6KE8.2	6.63	600	2.1	5.3
P6KE15	12.9	600	3.6	9.6
P6KE33	26.8	600	6.5	18.7
1N6037	6.63	1,500	0.83	2.1
1N6044	12.9	1,500	1.45	3.9
1N6051	26.8	1,500	2.6	7.5
5KP7.0	7.0	5,000	0.22	0.58
5KP13	13.0	5,000	0.40	1.1
5KP28	28.0	5,000	0.69	2.1

Source: [24].

inductance of the diode L_d and the element L_s. Typically, such a circuit is designed so that $L_s > 1{,}000\ L_d$.

To model the behavior of a protected circuit with two protection devices attached in the configuration of Figure 7.15, a circuit analysis code SpiceAge [25] was used. This program, like its predecessor Spice, allows the user to conduct either a transient or time-harmonic analysis of complicated circuits, including the effects of nonlinear elements. In conducting the analysis, the various clamping diodes, MOVs, and gas tubes have been used to explore the operation of different protection configurations. To do this, suitable electrical models for the protection elements are needed. These are discussed in following sections.

Zener or Avalanche Diode Model

From the discussion in [24], it is possible to infer a suitable circuit model for a Zener or avalanche diode, which serves to clamp the voltage at a specified level across a protected load. This circuit illustrated in Figure 7.16, with the ideal diode in series with R_d in Figure 7.15, usually employed as the fine protection of a circuit.

The switching action of the protective diode is represented by an ideal varistor, a device having an infinite resistance until the voltage across its terminals reaches a prescribed firing value; then its resistance is zero. When the actual device is conducting, there is a resistance through the junction R_d, which is included in the model. As given previously, the values for the parameters in this circuit element are presented in Table 7.5.

MOVs

Details of the MOV type of protection device have been discussed by a number of authors, including Standler [26] and Chen [27]. This device is typically constructed from zinc oxide or bismuth oxide and has a circuit model that is the same as in Figure 7.16.

Usually, the clamping voltage of the MOV devices is higher than for the protective Zener and avalanche diodes, and consequently, the MOVs are often used as a coarse protection element.

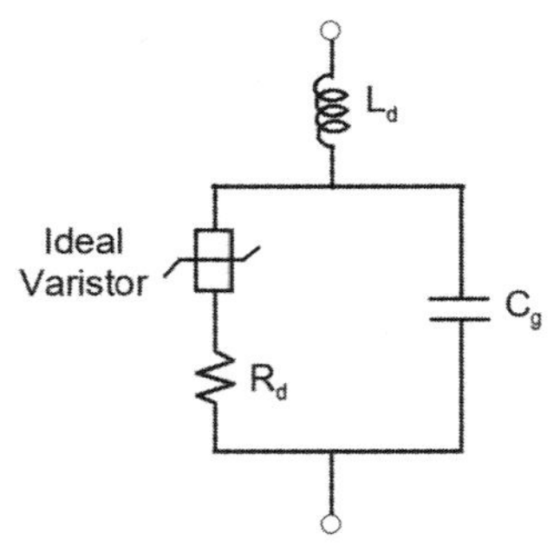

Figure 7.16 Simplified circuit model for a protection diode. (After: [24].)

Table 7.5
Typical Parameter Values for the Circuit of Figure 7.21 Representing a Zener or Avalanche Diode

Element Type	Clamping Voltage (V)	R_d (mΩ)	L_d (nH)	C_g (nF)
Clamping diode (Zener or avalanche)	20	40	14	10

Reference [27] presented typical circuit element values for MOV protection devices having clamping voltages of 200V and 270V, and these data are summarized in Table 7.6.

Gas Tube Devices

The model for a gas tube or other type of crowbar protection element is more complicated, because it does not use the standard diode elements available in circuit analysis programs. The fundamentals of operation and behavior of such a device were discussed in [15], where it was noted that its operation is considerably different from the clamping diode. In the gas tube, the voltage across its terminals increases until there is a breakdown within the tube, at which time the voltage abruptly drops to a much lower level. As a consequence, the waveform across the element can have a rather large spike just before the tube goes into conduction.

To develop a model for such a device, a standard Triac switch model may be used, together with two voltage-controlled switches S_{01} and S_{02}, as illustrated in Figure 7.17.

These switches are controlled by the voltage across terminals (*a-b*). Switch S_{01} is configured to fire when $V_{ab} > V_{threshold}$ and switch S_{02} closes when $V_{ab} < -V_{threshold}$. The 1Ω resistor models the constant resistance of the gas tube, when it is in a fired configuration. The inductance L in this model represents the effects of the leads of this device and is approximated by 50 nH.

7.3.1.4 Analysis of Conducted Protection: Standardized Waveforms

As has been variously discussed, a wide variety of transient HPEM environments, including lightning, HEMP, HPRF DE, and IEMI, can be applied to electrical systems. These external environments are modified by the interaction or coupling with the system. At the terminals of internal equipment, the transient voltage and currents can look considerably different from the original excitation waveform. This was discussed in Chapter 3.

To be able to provide a suitable analysis for the behavior of this equipment, or for the purpose of performing tests, several standard current or voltage

Table 7.6
Typical Parameter Values for the Circuit of Figure 7.16 Representing MOV Protection Devices

Element Type	Clamping Voltage (V)	R_d (mΩ)	L_d (nH)	C_g (nF)
MOV#1	200	16	10	1.9
MOV#2	270	8	50	3.5

Source: [27].

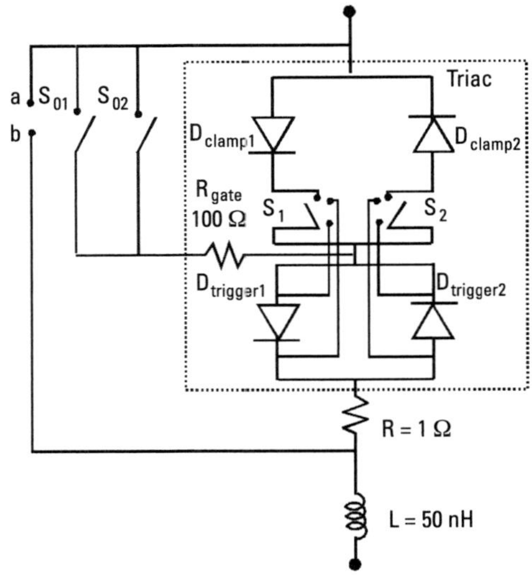

Figure 7.17 A nonlinear circuit model for representing the behavior of a gas tube protection element.

transient waveforms at the equipment level have been defined by cognizant standards organizations. Many of these waveforms were reviewed by Standler [15], and several of these environments are discussed here, with the intent of providing source drivers for the analysis that follows.

A summary of these environments and original HPEM environment to which they pertain is given in Table 7.7. Note that standardized waveforms for the conducted HPEM environments such as Hyperband, Mesoband, and Hypoband do not presently exist.

Table 7.7
A Summary of Some Conductive Transient Environments and Their Derivation

Environment Name	Environment Derivation
8 × 20 μs current waveform	Representative of the current produced in a direct lightning return stroke on an overhead power transmission or distribution line in a short-circuit load
1.2 × 50 μs voltage waveform	Representative of a surge voltage produced by a direct lightning return stroke in an open-circuit load
0.5 μs–100-kHz ring voltage waveform	Representative of a surge voltage produced by lightning inside a facility or building in an open-circuit load
Electrical fast transient (EFT)	Representative of fast transients that can be induced in power as well as data lines by the multiple reignitions occurring during the opening of a circuit by a contactor (i.e., switch bounce)
1.25 MHz ring voltage waveform	Representative of transient overvoltage on power lines resulting from the operation of switchgear in an electrical substation, for instance

The 8× 20-μs Current Waveform

The 8 × 20-μs current waveform was specified by the IEC [28] as one that is representative of a direct lightning return stroke on an overhead power transmission or distribution line. This represents the short-circuit current provided by the lightning source itself or in the laboratory by a simulated current source.

It is given by the analytical expression in (7.13):

$$I(t) = I_p A t^3 \exp(-t/\tau) \tag{7.13}$$

where t is the time in μs, $A = 0.01243$ (μs)$^{-3}$, and $\tau = 3.911$ (μs). The term I_p is the peak value of the transient pulse in amperes and typically ranges from 50A to more than 60 kA.

A commonly used measure of the effect that this surge has on equipment is the total charge contained in the waveform, which is the integral of the current. This is given in (7.14a) and (7.14b) as

$$Q_{total} = 6 \times 10^{-6} A \tau^4 \tag{7.14a}$$

or for the various waveform parameters just quoted,

$$Q_{total} = 1.746 \times 10^{-5} I_p \tag{7.14b}$$

where Q_{total} is the charge in coulombs.

This current waveform is plotted in Figure 7.18. Note that the waveform has been shifted slightly in time by 1 μs to illustrate more clearly the early-time behavior of the current. It is also normalized to provide a peak value of unity.

The 1.2× 50-μs Voltage Waveform

This waveform is also specified by the IEC [28] as a voltage surge that is suitable for testing insulation, especially for electric motors and transformers. Consequently, it is an open circuit test waveform, and as such, the input impedance of the source is not specified in the standard. According to Standler [15], commercial 1.2 × 50-μs surge generators have a source impedance ranging from 1 to 500Ω, and if a large current amplitude is desired, a source impedance of about 2Ω is recommended.

This waveform has a double exponential representation of the form described in:

$$V(t) = V_p A\left(1 - \exp(-t/\tau_1)\right)\exp(-t/\tau_2) \tag{7.15}$$

With t specified again in μs, $A = 1.037$, $\tau_1 = 0.4074$ (μs), and $\tau_2 = 68.22$ (μs). V_p is the peak value of the waveform.

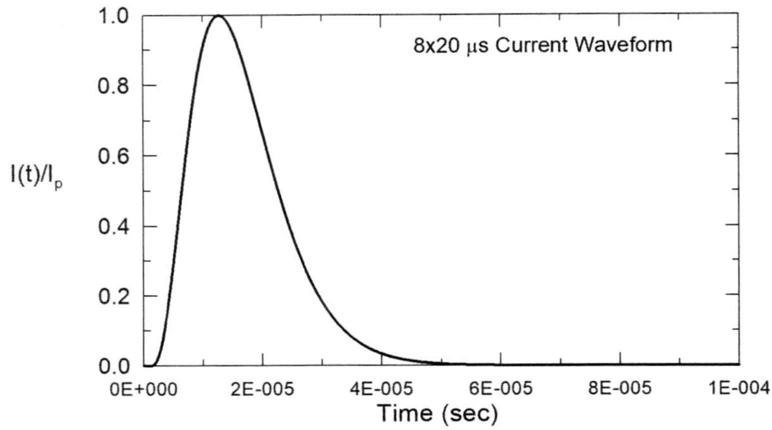

Figure 7.18 Plot of the normalized 8×20-μs short-circuit current waveform.

Figure 7.19 illustrates this voltage waveform, which also has been shifted by 1 μs to illustrate the early time start of the wave.

The 0.5 μs–100-kHz Ring Voltage Waveform

Observations of transient overvoltage on low-voltage power mains often show a rapidly rising peak, followed by a ringing waveform. This waveform was standardized by the IEEE [29] and is referred to as a ring wave.

This waveform is given by (7.16a) and (7.16b) as:

$$V(t) = \begin{Bmatrix} V_p B\left(1 + \eta\, y(t)\right) y(t) & 0 \le t \le 2.5\,\mu s \\ V_p B y(t) & t > 2.5\,\mu s \end{Bmatrix} \tag{7.16a}$$

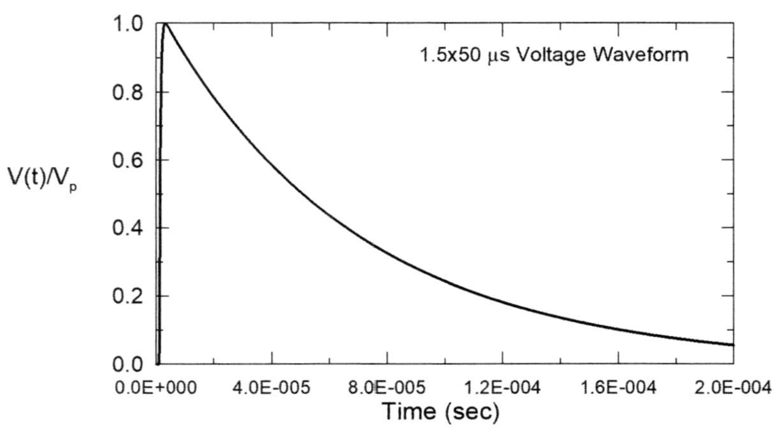

Figure 7.19 Plot of the normalized 1.2×50-μs open-circuit voltage waveform.

With

$$y(t) = A\left(1 - \exp\left(-t/\tau_1\right)\right)\exp\left(-t/\tau_2\right)\cos(\omega t) \quad (7.16b)$$

In these equations, t is in μs, $A = 1.590$, $B = 0.6205$, $\tau_1 = 0.4791$ (μs), $\tau_2 = 9.788$ (μs), $\eta = 0.523$, and $\omega = 2\pi \times 10^5$ (s). Again, V_p is the peak value of the waveform.

Standler [15] suggested that an appropriate source impedance for this source is on the order of 30Ω and that for power system testing a voltage amplitude of 6 kV is appropriate. This waveform is plotted in Figure 7.20 with the time axis shifted by 1 μs.

The 1.25-MHz Ring Voltage Waveform

Transient overvoltage on power lines resulting from the operation of switchgear were specified by ANSI [30]. These transients typically affect control circuit relays and form a part of the standard testing of the surge withstand capability of equipment connected to the power system.

This voltage waveform is described by (7.17):

$$V(t) = V_p A\left(1 - \exp\left(-t/\tau_1\right)\right)\exp\left(-t/\tau_2\right)\cos(\omega t) \quad (7.17)$$

With $A = 1.0096$, $\tau_1 = 2.35$ (ns), $\tau_2 = 15$ (μs), and $\omega = 2\pi \times 10^6$ (rad/s). Again, V_p is the peak value of the waveform.

Figure 7.21 plots this waveform, which is seen to be much more oscillatory than the previous ring wave. For this waveform, the source impedance of the pulser producing this overvoltage is specified at 150Ω.

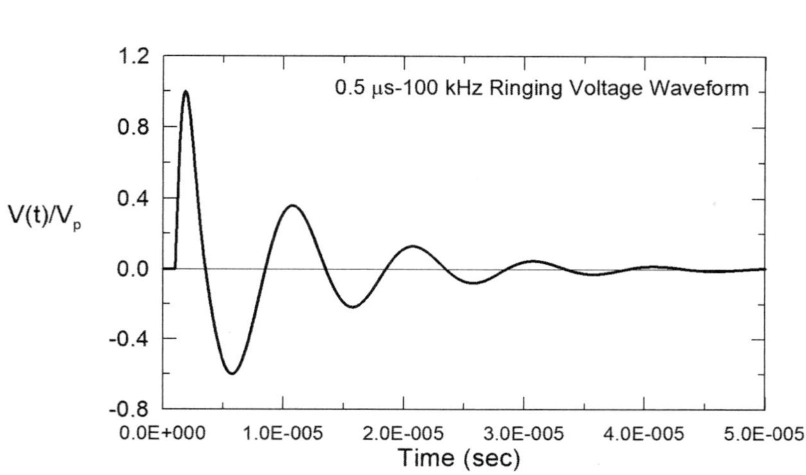

Figure 7.20 Plot of the normalized 0.5-μs–100-kHz ring voltage open-circuit waveform.

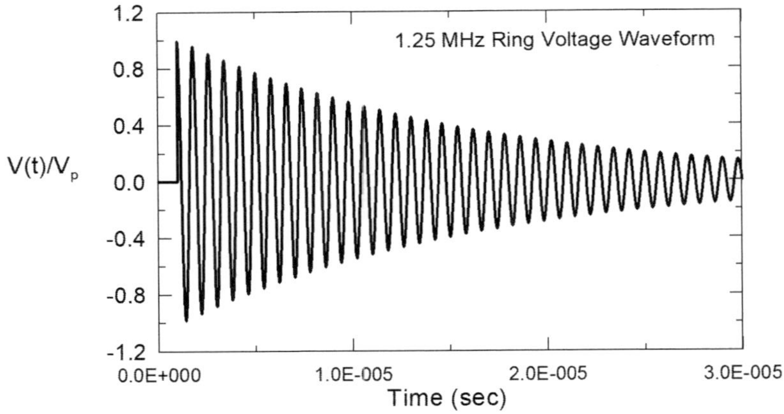

Figure 7.21 Plot of the normalized 1.25-MHz ring voltage open circuit waveform.

The Electrical Fast Transient (EFT)

The final waveform to be considered in this report is the EFT, which produces a waveform with roughly a 5-ns rise time and a 50-ns duration [31]. It is described mathematically by the expression in (7.18).

$$V(t) = V_p A\left(1 - \exp\left(-t / \tau_1\right)\right)\exp\left(-t / \tau_2\right) \tag{7.18}$$

with $A = 1.270$, $\tau_1 = 3.5$ (ns), and $\tau_2 = 55.6$ (ns). V_p is peak value of the waveform. Figure 7.22 plots this waveform, which is much faster than any of the other waveforms shown. For this waveform, the source impedance of the generator is specified to be 50Ω.

7.3.1.5 Analysis of Conducted Protection: Performance

Using the protection models and the standard surges discussed earlier, calculations of the transient responses of the sample circuit of Figure 7.23 have been performed and are discussed in this section. This circuit is modeled after the two-stage protection concept of Figure 7.23, with a gas tube device in the place of element #1 (serving as the coarse protection) and a Zener diode in place of protection element #2 (the fine protection). The coupling impedance, Z_s, connects these two parallel protection devices, as indicated in the figure. This circuit is denoted as configuration A.

The circuit of Figure 7.23 is excited by a Thevenin voltage source, consisting of a transient voltage source, $V_s(t)$, and a source impedance, Z_{source}. The load being protected by the two-stage circuit is denoted as R_L and is assumed to be a simple resistive load. In this circuit, we are interested primarily in the voltage response at the load, V_{Load}. Note that this response is also the voltage

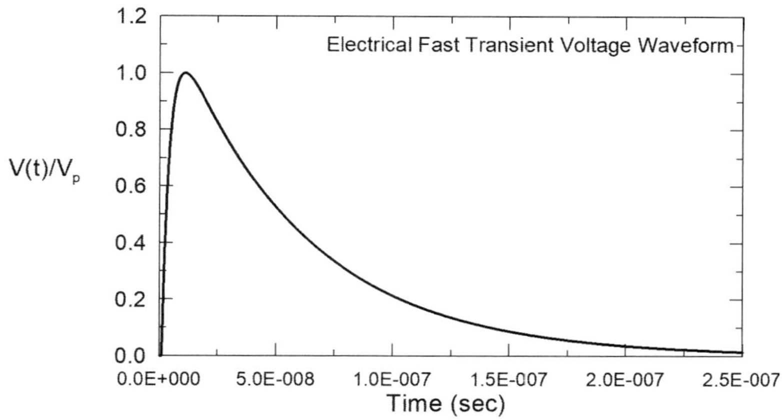

Figure 7.22 Plot of the normalized EFT open-circuit voltage waveform.

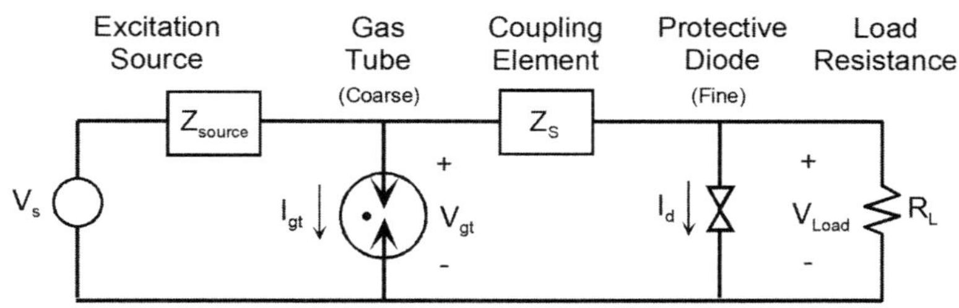

Figure 7.23 Circuit used for the analysis of two protection elements (configuration A).

that is across the protective diode. In addition, the currents through the diode and the gas tube, I_d and I_{gt}, respectively, as well as the voltage across the gas tube element, are of interest, and these will be calculated and plotted.

7.3.1.6 Responses for the 8 × 20-μs Current Waveform (Configuration A)

For computing the responses of the circuit of Figure 7.23 for the 8 × 20-μs current waveform described by (7.13), a Thevenin voltage source is used instead of the corresponding Norton source. In this manner, the equivalent excitation current source level, I_s, is given in terms of the voltage excitation, V_s, by the expression in

$$I_s = \frac{V_s}{Z_{source}} \tag{7.19}$$

For this calculation, the following parameters for the circuit have been chosen:

- Source impedance: resistive with the value *Zsource* = *Rsource* = 2Ω;
- Load impedance: resistive with the value *RL* = 50Ω;
- Coupling impedance: resistive with the value of *Zs* = *Rs* = 2Ω;
- Gas tube and diode protection devices installed in configuration A.

Responses with Nonlinear Elements Removed

Using the SpiceAge analysis package [25], a nonlinear model for the behavior of the circuit in Figure 7.23 was developed. However, the first response of interest is the linear response of the load voltage with the nonlinear protective devices removed. This gives an indication of how large the load voltage (and current) can be if there is no protection to the circuit.

The load voltage for this case can be expressed as shown in

$$\begin{aligned} V_{Load}(t) &= \frac{R_L}{R_L + R_S + R_{Source}} V_S(t) \\ &= 0.926 \times V_s(t) \end{aligned} \tag{7.20}$$

For the specific 8 × 20-μs voltage waveform having a peak amplitude of V_s = 700V, Figure 7.24 presents the transient load voltage, V_{Load}, together with the voltage across the terminals of the circuit where the gas tube would have been located. From both (7.20) and Figure 7.23, it is clear that the voltage across the unprotected load resistance is very close to the open circuit voltage of the 8 × 20-μs source.

Responses with Both Nonlinear Elements Present

Locating the protection devices in the circuit of Figure 7.23 will protect the circuit by limiting the voltage induced across the load resistance. Figure 7.25 presents a series of plots showing the transient behavior of the load voltage V_{Load} (which is also the voltage across the protection diode), together with the gas tube voltage, V_{gt}, and the protective element currents, I_d and I_{gt}. In these plots, the amplitude of the 8 × 20-μs voltage source is allowed to vary from 25V to 10,000V.

Even for the lowest level of 25V, the fine protection element (diode) clamps the load voltage at its 20-V level, as noted in Figure 7.25(a). As the excitation voltage increases in amplitude, the current through the diode increases, flowing only when the diode is conducting. Additionally, the voltage across the

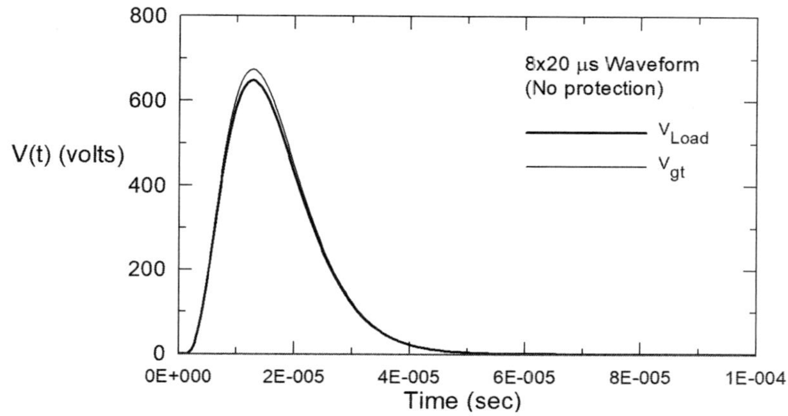

Figure 7.24 Plot of the transient load voltage (thick line) and the voltage across the Thevenin source (thin line) in the circuit of Figure 7.23 with the protection elements removed, for the 8 × 20-μs excitation waveform having a peak amplitude of $V_s = 700$V.

gas tube increases, but with no conduction, until the source voltage reaches approximately 600V (see Figure 7.25(c)). At this point, the gas tube begins to conduct. When this happens, there is only a small change in the voltage across the protected load, but the current flowing through the diode element is reduced markedly. Thus, the gas tube is now protecting the protective diode.

As the source voltage is increased even further, the currents flowing in the circuit also increase, including the current in the protective diode. Due to its small, but non-negligible resistance, the voltage across the diode increases past the 20-V clamping level, and at very high source voltages, the voltage across the protected load is higher than 20V (Figure 7.25(e)). Eventually, the voltage of the source can be set so high that the current through the diode is sufficient to cause its failure, even though it is protected by the gas tube.

One way of understanding the possibility of failure of the protective diode is to examine the cumulative charge that flows through the device. This charge is calculated by (7.21) as:

$$Q(t) = \int_{-\infty}^{t} I_d(\tau)d\tau \tag{7.21}$$

where $I_d(t)$ is the current through the diode.

Figure 7.26 presents the computed cumulative charge passing through the diode for different values of the source voltage. The clamping effect of the diode is shown on the plot, and occurs for all values of the source voltage. The gas tube firing occurs only for source voltages greater than 600V, and this also causes a change in slope in the curves.

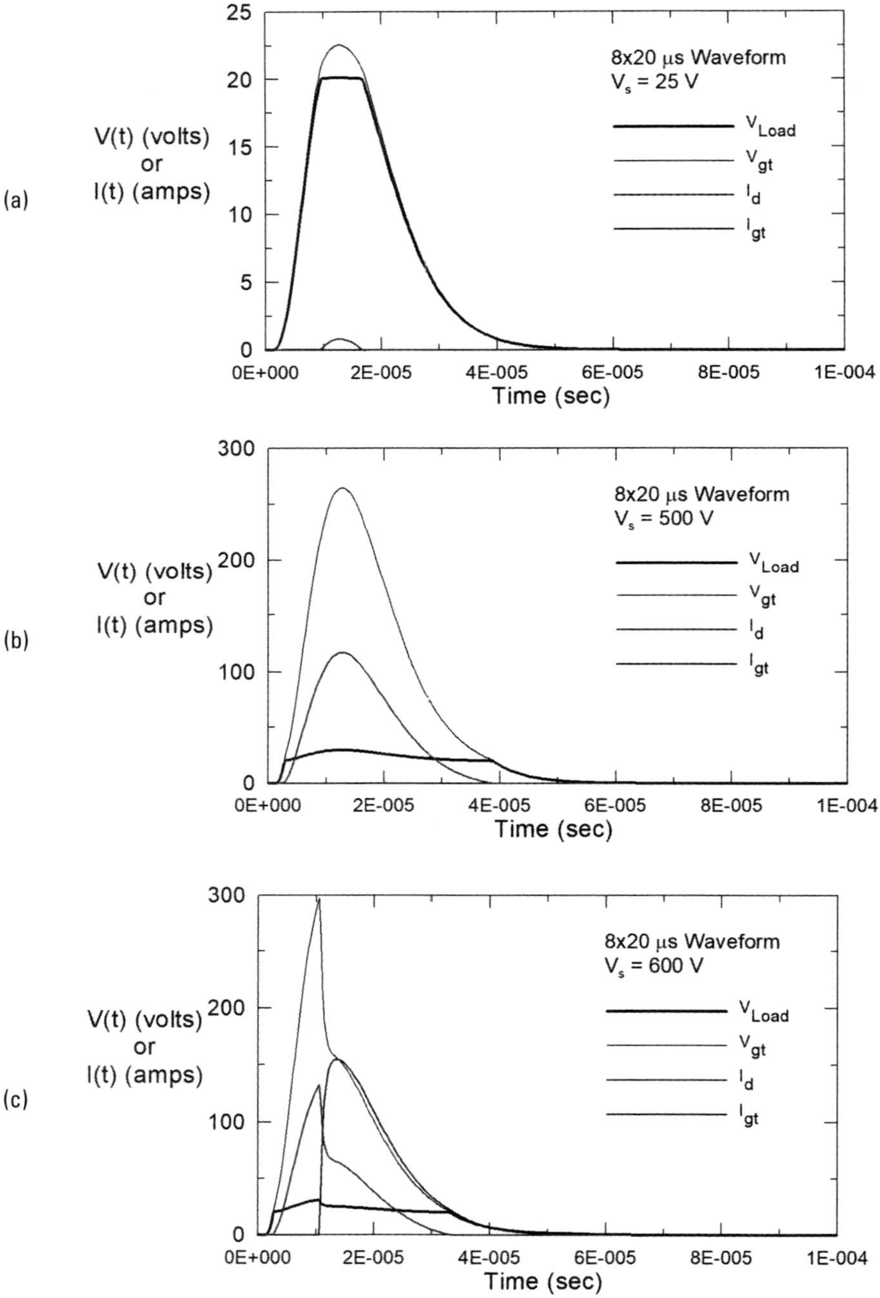

Figure 7.25 (a–e) Transient behavior of various currents and voltages in the circuit of Figure 7.23, for different values of peak voltage of the 8 × 20-μs excitation waveform: (a) $V_s = 25V$, (b) $V_s = 500V$, (c) $V_s = 600V$, (d) $V_s = 1{,}000V$, and (e) $V_s = 10{,}000V$.

(d)

(e)

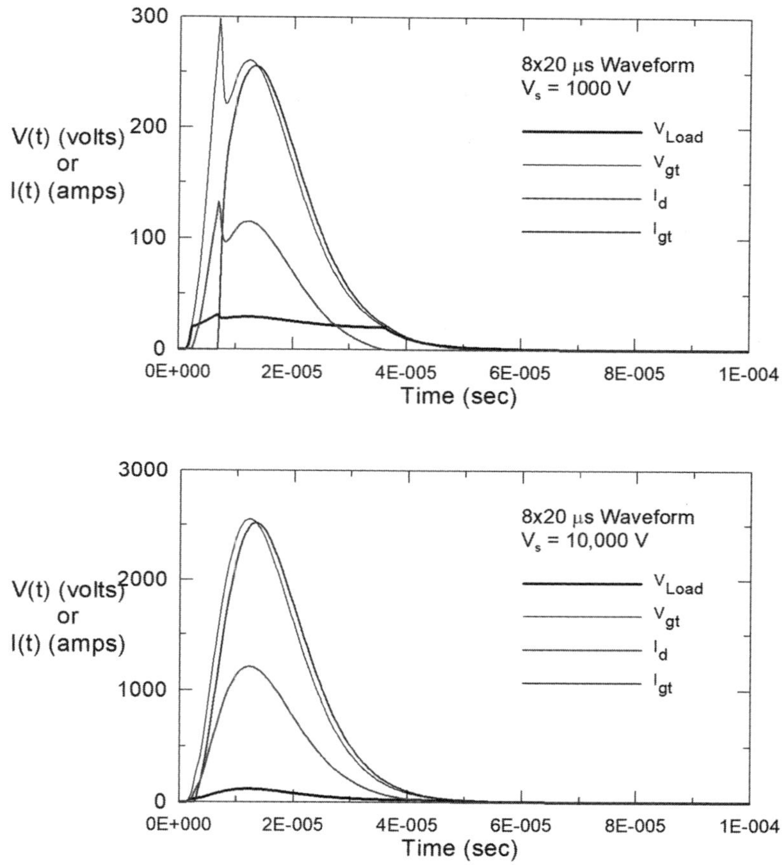

Figure 7.25 (continued)

The data presented in Figure 7.26 can be useful in trying to coordinate the properties of the gas tube and diode protection elements. Generally, the diode will fail at a specified level of the total charge. Knowing this value and its clamping voltage, the gas tube properties and the value of the coupling impedance can be adjusted to try to limit the total charge passing through the protective device.

Responses with Gas Tube Removed

It is clear from Figure 7.26 that at low source voltages (V_s < 600V) the gas tube has no effect on the response of the circuit. Above 600V, however, the gas tube fires and serves to protect the diode. It is instructive to examine the responses of the circuit at the higher voltages with the gas tube removed.

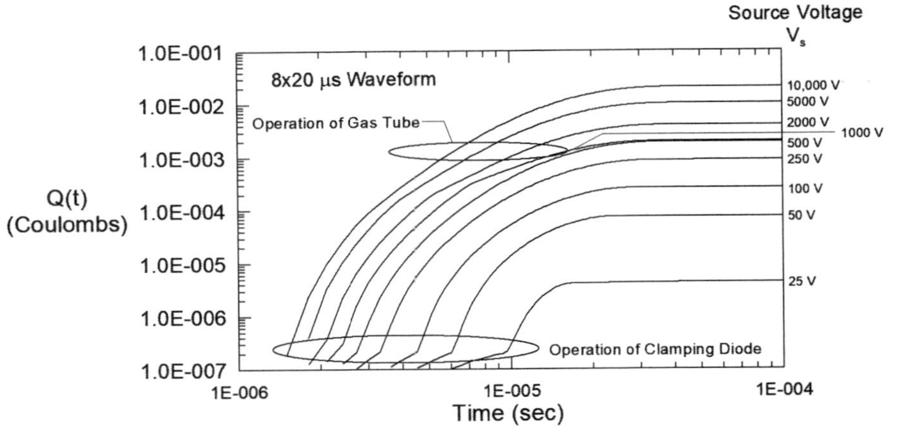

Figure 7.26 Plot of the cumulative charge passing through the protection diode (fine protection) of the circuit in Figure 7.23 for different values of the excitation amplitude, V_S.

For a source voltage of $V_s = 700$V, Figure 7.27 plots the transient behavior of various currents and voltages in the circuit of Figure 7.23 with the 8 × 20 μs excitation waveform. Figure 7.27(a) shows the responses for both the gas tube and diode protection elements present in the circuit. This figure is to be contrasted with Figure 7.27(b), which shows the same response with only the diode protection present. From this plot, it is clearly evident that the gas tube limits the current through the diode and consequently reduces the total charge flowing through the diode.

Figure 7.28 presents the total charge flowing through the diode for the case of the gas tube removed from the circuit. As earlier, the clamping action of the diode is clearly evident; however, for the higher source voltages, the curves are all linear and roughly parallel, indicating a linear response. A careful comparison shows that, at the higher voltage levels, the total charge passing through the diode element is larger than in the case with the gas tube protection present.

Responses with the Gas Tube Replaced by MOV Elements

Another variation of the protection circuit in Figure 7.23 is to use one of the MOV elements of Table 7.5 as the coarse protection element instead of the gas tube. For the case of the source excitation of $V_s = 700$V, Figure 7.29(a) shows the circuit responses for MOV#1 (having a 200-V clamping voltage), and Figure 7.29(b) shows the same responses for MOV#2 (with a 270-V clamping level). These responses are similar to those for the gas tube, with the notable exception that the initial spike from the gas tube switching is absent with the MOV devices.

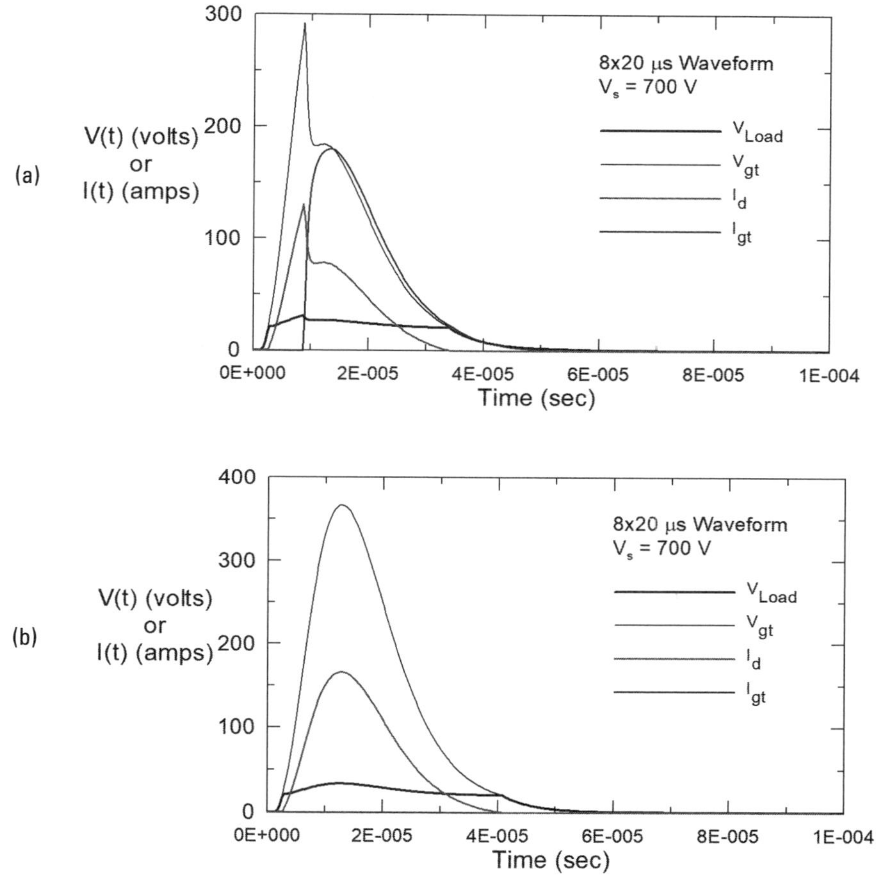

Figure 7.27 (a, b) Transient behavior of various currents and voltages in the circuit of Figure 7.23 with the 8×20-μs excitation waveform. (a) Gas tube and diode protection present. (b) Only diode protection present.

7.3.1.7 Responses for the 8×20-μs Current Waveform (Configuration B)

An alternative to the configuration of the protective devices shown in Figure 7.23 is to interchange the locations of the fine and coarse protection elements, as indicated in Figure 7.30. This is subsequently referred to as configuration B. In this configuration, the load voltage is the same as the voltage across the gas tube.

Figure 7.31 plots the transient currents and voltages in the circuit of Figure 7.30, for different values of peak voltage for the 8 20-μs excitation waveform. As in configuration A, the protection diode clamps the load voltage at around 20V for the lower values of the source voltage V_s, but as V_s increases, the resistance of this protection element becomes important and the load voltage

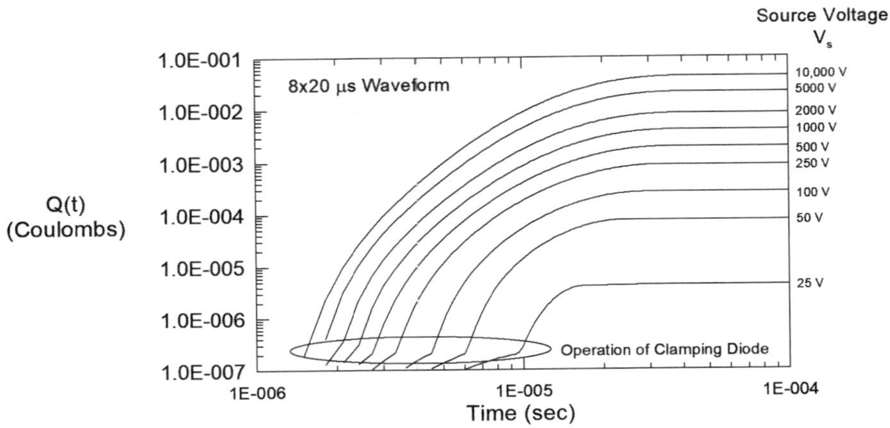

Figure 7.28 Plot of the cumulative charge passing through the protection diode of the circuit in Figure 7.23 with the gas tube removed, for different values of the excitation amplitude, V_S.

increases. However, the firing of the gas tube occurs at a much higher value of V_s, namely, in the range of 5,000V to 10,000V.

What happens in this configuration B is that the protective diode tries to clamp the voltage across the gas tube, preventing its firing, until the external voltage excitation reaches such a large value that the gas tube voltage fires. This permits much larger currents to flow in the diode, with the possibility of damage to the diode element. Figure 7.32 plots the cumulative charge passing through the protection diode in Figure 7.30 (configuration B) for different values of the excitation amplitude, V_s. From this plot, it is clear that significantly larger total charges are passed through the diode element than in configuration A.

To better illustrate the differences between these two protection configurations, Figures 7.32 and 7.33 plot the total charge passing through the diode protection elements, as a function of the applied voltage V_s for the 8×20-μs excitation waveform. It is evident that configuration B permits a much larger charge to enter the diode, with the consequence that it is potentially more damaging to the diode element.

7.3.1.8 Responses for Other Excitation Waveforms

Waveforms other than the 8×20-μs pulse are also of interest in examining the behavior of circuits protected by the two-element circuit. In this section, the other waveforms described earlier are applied to configuration A (Figure 7.23) of the protection circuit. Configuration B is not considered further in this chapter, as it appears to shunt more current onto the protective diode, possibly causing damage to the diode.

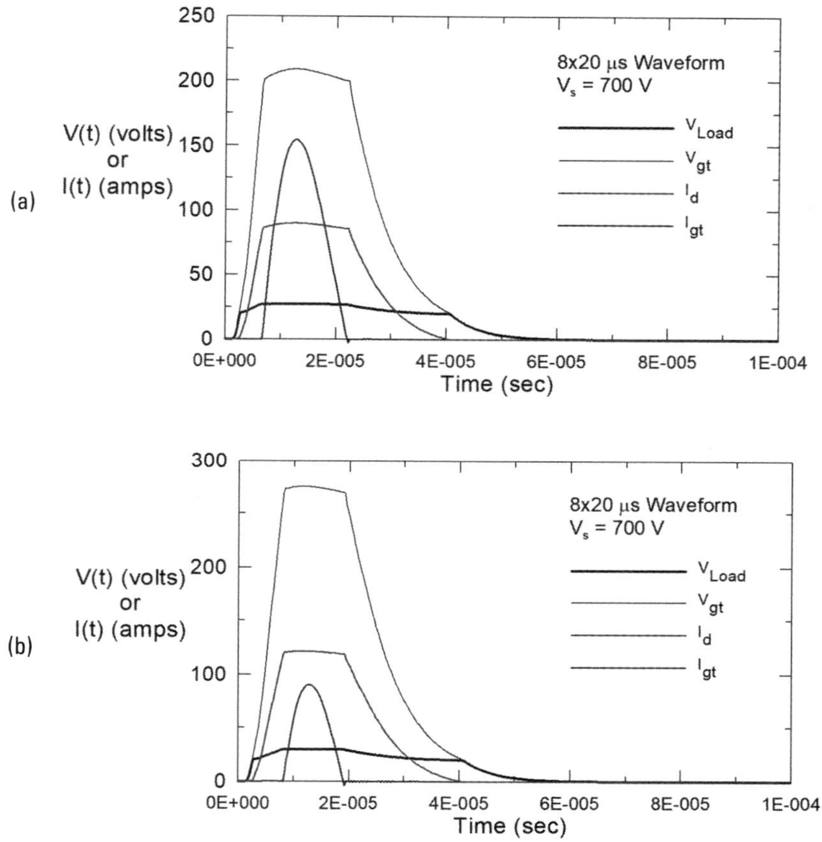

Figure 7.29 (a, b) Transient behavior of various currents and voltages in the circuit of Figure 7.23 with the 8 × 20-μs excitation waveform, and coarse protection provided by a MOV and fine protection by the clamping diode. (a) MOV#1 and protective diode. (b) MOV#2 and protective diode.

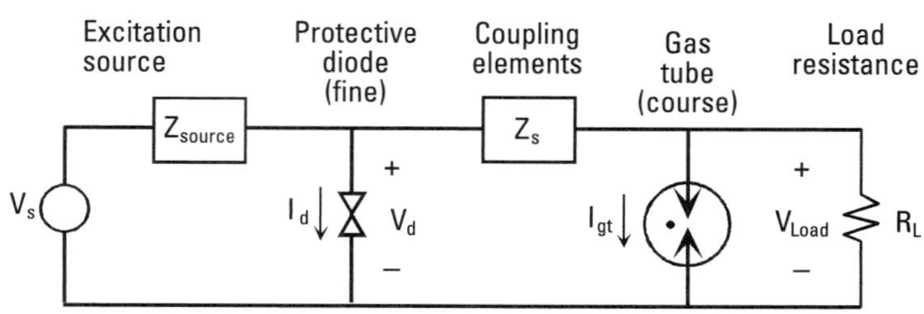

Figure 7.30 Revised circuit for the analysis of two protection elements with reversed positions (configuration B).

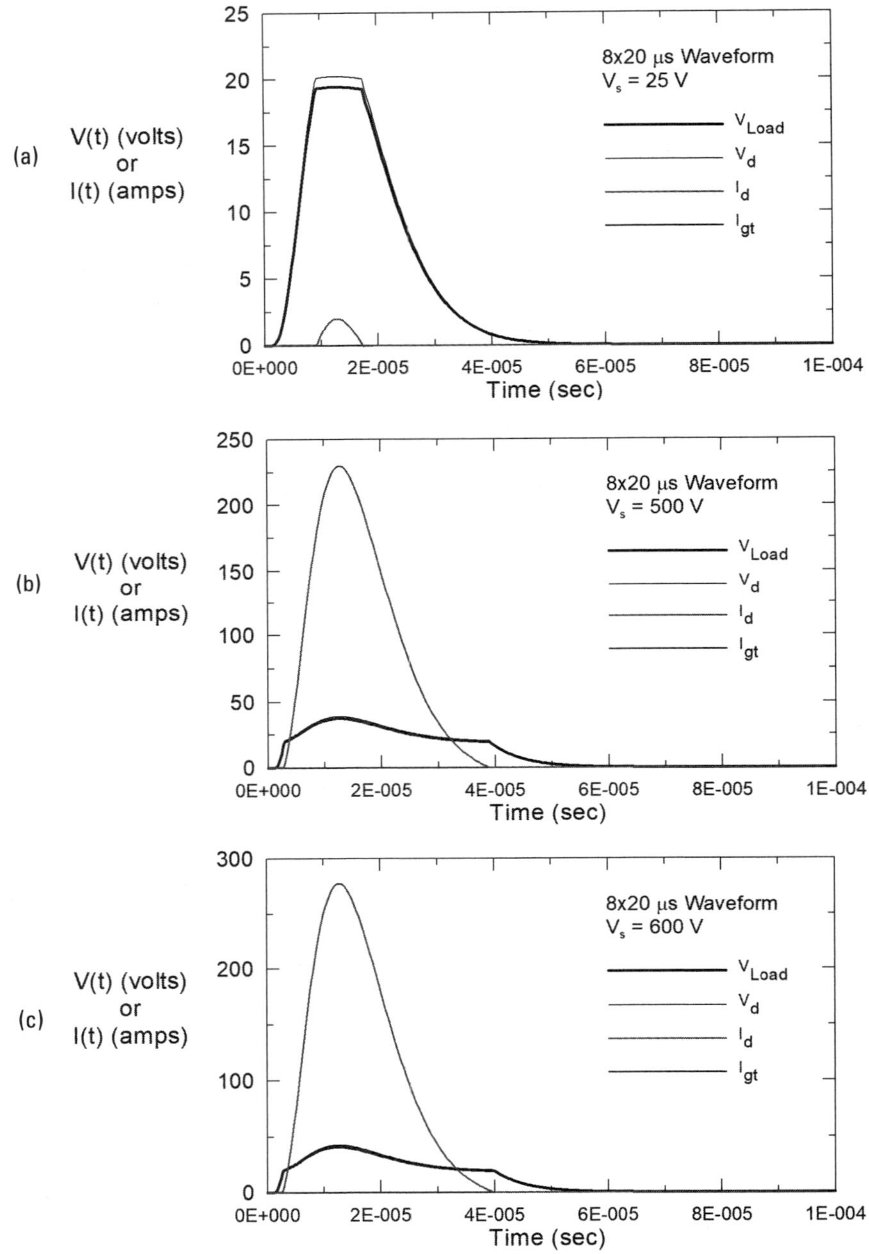

Figure 7.31 (a–e) Transient behavior of various currents and voltages in the circuit of Figure 7.30 (configuration B), for different values of peak voltage of an 8×20-μs excitation waveform. (a) $V_S = 25V$, (b) $V_S = 500V$, (c) $V_S = 600V$, (d) $V_S = 1{,}000V$, and (e) $V_S = 10{,}000V$.

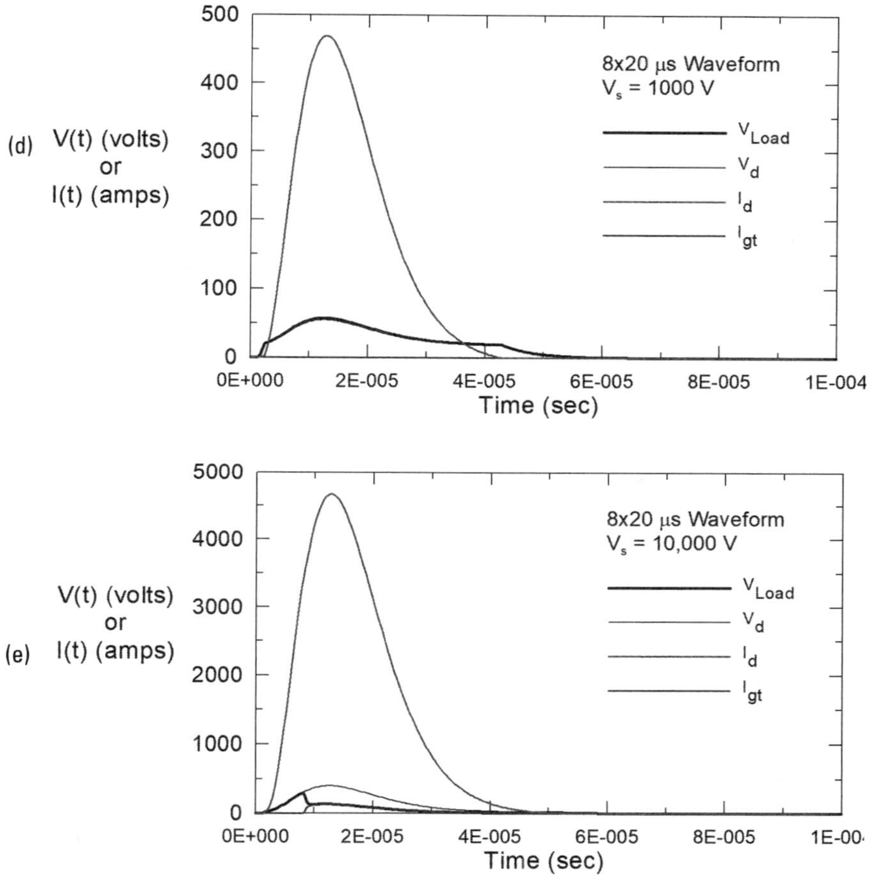

Figure 7.31 (continued)

1.5 × 50-μs Voltage Waveform

The 1.5 × 50-μs voltage waveform has been used to calculate the responses of the circuit in Figure 7.23 with the following parameters:

- Source impedance: resistive with the value $Rsource = 2\Omega$;
- Load impedance: resistive with the value $RL = 50\Omega$;
- Coupling impedance: resistive with the value of $Rs = 2\Omega$;
- Gas tube and diode protection devices installed in configuration A.

Figure 7.34 presents the plots of the transient load voltage V_{Load}, the gas tube voltage, V_{gt}, and the protective element currents, I_d and I_{gt}. Figure 7.34(a)

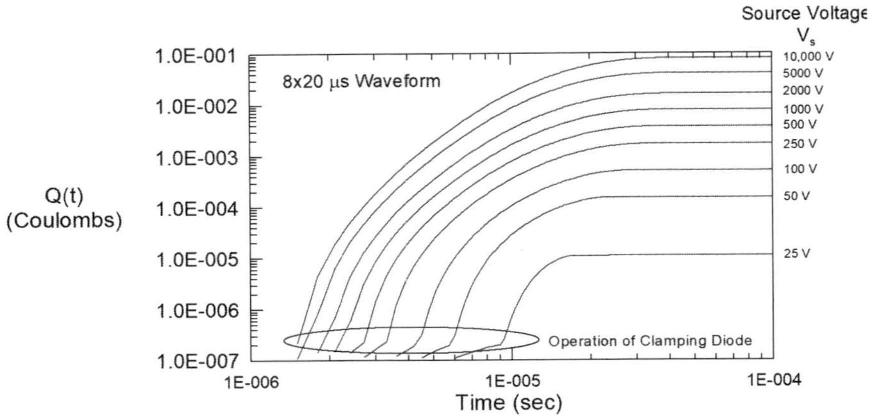

Figure 7.32 Plot of the cumulative charge passing through the protection diode (fine protection) of the circuit in Figure 7.30 for different values of the excitation amplitude, V_S.

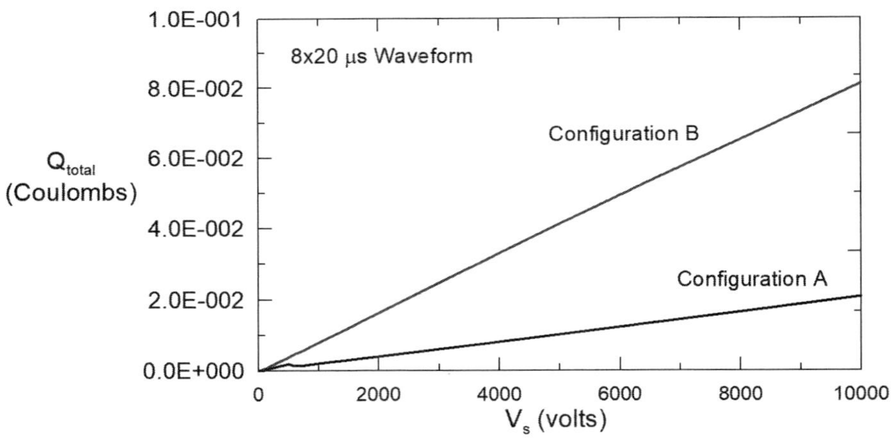

Figure 7.33 Plot of the total charge passing through the diode protection elements in configurations A and B, as a function of the applied voltage V_S for the 8×20-μs excitation waveform.

is for an excitation waveform amplitude of $V_s = 500$V. At this level, the gas tube does not have sufficient voltage across it to fire, and, as a consequence, the current I_{gt} is zero. However, for the higher excitation voltage peak of $V_s = 700$V (Figure 7.34(b)), the gas tube fires and we again note that the current flowing through the diode element is reduced. Notice that the gas tube voltage V_{gt} and current I_{gt} coalesce at late times on this plot. This is because the gas tube model assumes a 1Ω resistance, and at late times, the voltage and current are numerically equal.

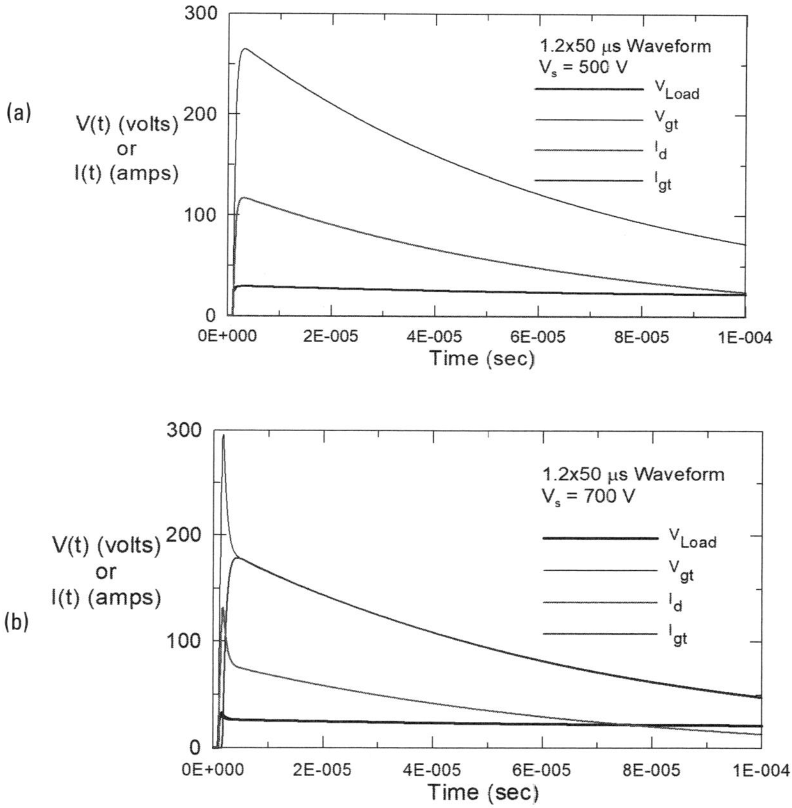

Figure 7.34 Plot of the voltage and current responses of the circuit in Figure 7.23 for the 1.5 × 50-μs voltage waveform with peak values of $V_S = 500$V (a) and 700V (b).

0.5 μs–100-kHz Ring Waveform

The 0.5 μs–100-kHz ring waveform has been used with the circuit of Figure 7.23 with the following parameters:

- Source impedance: resistive with the value *Rsource* = 30Ω;
- Load impedance: resistive with the value *RL* = 50Ω;
- Coupling impedance: resistive with the value of *Rs* = 2Ω;
- Gas tube and diode protection devices installed in configuration A.

Because the source impedance is so high in this case, the source voltage level that causes the gas tube to conduct is much higher than in the previous examples. Figure 7.35 presents the circuit responses for the ring wave for source voltages of $V_s = 700$V, 1,000V, and 2,000V in Figures 7.35(a–c), respectively.

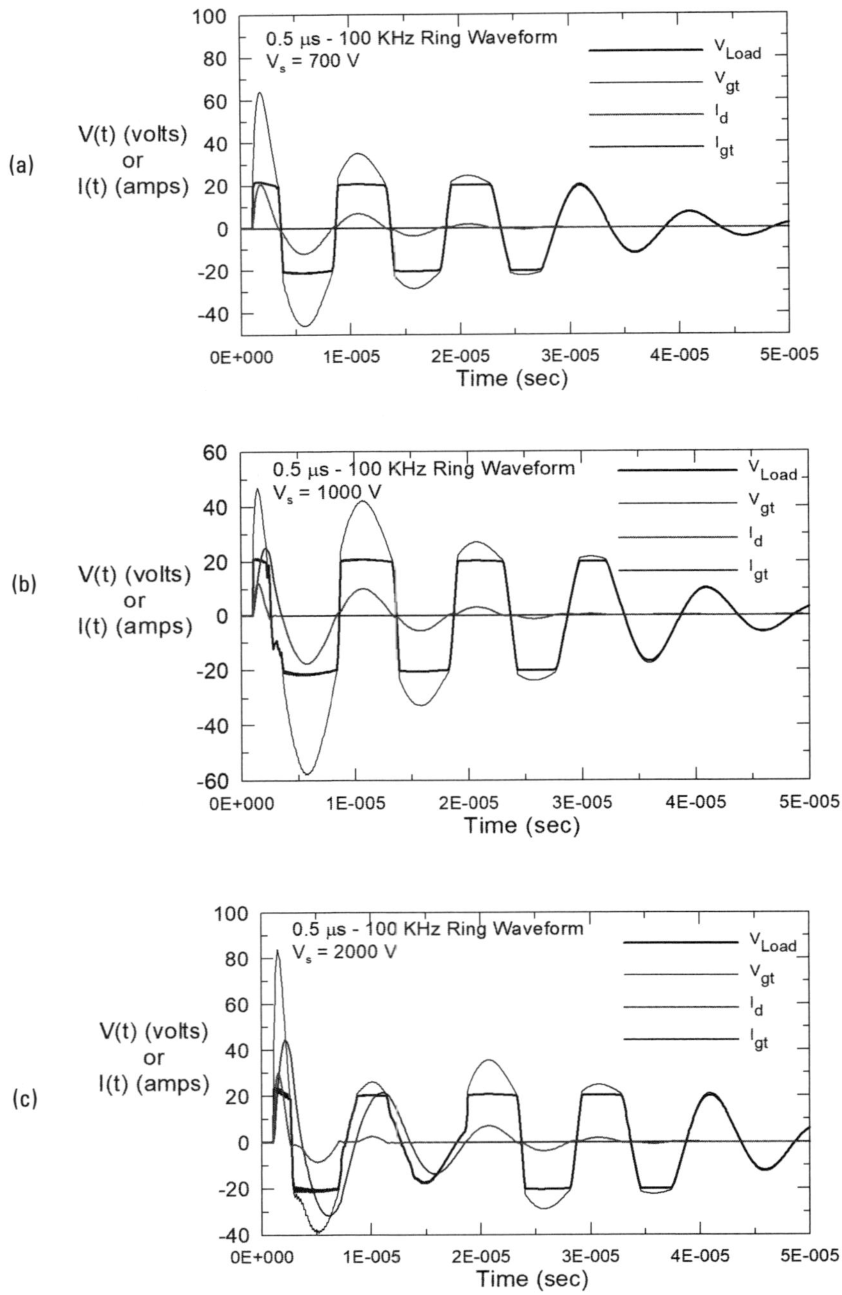

Figure 7.35 Plot of the voltage and current responses of the circuit in Figure 7.23 for the 0.5 μs–100-kHz ring waveform with peak values of $V_s = 700$V (a), 1,000V (b), and 2,000V (c).

For the 700-V source, which was sufficient to cause the gas tube to actuate in the previous examples, we note that it has not yet fired in this circuit. However, at the 1,000-V level, the gas tube conducts for a portion of the waveform. At the higher level of 2,000V, we see that the tube conducts for several cycles of the waveform. As in the other protective circuit examples, this waveform is adequately clamped to the 20-V level at the load resistance.

1.25-MHz Ring Waveform

We now consider the 1.25-MHz ring waveform. For this high Q waveform, the following parameters have been used in the circuit of Figure 7.23:

- Source impedance: resistive with the value *Rsource* = 150Ω;
- Load impedance: resistive with the value *RL* = 50Ω;
- Coupling impedance: resistive with the value of *Rs* = 2Ω;
- Gas tube and diode protection devices installed in configuration A.

Again, the source impedance is high and the required source voltage needed to cause the gas tube to conduct is much higher, between 2,000V and 5,000V. Figure 7.36 plots the various circuit responses for this 1.25 MHz ring wave for source voltages of V_s = 2,000V, 5,000V, and 7,000V in Figures 7.36(a–c), respectively.

As noted in Figure 7.36(a), the 1,000-V source is not large enough to fire the gas tube. However, for V_s = 5,000V, the tube has fired and there is a small gas tube current I_{gt} flowing.

Electrical Fast Transient Waveform

For the EFT waveform given by the double exponential form, circuit response calculations also were made for the following parameters:

- Source impedance: resistive with the value *Rsource* = 50Ω;
- Load impedance: resistive with the value *RL* = 50Ω;
- Coupling impedance: resistive with the value of *Rs* = 2Ω;
- Gas tube and diode protection devices installed in configuration A.

Because this waveform is so much faster than the others (i.e., it contains higher-frequency components), the parasitic inductance and capacitance elements of the nonlinear protection elements become very important. Figure 7.37 presents the calculated responses of the circuit in Figure 7.23 for this waveform with an amplitude of 1,000V and 5,000V. In these plots, we note that for early

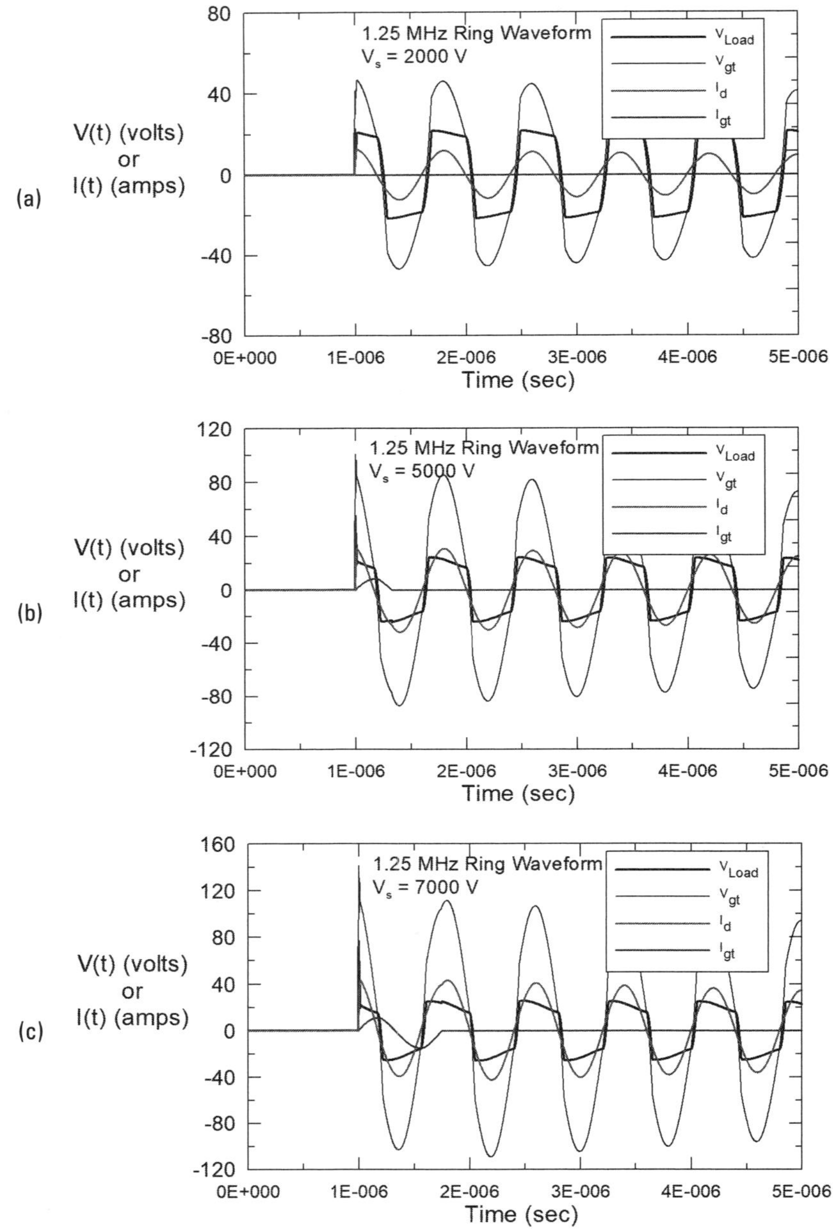

Figure 7.36 (a–c) Plot of the voltage and current responses of the circuit in Figure 7.23 for the 1.25-MHz ring waveform with peak values of V_S = 2,000V (a), 5,000V (b), and 7,000V (c).

times, the voltage across the load (V_{Load} in the plots) exceeds the 20-V diode clamping level by a considerable margin. This is attributable to the parasitic

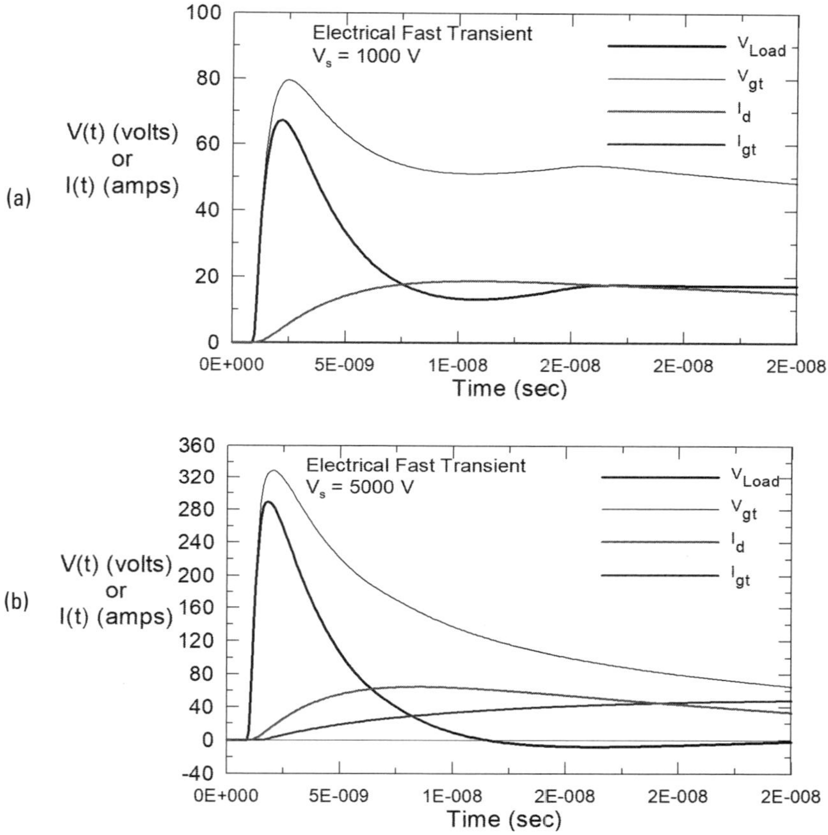

Figure 7.37 (a, b) Plot of the voltage and current responses of the circuit in Figure 7.23 for EFT with peak values of $V_s = 1{,}000V$ (a) and 5,000V (b).

elements of the diode. Moreover, we note that the current flowing through the gas tube in Figure 7.37(b) has a rather slow rate of increase, which is attributable to the inductance of the tube.

7.3.2 Summary of Analysis of Nonlinear Elements

Using appropriate computer models for nonlinear protective elements, clamping diodes, MOVs, and GDTs, calculations for a simple circuit have been conducted to illustrate how these devices function together and alone. From these simulations, several observations about these devices can be made, and these are summarized as follows:

- The design of a multistage nonlinear protection circuit may seem straightforward, but in reality it is quite difficult, due to the many interrelated and interacting parameters of the design: the source and load impedances, the clamping and/or crowbar voltage levels, and the coupling impedance between the two nonlinear devices.

- Even if a design of a protection circuit is realized, the actual device performance may be considerably different due to the parasitic capacitance and inductance in the protection elements. These parasitic elements become increasingly important at high frequencies (over 100 MHz) and they must be estimated in many cases, because manufacturer data sheets may not contain their values.
- Arrangement of the fine and coarse protection elements in a circuit is critical. The usual approach is to locate the coarse element (the one with the highest firing voltage) closest to the interference source and the fine element (the one with the lowest firing voltage) closest to the load being protected. In this manner, the fine element fires first, and if the HPEM environment is intense enough, the coarse protection device actuates and effectively protects the fine element. If the locations of these devices are reversed in the circuit, then this protection aspect of the coarse device is lost, and as has been shown here, there is a much larger current passing through the fine protection element, with the possibility of damage.
- It is often difficult to judge the effectiveness of a particular protective circuit by examining a transient waveform. One useful measure of the response is the total charge passing through a component.

7.4 HPEM Resilience and Detection

The concept of HPEM protection described previously is known to be very effective as long as the as-built protection is maintained and continuously assured. However, at least for HEMP, the protection concepts were developed by and for the military and indeed only for a subset of military systems that they described as time-urgent and mission-critical [32]. The military HEMP hardening and protection approach described in standards [33] was developed in the Cold War era when certain systems absolutely had to survive and work through a HEMP event.

The approach used for natural threats such as lightning protection, for example, and for HEMP for the military time-urgent and mission-critical systems can be described by the following steps.

- Set requirement for protection;
- Identify the threat(s) environment(s);

- Carry out a risk assessment for the items to be protected;
- Specify protection performance parameters and implement protection;
- Continuously assure the protection design against the performance parameters;

Here the protection design has a defined end-point (i.e., after the design is implemented and verified) because protection is mandated by some government or military authority. To comply with the mandate, the level of protection design in to the system is high enough to mitigate all risk to the system function. Put simply, the protection scheme has a very high probability of success if everything is well understood and controlled and the protection level can be fully assured. Occasionally, it may be necessary or at least prudent to revisit and update the threat and assess risks, but overall the approach is tractable. An advantage with this approach for HEMP, HPRF DE, and IEMI is that it can be expected that the installed protection will protect the system and is likely to deter malicious intent (i.e., the threat actor will think twice or look for a softer target). However, this approach has a number of disadvantages.

Ultimately, a protection scheme has to be affordable compared to the cost of failure of the system (risk) during the system lifetime due to the occurrence of the threat. The Foundation for Resilient Societies estimated that, "the cost to protect a high-voltage commercial generating facility would be about $250,000 USD [U.S. dollars]" [34] and that protection of the U.S. electrical grid from HEMP and geomagnetic disturbance is "from a low end of about $10 billion dollars to a high end of about $30 billion dollars" [35].

The Foundation for Resilient Societies pointed out that: "The projected costs of a regional or nationwide electric grid outage lasting months or years vary widely but range from $1 trillion to over $10 trillion." Baker [36] estimated the cost of HEMP protection for the bulk power system to be in the $50 billion range.

These costs are the up-front or capitalization costs; another aspect to consider is the whole-life costs of an HPEM protection scheme, which includes the cost of continued assurance and the cost of remedial repairs to a damaged protection scheme. As an example the HPEM protection designers of the past could not predict the rate of change of technology upgrades that are now a fact of modern life. For example, Intel [37] recommended: "a review and refresh cycle that modernizes your IT infrastructure every two to four years, as you evolve new business requirements, retire hardware and software that can no longer cope cost-effectively, and integrate new solutions that better meet your needs." Other aspects such as new energy efficiency initiatives can also drive the requirement for regular upgrades [38]. This means that infrastructure owner/operators are in an almost constant cycle of upgrades to meet the latest requirements. Unless great care is taken during an upgrade, which will inevitably require new penetrations through the protected boundary, a protection scheme can very easily be compromised.

Finally, when considering wider military needs and critical infrastructure protection, the vast majority of systems are not required to remain functional during (also known as work-through) an HPEM event but yet have to be protected sufficiently so that they can be restored after the event has subsided.

In summary, due to the factors discussed, work-through protection schemes are not widely adopted for man-made HPEM environments and threats. Ultimately, this is because of uncertainty over the risk and uncertainty of the likelihood of occurrence of the threat and it means that protection for HEMP, HPRF DE, and IEMI is only applied in very limited cases where:

- It is mandated by the authority;
- There is a belief that the threat likelihood of occurrence is high;
- A specific risk or threat is identified, and the risk can be fully mitigated by the implementation of a protection scheme.

The as-built and whole-life costs of HPEM protection and the difficulty in managing upgrades discourage many from applying protection at all.

7.4.1 A Risk-Based Approach

In more recent times, due in part to the realization that the military function is so intertwined with the civilian infrastructure, consideration has been given to applying a risk-based approach to HPEM protection.

Civilian standards organizations such as the IEC, the International Telecommunications Union (ITU), and CIGRE have been active in developing advice and guidelines on the protection of civilian infrastructure [39]. The approach for HPEM in the civilian sector can be broadly categorized as a risk-based approach, in that the threat environment is described, an assessment of risk is carried out (partly by understanding inherent protection and partly through an understanding of effects or immunity), and proportionate protection is defined.

Indeed, the U.S. Department of Homeland Security (DHS) has just published a four-level scheme [40], summarized in Table 7.8.

In the scheme in Table 7.8, only level 4 pertains to the work-through approach described previously. Still, the application of a risk-based approach is problematic because a requirement of any risk-based approach is to qualify or better to quantify the likelihood of several factors, such as the likelihood of oc-

Table 7.8
Protection Levels Based on Operational Criticality

Protection Level	Operational Criticality
Level 1	Lowest cost; longer mission outages permitted
Level 2	Only hours of mission outages are permitted
Level 3	Only minutes of mission outages are permitted
Level 4	Only seconds of mission outages permitted

currence of the threat environment and the probability of failure or degradation of the system.

The likelihood of occurrence of a man-made HPEM threat environment can be subdivided into two parts: (1) the motivation of the threat actor, which is highly subjective and (2) the technical capability of the threat system. This latter part has been discussed in Chapter 2.

The likelihood of occurrence of a lightning strike threat can be estimated through the evaluation of historical data and examination of lightning strike maps and forecasts, and live lightning strike data are available online. Lightning risk assessment tools are also available and even mandated in standards [41]. However, for man-made HPEM threats such as HEMP, HPRF DE, and IEMI, there are simply very limited data on which to base a likelihood assessment, although risk assessment techniques for IEMI are beginning to emerge [41–46]. Indeed, our understanding of HPRF DE and IEMI environments are still evolving and are likely to continue to do so.

For a civilian infrastructure owner or operator, HEMP, HPRF DE, and IEMI are several man-made threats to be considered in any protection scheme. Man-made threats are less predictable than natural events such as fire, flood, and extreme weather, which have a degree of inevitability. Man-made HPEM threat environments therefore falls a long way down the priority list for protection budget because of the low likelihood of occurrence compared to other known threats.

7.4.2 A Resilience Approach

For the reasons discussed, the concept of resilience of systems to HPEM environments, rather than or in addition to hardening and protection to ensure that functionality is coming to be of great interest.

A definition of resilience is: "The ability of a system, community or society exposed to a threat to resist, absorb, accommodate, adapt to, transform and recover from the effects of the threat in a timely and efficient manner." The key change here is a shift in focus from high-level protection to an acceptance that function may be disrupted and therefore that prompt restoration and recovery after an event are required. This approach does not negate the need for some protection because damaged components and systems will slow recovery and restoration and therefore damage should be prevented. However, the number of systems that must be protected in a resilience scheme can be far less. A model for resilience is shown in Figure 7.38.

For this model, consider, for example, the risk of a fire to populated building (e.g., a school, hotel, or office). The residual risk that a fire will damage the building or, worse, claim lives, cannot be fully mitigated by protection alone, although protection such as fire-retardant materials, fire doors, and fire suppression systems are likely to be implemented. Because of this residual risk and because the people at risk do not need to work through the fire incident, fire and smoke detection are implemented and there will be occasional tests of the system and evacuation drills, essentially training and rehearsal.

The emphasis in this scheme is in reducing the consequential impact, restoring function and returning to normal operations as quickly as possible.

This resilience approach is also followed for cyber threats as it is not possible for cyber security protection designers to keep pace with the speed of evolution of the cyber threat.

The resilience scheme shown in Figure 7.38 includes protection but in this case the level of protection is not required to mitigate all risks. One could consider having no specific protection whatsoever and merely focus all of the effort on detection and recovery, but this approach does not provide deterrence, which is often a vital requirement. Inevitably, some parts of the system have to be protected in some way; for example, if spares or replacement parts

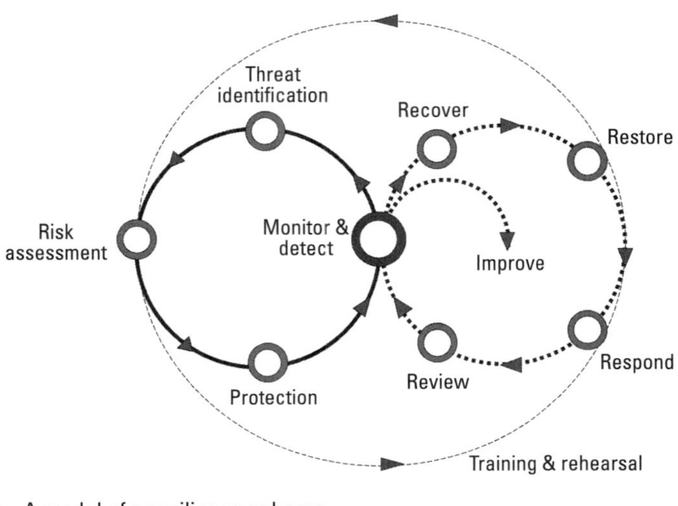

Figure 7.38 A model of a resilience scheme.

for damaged electronics are part of the recovery strategy, then the spares must be stored in such a way as to be protected from the HPEM event. The protection scheme in this instance could be as simple as storing spares in a shielded container in a basement.

This post-event recovery scheme has five stages: detect, respond, restore, recover, and improve. Detection is discussed at length later.

- *Respond:* The respond node refers to the immediate actions taken after an HPEM event is detected. Immediate actions may include, for example, confirming that there has been degradation or damage of the affected system function, reducing functionality so that functional safety is presevered, identifying other confirmations of an HPEM event, reverting to other backup modes of operation, alerting first responders, which in this case may be a field service engineer, and prioritizing restoration activities. These steps are vital first response or triage types of action.
- *Restore:* The restoration stage is about actions in the short term that restore the core functionality of the system. It may be possible to skip this stage and recover to full operating capability straight away, but it may be desirable to only restore the core functionally if other events could still occur, or the full extent of degradation and damage is unknown. This step could prevent cascading or second-order to n^{th}-order effects.
- *Recover:* The recovery stage is about actions that establish the full capability of the system back to normal operations. Progressing to full recovery will require confirmation that the threat has abated or has been mitigated and that cascading effects have been identified and resolved. Some parts of the recovery process may take considerable time, for example, if the affected process is safety-critical requiring certain protocols to be followed or if the affected system is a business system where it may take months or years to recover financial stability or reputation.
- *Improve:* Vitally, all of the stages provide information that can be used to improve the process, reduce the impact, and improve response and recovery times. The approach enables learning and improvement in overall resilience.

In summary, the overall aim of a resilience scheme is:

- To prepare for events that either have an unknown likelihood of occurrence or cannot be fully mitigated by a pragmatic level of protection;
- To recover from the event in a timely and efficient manner;
- To learn from events to improve response and recovery time.

This type of resilience scheme is just starting to be explored for man-made HPEM threats including HEMP, HPRF DE, and IEMI. One of the key requirements that enable a resilience, restoration, and recovery approach is HPEM detection.

7.4.3 HPEM Detection

Very simply, steps to recover in a timely and effective manner from the effects of the HPEM environment cannot be taken if there is no awareness that an HPEM event has been or is currently taking place. It is not possible in most cases for human senses to detect HPEM and an HPEM effect on a system leaves no easily discernible telltale sign or witness [47]. Detection technologies are therefore required to provide confirmation that a disruption in the system function is the result of an HPEM environment.

Warning of HPEM events well before they threaten a system is clearly preferable because if warning can be provided, then steps can be taken to mitigate the threat before the event reaches the system and actions can be taken to minimize effects. For natural HPEM threats such as lightning, storm forecasting is now mature enough to provide at least several hours of warning. However, threat warning for man-made HPEM cannot be forecast. It may be possible to detect a threat actor preparing to use an HPEM effector, but this is very challenging because there will be a need to be in the right place at the right time. This pre-deployment detection of man-made HPEM is an intelligence-gathering exercise.

An HEMP event requires detonation of a nuclear device in space. Systems to detect pre-launch and launch of rocket or missile-based systems capable of delivering a nuclear payload into space no doubt exist.

For HPRF DE and IEMI, the delivery mode of the effector could be in a variety of guises; however, it must be anticipated that effector systems will be tested and trialed before they are deployed and it may be possible to identify this activity [45–49]. Still, providing early warning of a man-made HPEM threat is very difficult. To date, most of the research and development on HPEM detection has been for detection of HPEM events as they occur at the location of some critical asset.

7.4.3.1 Requirements and Technology for HPEM Detection

The technical requirements for HPEM detection technologies are challenging. In addition to these technical challenges are nontechnical challenges, the information provided by the detector must be in a format that is understandable to the user but perhaps more importantly the detection technology must be affordable, since it must be lower than the whole life cost of building and maintaining protection.

Technology types that could provide the basis for a man-made HPEM environment detector include:

- *RF receivers:* There are very many types of RF receivers commercially available that vary from handheld scanners to full-scale electronic surveillance measure (ESM) receivers. Most RF receivers are narrowband receivers and step or sweep through a broad frequency range. Most RF receivers are designed for low-power, small-signal applications.
- *Advantageous use of ubiquitous receiver technology:* Lopes-Esteves et al. [48] proposed that generally available ubiquitous receiver technologies such as those in everyday cell phones, laptop computers, and even drones (i.e., GSM, Wi-Fi, and GPS) will naturally detect HPEM disturbances.
- *Use of software to correlate effect to an HPEM disturbance:* Lopes-Esteves et al. also proposed that software tools or applications can be specially developed to interrogate status information of the semiconductor device inside an item of victim equipment [50]. In theory, readily available device parameters such as temperature and error status may change due to the HPEM disturbance and this change can be detected.
- *EM field (EMF) detectors:* EMF detectors are commercially available for the purpose of detecting and monitoring EM environments, which are hazardous for human health. There are examples that are very broadband and some have data-logging features.
- *Electro-optic devices:* This technology is highly promising for the HPEM detection application. Examples in the literature [51–56] employ devices that change some property of light (modulation, intensity, wavelength) when an external EM field is applied. The light source is commonly a laser and the electro-optic device is usually a crystal, although polar liquids are also found to exhibit the phenomenon. Electric field detectors based around birefringent crystals, which exhibit the Pockel's effect, appear to be the most common.
- *Other RF transducers:* This is a group of front-end sensor options including neon tubes [56, 57], liquid crystals [58, 59], fuses [60, 61], resistors [62], and rectennas [63].

  Neon and fluorescent lamps (both examples of GDTs) are known to fluoresce or produce a glow discharge in the presence of high electric fields. The lamps respond to energy deposition (heating) to initiate the glow discharge.

  Liquid crystal detectors can change color when heating is induced in them by an incident EM field. This is an average power phenomenon.

Fusing (vaporization of a thin conductor) can be shown to be initiated by high-intensity electric fields and this has been proposed as a means of HPEM detection. In a very simple design, fuses are formed on thermal paper or micron thin printed circuit boards and coupled to a planar antenna structure which is also printed such as a Vivaldi antenna.

Resistive sensors for narrowband HPM have been developed by Dagys et al. [62]. A resistive sensor is a device primarily designed for HPM pulse measurement within a transmission line. The performance of the resistive sensor is based on the electron heating effect in semiconductors. A photograph of some resistive sensors is shown in Figure 7.39.

The term rectenna is a concatenation of the terms rectification and antenna. A typical rectenna design incorporates a rectifying semiconductor device directly into an antenna structure so that a quasi-direct current signal proportional to the magnitude of the HPEM exposure is produced. By fusing the rectifying element directly with the antenna, the efficiency of the transfer of EM energy to electrical energy is optimized. To date, most rectenna designs have been narrowband. However, the Canary HPEM detector [49] utilized a nested periodic antenna.

Table 7.9 describes the advantages and limitations of each technology type for the HPEM detection application.

7.4.3.2 Examples of HPEM Detection Solutions

HPEM detection technologies are still in their infancy; while mature prototypes exist, commercially proven technologies were not widely available at the time of writing.

Figure 7.39 Resistive sensors for HPEM measurement or detection. (Courtesy of Dagys et al. [62].)

Table 7.9
Some Advantages and Limitations of Different Technologies for HPEM Detection Applications

Technology	Advantage(s)	Limitation(s)
RF receiver	Available technology; logging/data retrieval function; waveform analysis	Most receiver technologies have insufficient instantaneous bandwidth for the detection of Hyperband environments; possibility of damage; high cost
Advantageous use of ubiquitous receivers	Widely available technology; receivers are already deployed; low cost	Receivers are narrowband by design, therefore not suitable for detection of Hyperband waveforms; requires development and embedding of an application on to device hardware; likely to be difficult to set meaningful detection thresholds
Software correlation	No extra hardware required; easily deployed; low cost	Requires development and embedding of an application on to device hardware; if the HPEM disturbance is very intense or the host device loses power, then detection may not be recorded
EMF detector	Available technology; low cost; logging/data retrieval function	Sensor element of detection is based on human body heating model (thermal) and therefore not suitable for detection of Hyperband waveforms
Electro-optic devices	Non-semiconductor sensor, so robust to HPEM environment; wide instantaneous bandwidth	High cost; mechanical robustness
Neon: glow discharge transient detector	Nonsemiconductor sensor, so robust to HPEM environment; compact; very low cost	Requires high average power for detection, not suitable for detection of Hyperband waveforms
Liquid crystal transient detector	Nonsemiconductor sensor: so robust to HPEM environment; compact; very low cost	Requires high average power for detection
Fuse	Available technology; very simple; compact; low cost	Requires tuning; difficult to interrogate; not broadband so not suitable for detection of Hyperband waveforms
Resistive sensor	Available technology; very simple; wide bandwidth; compact; low cost	Generally relies on thermal/heating effect, not suitable for detection of Hyperband waveforms
Rectenna	Very simple; compact; low cost	Most examples are not broadband so not suitable for detection of Hyperband waveforms

HPM Detector, TNO, The Netherlands

The high-power microwave detector prototype developed by TNO, The Netherlands [64], has been designed to be a low-cost device and meet specifications of most HPEM signal types, flat frequency response, omnidirectionality, and response time. The characteristics of the detector are as follows:

- Signal types: carrier-based pulses, UWB, and CW;
- Frequency range: 100 MHz–8 GHz (flat response);

- Sensitivity threshold:
    - Low level: 3 V/m;
    - Medium level: 10 V/m;
    - High level: 40 V/m.
- Visual indication via 3 indicator lights, separate indication CW and carrier-based pulses;
- Maximum input level: >1,000 V/m (carrier based), >3 kV/m (UWB).

EM Inferential-Detector system, Emprimus, USA

The detector developed by Emprimus, USA [65], detects a wide range of EM threat intensities, ranging from nuisance levels to those that damage electronics and cause data integrity issues. Emprimus proposes a proprietary inferential approach to high EM field detection. Inferential refers to the fact that one uses the far-field measurement of the magnetic field, which is directly proportional to the high electric field of interest in air. The high electric field is shielded from the measurement [66].

The characteristics of the detection system are as follows:

- Frequency range: 100 MHz–10 GHz;
- Field strength: 100 V/m–100 kV/m;
- Minimum pulse width: 10 ns;
- Rise time: <10 ns;
- Measurement interval: every 1 ms.

IEMI Detector system, Fraunhofer INT, Germany

Fraunhofer INT as part of the HIPOW project [67, 68] has developed an IEMI detection system. The system shown in Figure 7.40 is composed of a 4-channel antenna system to enable direction finding, detection electronics based on a logarithmic amplifier and a laptop linked to the detection electronics. The laptop includes a software application that can be used to collect and visualize detection data.

The characteristics of this detection system are as follows:

- Detection of signals (pulse, UWB, CW) with field strength >1 kV/m;
- Damage immunity against field strengths of up to 10 kV/m;

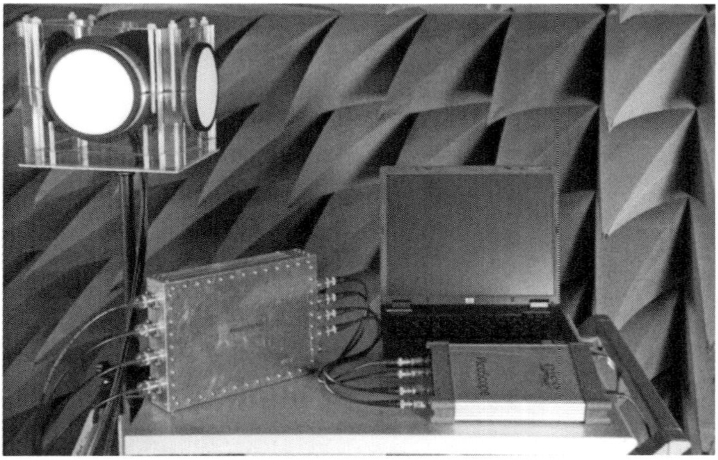

Figure 7.40 IEMI detector developed by Fraunhofer INT, Germany. (Courtesy of Adami, et al. [68].)

- Frequency-independent detection (500 MHz–8 GHz) of HPM sources in medium distances (at least detection of field strengths >100 V/m) for warning and searching;
- Measuring dynamics >60 dB;
- Polarization independence: directional independence (at least in the horizontal plane) or in a defined sector (e.g., 90°);
- Classification of the detected events by amplitude, pulse duration, pulse repetition frequency (*prf*) or pulse number, and pulse form;
- Threat direction identification (four-channel system).

Norms Detector, China

Kong and Xie [69] developed a detection system that uses an integrated asymptotic conical antenna (d-dot), followed by an impedance matching and hardware integrator and fiber-optic link for the transmission of detection data. The detector can also store data [69]. A photograph of the system is shown in Figure 7.41.

The characteristics of this detection system are as follows:

- Operating bandwidth: 1 kHz to 460 MHz;
- Field strength: 0.1 to 55 kV/m by adjusting the antenna's height;
- Maximum sampling rate: 2 G samples per second;
- Memory depth: 8 kB.

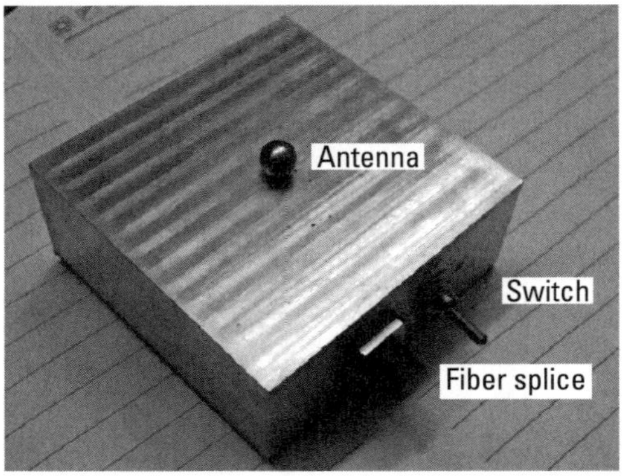

Figure 7.41 Norms detector developed by Kong and Xie, China. (Courtesy of Xie et al. [70].)

HPM Attack Detector System, Fraunhofer EMI, Germany

Fraunhofer EMI as part of the PROGRESS European framework project [71] developed an HPM attack detection system. The system is integrated with other detectors such as physical (blast), cyber, and Global Navigation Satellite System (GNSSS) jamming and spoofing detection. The detector was developed specifically to improve the resilience of satellite ground stations.

The characteristics of this detection system are as follows:

- Frequency range 500 MHz–8 GHz;
- Polarization dependence independent;
- Direction finding accuracy ± 15°;
- Dynamic range 60 dB;
- Minimum pulse width 20 ns;
- Maximum pulse repetition rate 1 kHz.

TOTEM Detector System, QinetiQ, UK

The TOTEM detector system has been developed by QinetiQ in the United Kingdom. It is designed to detect HEMP E1, HPRF DE, and IEMI environments, which have the possibility to threaten electronic systems [72]. The detector has a novel broadband spiral antenna and uses a logarithmic receiver [73]. The detector has on-board flash memory for storing event data, an integral uninterruptable power supply, and fiber-optic connections for event data transmission and can be configured to interface with a 24/7 web service. A photograph of the TOTEM detector is shown in Figure 7.42.

The characteristics of this detection system are as follows:

Figure 7.42 TOTEM detector developed by QinetiQ UK. (Courtesy of QinetiQ Ltd.)

- Frequency range: 10 MHz to 10 GHz;
- Instantaneous bandwidth: ~3 GHz, proven to detect Hyperband environments (200 ps pulse width);
- GPS/GNSS interference detection;
- Event log: IEMI-hardened event logger recording event time, date, and magnitude;
- UPS: includes integrated battery backup in the event of power outage in normal operating conditions;
- Fiber-optic ports for data transmission.

7.4.3.3 Challenges for HPEM Detection

There are several challenges that remain for the deployment of HPEM detection systems, which are discussed here.

- *False alarms:* One of the biggest problems with any sort of detection is in differentiating between truly malicious or unacceptable behavior and accidental or acceptable behavior. Any alarm type system quickly loses credibility if too many false alarms are produced [74]. Four outcomes are considered possible from a detection system that can be described by Figure 7.43.

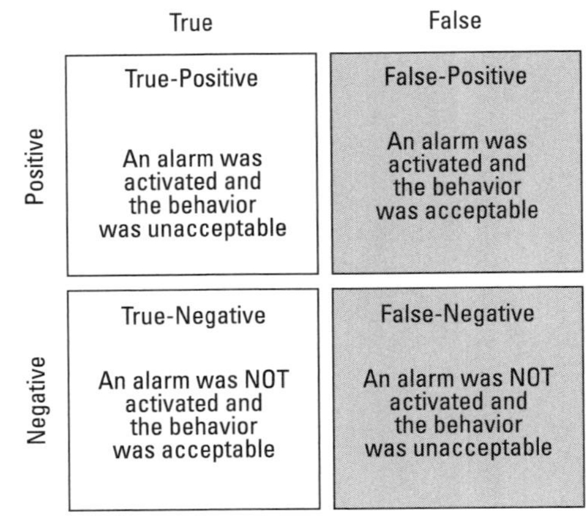

Figure 7.43 True or false negative or positive diagram.

False-positives are seen to be the most problematic because alarms require analysis, and therefore a false alarm requires nugatory activity, which has an inherent time penalty that may slow down processing functions. False-negative alarms are perhaps the most serious because unacceptable behavior was allowed and was not detected. However, false-positives are the most likely to cause degradation of user confidence of the detection system.

- *Alarm threshold setting:* As discussed throughout this book, HPEM effects are imbued with a high degree of uncertainty and therefore it may be challenging to set event detection thresholds that are meaningful and yet do not produce false alarms. Fortunately, the understanding and definition of the HPEM environment are now fairly mature and there are published environment definitions. A detection solution must have the ability to rapidly and remotely adjust threshold settings as the operational understanding of the deployment environment evolves.
- *Detector location and orientation:* The installed location of an HPEM detector is important as the appropriate selection of the location can be key to minimizing uncertainty with respect to the detection threshold. For example, at first, it may seem desirable to place the HPEM detector on the external façade of a building to provide detection for the building. However, unless the coupling into the building structure is known for all HPEM environments of concern, then the threshold cannot be set with sufficient fidelity to reduce false alarms.

7.5 HPEM Standards

In the past and particularly in the Cold War era, the primary focus on HEMP standards was on the protection of military systems. Many HEMP standards in the military domain remain classified and cannot be discussed in detail within this chapter. However, in recent years, it has been recognized that HEMP could severely impact the functionality of civilian infrastructures [75–77]. There also appears to be a new impetus in the civilian sector to consider unifying HEMP protection requirement standards for infrastructure assets with other HPEM threats such as IEMI and with natural phenomena such as geomagnetic disturbances caused by solar storms [78, 79].

Recently, much effort has been expended on producing HPEM standards for guidance to the civilian community on HEMP environments and impacts and for protection of civilian critical infrastructure (CI) assets. This section discusses the progress in HPEM protection standardization efforts.

7.5.1 HPEM Standards Organization

There are several organizations that are presently active in producing or maintaining HPEM standards. In many cases and particularly for the military standards organizations, HPEM is often treated as a component of the combined or unified EM environment (EME) along with other phenomena such as lightning, electrostatic discharge (ESD), and EM interference (EMI) [80–83]. Alternatively, HEMP in particular can be treated as a component of nuclear weapons effect requirements [84, 85]. However, there are always exceptions and for some specific military applications HEMP is treated separately such as in MIL-STD-188-125 Part 1, for example [86].

Often the military organization of standards is based on the ad hoc grouping of experts into a Working Group under a steering committee. An example is MIL-STD-188-125, which was organized by a special Working Group under the U.S. Defense Special Weapons Agency (DSWA), now known as the Defense Threat Reduction Agency (DTRA).

Within the civilian domain, arguably the most active and prolific group is the IEC Subcommittee 77C (IEC SC 77C).

7.5.1.1 IEC SC 77C: High-Power Transient Phenomena

Since 1989, the IEC, headquartered in Geneva, Switzerland, has been developing and publishing standards and reports dealing with the HEMP environments and methods to protect civilian systems from these environments under the leadership of IEC SC 77C. As HEMP is an EM disturbance, it was decided from the beginning that this work would be closely integrated with the EMC work being performed by the IEC and other organizations throughout the world. The IEC Technical Committee 77 (TC 77), which is the parent committee of SC 77C, has the title "EMC."

There have been several articles [87–89] that provided details of the 21 published IEC SC 77C documents that can be applied to the definition of the environments, the coupling to systems, and the protection of systems. The document set is shown in Figure 7.44. Those standards in black or bold text relate directly to HEMP, whereas those in lighter text relate to the IEMI.

7.6 Summary

Protection techniques for HPEM environments have been rigorously studied and have been applied for more than 50 years. Protection concepts are very mature, are tried and tested, and can provide a high degree of protection for devices, equipment, systems, and installations. Still, HPEM protection design can be complex and requires skill to implement.

Many of the well-established protection concepts are designed to provide a work-through protection solution such that the function performed by the installation or system is unperturbed by HPEM exposure. However, a work-through protection design requires the installation of shielding, filtering, and transient protection, all of which require continuous surveillance and maintenance to provide work-through protection throughout the lifetime of the

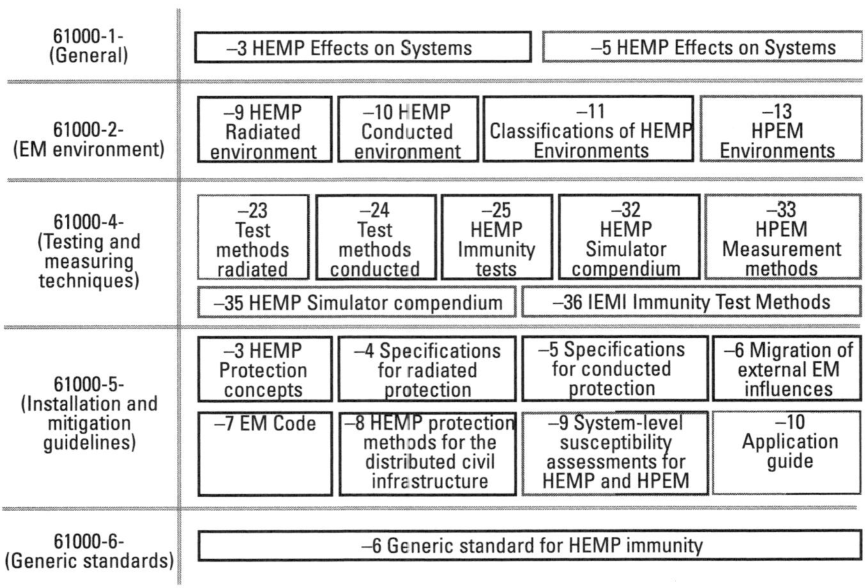

Figure 7.44 The body of HPEM standards produced by the IEC SC 77C.

protected asset. These protection schemes can be financially difficult to justify and expensive to implement and maintain.

The vast majority of military and civilian systems do not require to work through an HPEM exposure or event but must recover from HPEM exposure in an acceptable manner. Man-made HPEM environments are continually evolving, and the pace of evolution is not expected to abate. Traditional work-through protection schemes can be inflexible, in that they are easily compromised due to the rate of change of the asset that they are designed to protect. The cost of a full work-through protection scheme is prohibitive, especially given the likelihood of occurrence of an HPEM environment.

A recent trend is to consider a risk-based approach and increasingly a resilience model for protection. The resilience model relies on detection of HPEM environments and commercially available HPEM detection solutions are beginning to emerge.

The deployment of HPEM detectors will greatly assist in quantifying the occurrence of man-made and natural HPEM impacts. Detection in itself provides deterrence because an important aspect of deterrence is the ability to attribute events to the cause of the event.

Finally, a resilience approach, informed by detection, actually helps to quantify the likelihood of occurrence of an HPEM threat. These data are vital as they help to identify the scale of the protection challenge.

References

[1] Bell Laboratories, *EMP Engineering and Design Principles,* Technical Publications Department, Bell Laboratories, Whippany, NJ, 1975.

[2] Lee, K. S. H., (ed.), *EMP Interaction: Principles, Techniques, and Reference Data,* New York: Hemisphere Publishing Co., 1986.

[3] Radasky, W. A., and M. W. Wik, *IEC Standardization: Immunity to High Altitude Nuclear Electromagnetic Pulse (HEMP),* Report on the Status of Work, June 1998.

[4] Vance, E. F., "EMP Hardening of Systems," *Proceeding of the 4th Symposium and Technical Exhibition on Electromagnetic Compatibility,* Zurich, March 10–12, 1981.

[5] U.S. Army Corps of Engineers, "Electromagnetic Pulse (EMP) and Tempest Protection for Facilities," Publication 1110-3-2, December 31, 1990.

[6] Clayborne, D., D. Taylor, and V. Giri, *High Power Microwave Systems and Effects,* Taylor & Francis, 1994.

[7] Giri, D. V., and A. W. Kaelin, "Many Faces of High-Power Electromagnetics (HPEM) and Associated Problems in Standardization," *AMEREM,* Albuquerque, NM, 1996.

[8] Baum, C. E., "Electromagnetic Topology for the Analysis and Design of Complex Electromagnetic Systems," in *Fast Electrical and Optical Measurements, Vol. I,* I. E. Thompson and L. H. Luessen, (eds.), Dordrecht: Martinus Nijhoff, 1986, pp. 467–547.

[9] Baum, C. E., "How to Think About EMP Interaction," *Proceedings of the 1974 Spring FULMEN Meeting*, Kirtland AFB, April 1974.

[10] Tesche, F. M., et al., "Internal Interaction Analysis: Topological Concepts and Needed Model Improvements," *Interaction Note Series*, IN-248, October 1975.

[11] Tesche, F. M., "Topological Concepts for Internal EMP Interaction," *IEEE Transactions on Antennas and Propagation*, Vol. AP-26, No. 1, January 1978.

[12] Schulz, R. B., "Electromagnetic Shielding," Ch. 9 in *The Handbook of EMC*, R. Perez, (ed.), New York: Academic Press, 1995.

[13] Archambeault, B., and R. Thibeau, "Effects of Corrosion on the Electrical Properties of Conducted Finishes for EMI Shielding," *IEEE EMC Conference*, May 23–25, 1989.

[14] Das, S. K., et al., "The Effect of Corrosion on Shielding Effectiveness of a Zinc-Coated Steel Enclosure," *IEEE EMC Conference*, 1999.

[15] Standler, R. B., *Protection of Electronic Circuits from Over Voltages*, 1st ed., New York: Wiley-Interscience, 1989.

[16] *Surge Protection and Applications Guide*, Powerforce Systems, Inc., Alcoa, TN.

[17] *A Guide to the Use of Spark Gaps for Electromagnetic Pulse (EMP) Protection,* Joslyn Electronic Systems, Goleta, CA, 1973.

[18] *Electrical Distribution and System Protection: A Textbook and Practical Reference on Overcurrent and Overvoltage Fundamentals, Protective Equipment and Apparatus,* Cooper Power Systems, 1990.

[19] Technical Staff of Keytek Instrument Corp., *Surge Protection Test Handbook: A Practical Guide for Designers, Builders and Users of Modern Electronic Equipment*, 2nd ed., 1982.

[20] *A Comparison Report of Transzorbs Versus Metal Oxide Varistors,* General Semiconductor Co., Tempe, AZ.

[21] Sweetana, A., "Characteristics, Application and Field Testing of Westinghouse Gapless Metal Oxide Surge Arresters," *Proceedings of the 1982 Doble Conference*, Boston, April 19–23, 1982.

[22] "Citel Surge Arrester Gas Tubes," Product brochure from Citel Inc., www.citelprotection.com.

[23] *Reference Data for Radio Engineers Handbook*, New York: International Telephone and Telegraph Co., 1956.

[24] Büchler, W., "Overvoltage Protection Circuits," Meteolabor AC, Wettikon, Switzerland, *Proceedings of the 1985 Zurich Symposium on EMC*, 1985.

[25] *Spice Age, Nonlinear Analog Circuit Simulator*, Tatum Labs, Inc., Ann Arbor, MI.

[26] Standler, R. B., "Transmission Line Models for Coordination of Surge-Protection Devices," *Proceedings of the 1993 IEEE EMC Symposium*, 1993.

[27] Chen, C. -L., "Transient Protection Devices," *Proceedings of the 1975 IEEE EMC Symposium*, 1975.

[28] International Electrotechnical Commission Standard IEC 60-2, International Electrotechnical Commission Standard. (Also see ANSI/IEEE Std 4-1798 and ANSI C62.1-1984.)

[29] Institute of Electrical and Electronics Engineers Standard, IEEE Standard 587. (Also see ANSI Std C62.41-1980.)

[30] American National Standards Institute, ANSI C37.90a-1974.

[31] International Electrotechnical Commission Standard IEC 801-4, International Electrotechnical Commission Standard, 1998.

[32] MIL-STD-188-125, "HEMP Protection for Ground-Based C4I Facilities Performing Critical, Time-Urgent Missions," 1998.

[33] Hoad, R., and W. Radasky, "Progress in High Altitude Electromagnetic Pulse (HEMP) Standardization," *IEEE Transactions on Electromagnetic Compatibility*, Special Issue on HEMP, Vol. 55, No. 3, June 2013.

[34] Public Submission to the U.S. National Reliability Council, "Comments of the Foundation for Resilient Societies, Inc. on Mitigation Strategies for Beyond-Design-Basis Events," NRC Docket No. NRC-2014-0240, 80 FR 70609, August 8, 2019.

[35] Fact Sheet, "On Preliminary Costing Model of the Foundation for Resilient Societies to Protect the U.S. Electric Grid from Man-Made Electromagnetic Pulse (EMP) Hazards and Solar Geomagnetic Disturbances (GMD)," Foundation for Resilient Societies, May 13, 2015.

[36] Baker, G. H., "Testimony Before the Senate Committee on Homeland Security and Governmental Affairs," February 27, 2019.

[37] Intel IT Center planning guide, "Updating IT Infrastructure Four Steps to Better Performance and Lower Costs for IT Managers in Midsize Businesses," December 2013.

[38] Huang, R., and E. Masanet, "Data Center IT Efficiency Measures - The Uniform Methods Project: Methods for Determining Energy Efficiency Savings for Specific Measures," NREL/SR-7A40-63181, January 2015.

[39] Hoad, R., E. Schamiloglu, and W. Radasky, "An Update on HPEM Standards and the Work of IEC SC 77C in 2018," *AMEREM 2018*, Santa Barbara, CA, 2018.

[40] Baker, G. H., et al., "Electromagnetic Pulse (EMP) Protection and Resilience Guidelines for Critical Infrastructure and Equipment, Version 2.2," National Coordinating Center for Communications (NCC), February 5, 2019.

[41] Mata, C. T., and T. Bonilla, "Lightning Risk Assessment Tool, Implementation of the IEC 62305-2 Standard on Lightning Protection," *2012 International Conference on Lightning Protection (ICLP)*, Vienna, Austria, 2012.

[42] Petit, B., R. Hoad, and A. Fernandes, "An Overview of Some Site Specific IEMI Risk Assessment Tools," *AMEREM 2014*, Albuquerque, NM, July 27–August 1, 2014.

[43] Chatt, L., B. Petit, and R. Hoad, "High Power Radio Frequency Risk/Hazard Assessment Tool," *EUROEM 2016*, Imperial College, London, U.K., July 2016.

[44] Sabath, F., and H. Garbe, "Assessing the Likelihood of Various Intentional Electromagnetic Environments the Initial Step of an IEMI Risk Analysis," *IEEE International Symposium on Electromagnetic Compatibility (EMC)*, 2015, pp. 1083–1088.

[45] Genender, E., H. Garbe, and F. Sabath, "Probabilistic Risk Analysis Technique of Intentional Electromagnetic Interference at System Level," *IEEE Transactions on Electromagnetic Compatibility*, Vol. 56, No. 1, February 2014, pp. 200–207.

[46] Lanzrath, M., M. Suhrke, and H. Hirsch, "HPEM-Based Risk Assessment of Substations Enabled for the Smart Grid," *IEEE Transactions on Electromagnetic Compatibility*, 2019, pp. 1–13.

[47] Hoad, R., and I. Sutherland, 'Malicious Electromagnetic (EM) Threats to Information Processing Installations – How Do We Respond and Protect?" University of Plymouth, Plymouth, U.K., June 30–July 1, 2008.

[48] Lopes-Esteves, J., E. Cottais, and C. Kasmi, "Software Instrumentation of an Unmanned Aerial Vehicle for HPEM Effects Detection," *2nd URSI Atlantic Radio Science Meeting (AT-RASC)*, 2018, pp. 1–4.

[49] Hoad, R., "The Utility of Electromagnetic Attack Detection to Information Security," Ph.D. Thesis, University of Glamorgan, December 2007.

[50] Kasmi, C., et al., "Event Logs Generated by an Operating System Running on a COTS Computer During IEMI Exposure," *IEEE Transactions on Electromagnetic Compatibility*, Vol. 56, No. 6, 2014, pp. 1723–1726.

[51] Kanda, M., and K. D. Masterson, "Optically Sensed EM-Field Probes for Pulsed Fields," *Proceedings of the IEEE*, Vol. 80, No. 1, January 1992, pp. 209–215.

[52] Zaldivar-Huerta, L., and J. Rodriguez-Asomoza, "Electro-Optic E-Field Sensor Using an Optical Modulator," *14th International Conference on Electronics, Communications and Computers, 2004, CONIELECOMP 2004*, February 16–18, 2004, pp. 220–222.

[53] Deibel, J. A., and J. F. Whitaker, "A Fiber-Mounted Polymer Electro-Optic-Sampling Field Sensor," *The 16th Annual Meeting of the IEEE Lasers and Electro-Optics Society, 2003 (LEOS 2003)*, Vol. 2, 2003, pp. 786–787.

[54] Torihata, S., and B. Loader, "The New Principle E-Field Sensor for Automotive Immunity Test," *Automotive EMC 2003*, Milton Keynes, NEC Tokin Corp., Japan, November 6, 2003.

[55] Gaborit, G., et al., "Packaged Optical Sensors for the Electric Field Characterization in Harsh Environments," *2015 International Conference on Electromagnetics in Advanced Applications (ICEAA)*, 2015, pp. 1468–1471.

[56] Thickpenny, J., "The Measurement of Electric and Magnetic Fields for Prediction," Royal Military College of Science Technical Note RT 56, 1971.

[57] Yamamoto, Y., et al., "Measurement of Intense Microwave Field Patterns Using a Neon Glow Indicator Lamp," *International Journal of Infrared and Millimeter Waves*, Vol. 16, No. 3, March 1995.

[58] Giannini, F., P. Maltese, and R. Sorrentino, "Liquid Crystal Technique for Field Detection in Microwave Integrated Circuitry," *Alta Frequenza (English Edition)*, Vol. 46, April 1977, pp. 170–178.

[59] Sato, S., and T. Hara, "Application of a Ferroelectric Liquid Crystal Cell to an Electric Field Sensor," Institute of Scientific Instrumentation, 1998.

[60] Seregelyi, J. S., T. Lapohos, and C. Gardner, "Design and Characterisation of Broadband Dosimetry Plates," NATO presentation SCI-019, Tactical implications of HPM, June 8–10, 1999.

[61] Eriksson, K., et al., "Microwave Shielding Effectiveness of Large Mobile Military Systems and a Low Cost HPM Indicator," Reprint from RVK02, Stockholm, Sweden, June 10–13, 2002.

[62] Dagys, M., et al., "Resistive Sensor for High Power Microwave Pulse Measurement," *2011 International Conference on Electromagnetics in Advanced Applications*, 2011, pp. 43–46.

[63] McSpadden, J. O., L. Fan, and K. Chang, "Design and Experiments of a High-Conversion-Efficiency 5.8-GHz Rectenna," *IEEE Transactions on Microwave Theory and Techniques*, Vol. 46, No. 12, 1998, pp. 2053–2060.

[64] Arnesen, O. H., and R. Hoad, "Overview of the European Project 'HIPOW,'" *IEEE Electromagnetic Compatibility Magazine*, Vol. 3, No. 4, 2014, pp. 64–67.

[65] Jackson, D. B., T. R. Noe, and G. H. Baker III, "High Dynamic Range, Wide Bandwidth Electromagnetic Field Threat Detector," *Ultra-Wideband, Short-Pulse Electromagnetics 10*, November 28, 2013, pp. 355–368.

[66] Jackson, D. B., et al., "Electromagnetic Field Detection Systems and Methods," U.S. Patent 8860402, Application Number: 12/906912, October 14, 2014.

[67] Suhrke, M., "Threat and Detection of High Power Microwaves," *Progress in Electromagnetics Research Symposium Abstracts*, Stockholm, Sweden, August 12–15, 2013.

[68] Adami, C., et al., "HPM Detector with Extended Detection Features," *Ultra-Wideband, Short-Pulse Electromagnetics 10*, New York: Springer, 2014, pp. 345–353.

[69] Kong, X., and Y. Z. Xie, "An Active E-Field Sensor Based on Laser Diode for EMP Measurement," *XXXIth URSI General Assembly and Scientific Symposium (URSI GASS)*, 2014, pp. 1–4.

[70] Kong, X., et al., "Development of One-Dimensional Norm Detector for the Measurement of Nanosecond Transient Electrical Signals," *IEEE Transactions on Electromagnetic Compatibility*, Vol. 59, No. 4, 2017, pp. 1035–1040.

[71] Schopferer, S., et al., "PROGRESS Project: Improving the Resilience of Satellite Ground Station Infrastructures: High Power Microwaves Threat Detection System and Protection Strategies," *2016 International Symposium on Electromagnetic Compatibility (EMC EUROPE)*, 2016, pp. 746–749.

[72] Herke, D., et al., "Lessons Learnt from IEMI Detector Deployments," *EUROEM 2016*, Imperial College, London, U.K., July 2016.

[73] Hoad, R., and D. L. Herke, "Electromagnetic Interference Indicator and Related Method," International Patent Publication No. WO17/125465 A1, January 19, 2017.

[74] Proctor, P. E., *The Practical Intrusion Detection Handbook*, Upper Saddle River, NJ: Prentice Hall, 2001.

[75] Wik, M., "URSI Statement - Nuclear Electromagnetic Pulse [EMP] and Associated Effects," *IEEE Antennas and Propagation Society Newsletter*, Vol. 29, No. 3, 1987, pp. 19–23.

[76] *Report of the Commission to Assess the Threat to the United States from Electromagnetic Pulse (EMP) Attack, Vol. I: Executive Report*, April 7, 2004, www.empcommission.org.

[77] *Report of the Commission to Assess the Threat to the United States from Electromagnetic Pulse (EMP) Attack, Critical National Infrastructures*, April 2008, www.empcommission.org.

[78] Hoad, R., "Protection from High Power Electromagnetic Environments (HPEM) on Civilian Infrastructure," *Directed Energy Weapons 2010*, London, U.K., March 22–23, 2010.

[79] Radasky, W. A., "The Potential Impacts of Three High-Power Electromagnetic (HPEM) Threats on Smart Grids," *IEEE Electromagnetic Compatibility Magazine*, Vol. 1, No. 2, 2012.

[80] EP 1110-3-2, "Engineering and Design - Electromagnetic Pulse (EMP) and Tempest Protection for Facilities Proponent," December 31, 1990. http://everyspec.com/ARMY/ARMY-General/EP_1110-3-2_1990_18942.

[81] NATO AECTP 250, "Electrical and Electromagnetic Environmental Conditions - Leaflet 256 Nuclear Electromagnetic Pulse (NEMP/EMP)," 2nd ed., November 2010.

[82] United Kingdom, Defence Standard 59-411, Part 2, Issue 1, Amendment 1, "Electromagnetic Compatibility – Part 2 – The Electric, Magnetic & Electromagnetic Environment," January 31, 2008.

[83] USA, MIL-STD-464C, Military Standard, "Electromagnetic Environmental Effects, Requirements for Systems," December 1, 2010.

[84] NATO AEP 4: "Nuclear Survivability Criteria for Armed Forces Material and Installations," 1991.

[85] Defense Standard 08-4, "Nuclear Weapons Explosions Effects and Hardening – Part 5 Section 4, Nuclear Electromagnetic Pulse," 2010.

[86] USA, MIL-STD-188-125, Part 1, "High-Altitude Electromagnetic Pulse (HEMP) Protection for Ground-Based C4I Facilities Performing Critical, Time-Urgent Missions, Part 1 Fixed Facilities," July 17, 1998.

[87] Ni, L., "The Standards of High Power Transient Phenomena," *5th Asia-Pacific Conference on Environmental Electromagnetics (CEEM)*, 2009.

[88] Radasky, W. A., "2007 Status of the Development of High-Power Electromagnetic (HPEM) Standards in the IEC," *International Symposium on Electromagnetic Compatibility*, 2007.

[89] Hoad, R., and W. Radasky, "Progress in IEC SC77C High Power Electromagnetics Publications During 2009," *Asia Pacific EMC Conference*, Beijing, China, April 12–16, 2010.

8

Epilogue

This book has provided information on the broad topics of HPEM environments, HPEM interaction and coupling, HPEM effects test techniques, HPEM effects mechanisms, HPEM effects classification, and protection of systems from HPEM environments. This chapter looks at these topics with a view toward the future many years after the publication of this book.

We have shown that HPEM environments cover both natural and man-made phenomena. The natural phenomena such as lightning and geomagnetic disturbance have been part of the Earth's environment for millennia, yet our understanding of the particulars and peculiarities of these phenomena continues to improve. There are some suggestions that, with climate change, there will be more extreme weather events, which no doubt will lead to the possibility of more frequent lightning events. Indeed, at least one article [1] predicted that the number of lightning strikes in the United States will increase by about 12% for every degree of rise in global average air temperature. It is possible to speculate that, because of climate change, regions of the Earth that presently have a low incidence of lightning strikes may see more lightning.

Geomagnetic disturbances caused by geomagnetic storms, in turn, caused by energetic particles emitted from the Sun, are really only just beginning to be understood [2]. It is not really known if changes on the Earth will affect the likelihood or severity of geomagnetic disturbances. It is known that occasionally there are rare severe geomagnetic disturbances; sometimes these severe storms are referred to as 100-year storms. The last really severe event, known as the Carrington storm after the English astronomer who observed it, was in 1859. Efforts are underway to predict when the next big geomagnetic storm will hit [3], but there is no doubt that the severity of the effect on electrical and electronic systems, today and in the future, will be much larger than in 1859.

The man-made HPEM environments such as HEMP, HPRF DE, and IEMI have really only been with us for the last 60 years, and our experience and knowledge of them is still evolving.

HPRF DE is still a rapidly growing and evolving technology area. It is clear that the technologies that can produce HPRF DE systems are increasingly more capable. This is due in part to continued defense research and development efforts and in part due to developments in subsystem technologies developed in nondefense domains. Technologies such as compact high-energy density batteries from the electric vehicle industry, compact electrical generators from aerospace, lightweight meta-material antennas, and phased array antenna technologies from radar are being employed to increase the capability of HPRF DE systems.

Market intelligence reports [4, 5] suggested that the Global Directed Energy Weapons (DEW) market (which includes the vibrant field of laser DEW) is expected to reach $41.97 billion by 2023 with a compound annual growth rate (CAGR) of 26.4%.

Some recent work [6] has looked at combining the capabilities of HPRF/IEMI sources to deliver a cyber-type effect. In one experiment, the authors showed that they can modulate a high-power RF amplifier with a voice command and that voice command can couple to cables connected to smart devices, such as smart phones, and can unlock or access the device.

For the case of IEMI, it has already been identified that developments in the IEMI domain seem to track developments in HPRF DE domain, such that the technology to make more powerful IEMI sources will improve and become more readily available. An observation that many scientists and engineers make, those who have an understanding of HPEM, is that there seems to be more interest annually and more discussion on the possibility of IEMI in public forums and social media. Indeed, at least one author [7] believed that:

- Accounts of nonmalicious EM interference, such as the Forrestal disaster discussed previously;
- Warnings given by airlines to turn off personal electronic devices (PED) such as mobile phones and IT equipment on commercial flights;
- Military demonstrations of high-power RF technology, such as those discussed herein.

This could encourage adversarial groups to try out the IEMI technology. Still, to date, while the number of disruptive events on system functions seems to be growing, incontrovertible evidence that an IEMI or, even more broadly, an HPEM environment was the cause of at least some of the disruption has not been forthcoming.

As for the scientific understanding of HPEM interaction and effect mechanisms, it should be clear from this book that while a considerable amount of good work has been done, there remains a great deal of uncertainty in the effects data gathered to date which has a profound impact on the prediction of effects. The origins of this uncertainty have been discussed.

Computational EM modeling will continue to improve and, with more powerful computing capabilities, will no doubt be able to challenge uncertainties in coupling and interaction. However, for the reasons discussed in this book, it is unlikely that modeling by itself will be capable of closing the uncertainty gap for effects.

It seems clear that a major, systematic, and statistically credible study, probably involving many teams globally, employing strict test protocols and testing many systems to HPEM to gather effects data is the only viable way to reduce uncertainty in effects prediction. The problem of effect prediction probably has more in common with the data and enable effects prediction of the incidence and mortality rates of contracted diseases in humans, for instance, than it does with other science. It is likely that practiced and accepted statistical techniques used in human health studies and epidemiological studies will have to be used to improve our understanding of HPEM effects.

HPEM protection technologies arguably reached a pinnacle of development in the late 1980s when the Cold War was at its height. Protection technologies such as shielding, filtering, and transient protection are now well established and can be implemented to provide a work-through solution to HPEM disturbances. However, many modern industries and infrastructure providers are finding it increasingly challenging to implement work-through HPEM protection due to the up-front cost and need for continuous maintenance. The concept of providing resilience to HPEM disturbances, which is informed by detection and attribution of HPEM disturbances, is emerging. It is expected that interest in detection and prompt recovery will grow and, in turn, the employment of detection capability will inform the scale of protection to be implemented. It is expected that there will be a need to revive research in HPEM protection technologies, which are low cost, lightweight, and easily maintainable.

What can we say of the impacts of HPEM effects on modern and future society? Prognostications on the impact of a HEMP event on modern society are frankly terrifying [8] and have prompted drastic action in the United States [9]. With more nations around the world than ever possessing nuclear weapons capabilities and no clear sign of nuclear weapon de-escalation on the world political agenda, the threat of nuclear weapons use has not abated.

In an earlier book on effects published in 1994 [10], the authors asked: "Will high power microwave beams be used as a weapon?" The answer to the question is undoubtedly yes, but the more difficult question is when.

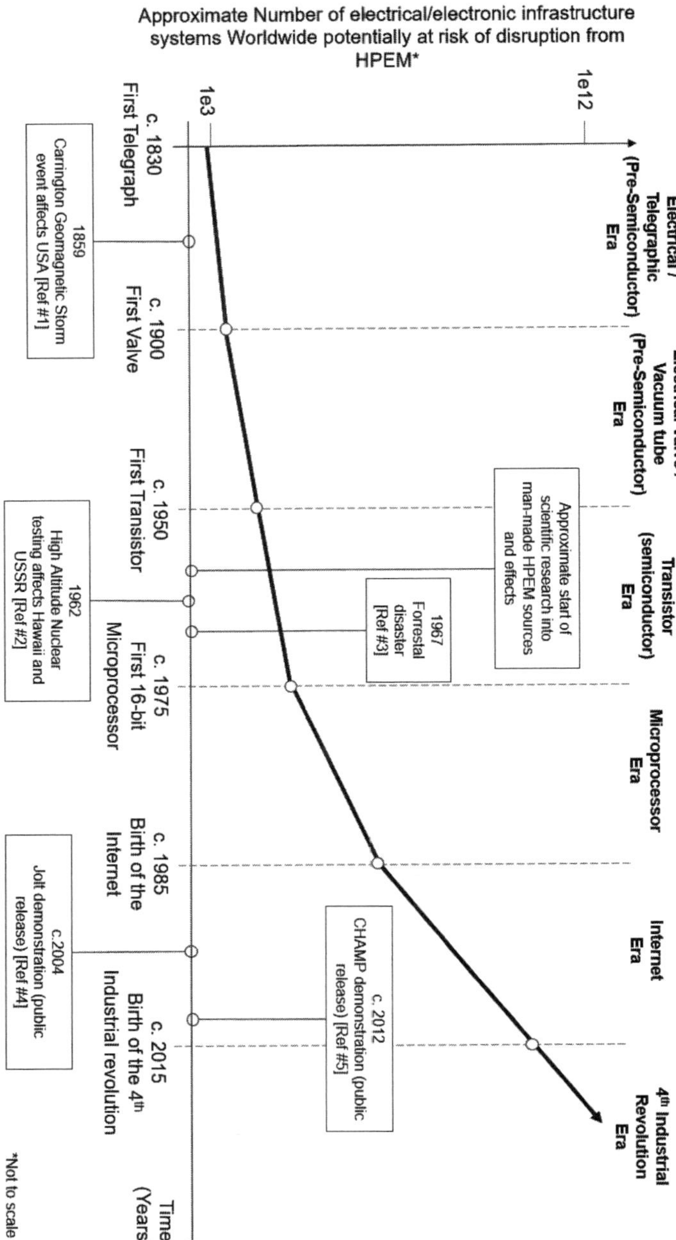

Figure 8.1 Technology eras, numbers of devices at risk, and important milestones in HPEM.

Ultimately and very simply, HPEM environments affect electronics. Civilized life today depends on complex electrical and electronic systems, from handheld devices and man-machine interfaces to space-based capabilities. We rely on widespread technological advancements in many fields such as law enforcement, firefighting, healthcare, financial institutions, civil aviation, communication, internet, and other forms of commerce.

To date and for the foreseeable future, the vast majority of advanced disruptive technology put forward to improve modern society, from autonomous transportation through robotics through to enabling the fourth Industrial Revolution (4IR) and mega-cities, is totally reliant on sensors and electronics, which have electronics at their core.

As modern society evolves and becomes more dependent on electronics-based technologies, even the ancient natural phenomena of lightning and geomagnetic disturbance pose serious new concerns, not entirely because the HPEM environment is becoming more severe, although there is evidence that it will, but because our modern society is more vulnerable to their effects.

Figure 8.1 attempts to articulate the underlying trends to enable the reader to visualize this issue.

The total number of electronic devices in use today or in the near future is difficult to estimate, but at least one prediction [11] is that 34 billion IoT devices will be in use by 2020. IoT devices form a subset of all electrical/electronic systems that perform important functions. Indeed it is sobering to reflect that the total number of electrical/electronic systems potentially at risk from HPEM is much greater than those at risk from cyber disruption. This is because, to deliver cyber disruption, the system affected must be connected to the internet or at least a network. This is not the case for HPEM since it can affect non-networked electronic systems.

We, the authors speculate that at some point, possibly in the very near future, society's vulnerability to HPEM will be at a tipping point where consideration of HPEM effects, protection, and resilience will become essential for civilization.

References

[1] Romps, D. M., et al., "Projected Increase in Lightning Strikes in the United States Due to Global Warming," *Science*, Vol. 346, No. 6211, November 14, 2014, pp. 851–885.

[2] Radasky, W., "Summary of the Cigré Study Committee C4 Project on Geomagnetic Storm Environments," *URSI General Assembly*, Montreal, Canada, August 22, 2017.

[3] Moriña, D., et al., "Probability Estimation of a Carrington-Like Geomagnetic Storm," *Nature, Scientific Reports*, Vol. 9, 2019.

[4] "Directed Energy Weapons (DEW) Market 2018 Global Trends, Market Share, Industry Size, Growth, Opportunities and Forecast to 2023," October 2018. https://www.marketwatch.com/press-release/directed-energy-weapons-dew-market-2018-global-trends-market-share-industry-size-growth-opportunities-and-forecast-to-2023-2018-10-10.

[5] Cision PR Newswire, Report Buyer "Global Directed Energy Weapons (DEW) Technologies and Market Forecast to 2025," London, U.K., September 2017, https://www.prnewswire.com/news-releases/global-directed-energy-weapons-dew-technologies-and-market-forecast-to-2025-300516162.html. Cision PR Newswire, Report B Buyer September 7, 2017, Accessed September 2019.

[6] Kasmi, C., and J. Lopes-Esteves, "'Smart' IEMI and RFDEW: Emerging Threats for Information Security," *Proceedings of ASIAEM 2017*, Bangalore, India, July 2017.

[7] Shahar, Y., "Directed Energy Hazards to Civil Aviation," *Workshop B, IQPC Directed Energy Weapons (DEW) 2009*, London, U.K., February 25, 2009.

[8] Graham, W. R., *Report of the Commission to Assess the Threat to the United States from Electromagnetic Pulse (EMP) Attack*, April 2008.

[9] White House, USA, "Executive Order on Coordinating National Resilience to Electromagnetic Pulses," March 26, 2019, https://www.whitehouse.gov/presidential-actions/executive-order-coordinating-national-resilience-electromagnetic-pulses/.

[10] Giri, D. V., and C. D. Taylor, *High-Power Microwave Systems and Effects*, Washington, D.C.: Taylor and Francis International Publishers, 1994.

[11] BI Intelligence, Business Insider Magazine, June 9, 2019, "There Will Be 34 Billion IoT Devices Installed on Earth by 2020," https://www.businessinsider.com/there-will-be-34-billion-iot-devices-installed-on-earth-by-2020-2016-5?r=US&IR=T. Accessed September 2019.

Glossary

4iR	Fourth Industrial Revolution
AC	alternating current
AGC	automatic gain control
AM	amplitude modulation
AMN	artificial mains network
ANSI	American National Standards Institute
BCI	bulk current injection
BER	bit error rate
BFR	breakdown failure rate
BJT	bipolar junction transistor
BWO	backward wave oscillator
C4I	command, control, communications, computers, and intelligence
CAD	conductive penetrations aperture penetrations and diffusion
CAGR	compound annual growth rate
CB	citizens band
CBRN	chemical biological radiation and nuclear
CDF	cumulative density function
CI	critical infrastructure

C-IED	counter improvised explosive device
CIGRE	International Council on Large Electric Systems (translated from French)
CIPA	Critical Infrastructure Protection Act
CISPR	Special Committee on Radio Interference (translated from French)
CL	confidence level
CMOS	complementary metal oxide semiconductor
CNI	critical national infrastructure
CPNI	Center for the Protection of the National Infrastructure
CPU	central processing unit
C-UAV	counter unmanned aerial vehicle
CW	continuous wave
DC	direct current
DCI	direct current injection
DEW	directed energy weapons
DEXP	double exponential
DFR	destruction failure rate
DFT	discrete fourier transform
DHS	Department of Homeland Security
DMA	Direct Memory Access
DoD	Department of Defense
DoS	denial of service
DQPSK	Differential Quadrature Phase Shift Keying
DSWA	Defense Special Weapons Agency
DTRA	Defence Threat Reduction Agency
DTV	digital television
EE	electronic exploits
EFT	electrical fast transient
EFTB	electrical fast transient burst

EIRP	effective isotropic radiated power
EKG	electrocardiogram
EM	electromagnetic
EMC	electromagnetic compatibility
EME	electromagnetic environment
EMF	electromagnetic field
EMI	electromagnetic interference
EMP	electromagnetic pulse
EMT	electromagnetic topology
ERTMS	European Rail Traffic Management System
ESD	electrostatic discharge
ESM	electronic surveillance measures
EW	electronic warfare
FET	field effect transistor
FIAC	fast incoming attack craft
FM	frequency modulation
FOL	fiber optic link
FT	Fourier transform
FWHM	full width half maximum
GBAD	ground based air defence
GDT	gas discharge tube
GIC	geomagnetically induced current
GMD	geomagnetic disturbance
GNSS	Global Navigation Satellite System
GPS	Global Positioning System
GSM	Global System for Mobiles
GTEM	gigahertz transverse electromagnetic mode
HARM	high speed anti radiation missile
HCDC	House of Commons Defence Committee

HDD	hard disk drive
HEMP	high altitude electromagnetic pulse
HERF	high energy radiated fields
HIPOW	high-power effects on critical infrastructure
HM/HS	hardness maintenance/hardness surveillance
HPD	horizontal polarised dipole
HPEM	high power electromagnetics
HPM	high power microwave
HPRF DE	high power radio frequency directed energy
IC	integrated circuit
ICT	Information Communication Technology
IEC	International Electrotechnical Commission
IEEE	Institute of Electrical Electronic Engineers
IEMI	intentional electromagnetic interference
IET	Institution of Engineering and Technology
IEV	International Electrotechnical Vocabulary
IFT	inverse Fourier transform
ILS	instrumented landing system
IoT	Internet of Things
IRA	impulse radiating antenna
ISD	interconnection sequence diagram
ISM	industrial scientific medical
IT	information technology
ITU	International Telecommunications Union
JEC	Joint Economic Committee
LAN	local area network
LEMP	lightning electromagnetic pulse
LISN	line impedance stabilising network
LLS	low level swept

LLSC	low level swept current
LNA	low noise amplifier
LTBT	Limited Test Ban Treaty
LUF	lowest usable frequency
LV	low voltage
MCG	magneto cumulative generator
MHD	magneto hydro dynamic
MOD	Ministry of Defence
MOSFET	metal oxide semiconductor field effect transistor
MOV	metal oxide varistor
MPA	minimum phase algorithm
MT	mega ton
MTF	microwave test facility
NASA	National Aeronautics and Space Administration
NATO	North Atlantic Treaty Organisation
NEMP	nuclear electromagnetic pulse
NLTL	non linear transmission line
NNEMP or N2EMP	non-nuclear electromagnetic pulse
NPL	National Physical Laboratory
PCI	peripheral component interconnect
PDF	probability density function
PED	portable electronic device
PIT	programmable interval timer
PLL	phase locked loop
POE	point or port of entry
prf	pulse repetition frequency
PWB	power balance
QEXP	quotient of exponentials
R2SPG	repetitive random square wave generator

RAM	radio or radar absorbing material
RAM	random access memory
RAS	Russian Academy of Sciences
RC	reverberation chamber
RF	radio frequency
RFDEW	radio frequency directed energy weapons
RFM	radio frequency munition
RMS	root mean square
SCR	silicon controlled rectifier
SE	shielding effectiveness
SEAD	Suppression of Enemy Air Defence
SECRET	Security of the Railway Network Against EM Attacks
SGEMP	system generated electromagnetic pulse
SNR	signal to noise ratio
SPD	Surge Protection Device
SREMP	source region electromagnetic pulse
SSPA	solid state power amplifier
STP	shielded twisted pair
STRUCTURES:	Strategies for the Improvement of Critical Infrastructure Resilience to Electromagnetic Attacks
SUT	system under test
TED	transient electromagnetic device
TEM	transverse electromagnetic mode
TNO	Netherlands Organisation for Applied Scientific Research (translated from Dutch)
TTL	transistor-transistor logic
TWT	travelling-wave tube
TWTA	travelling-wave tube amplifier
UAV	unmanned aerial vehicle
UHF	ultrahigh frequency

URSI	Union of Radio Science International
USS	United States Ship
USSR	Union of Soviet Socialist Republics
UWB	ultra wide band
VHF	very high frequency
VPD	vertical polarised dipole
WIS	Research Institute for Protective Technologies (translated from German)
WLAN	wireless local area network

About the Authors

D. V. Giri has 45 years of work experience in the general field of electromagnetic theory and its applications in nuclear electromagnetic pulse (NEMP), high-power microwaves (HPM), lightning, and ultrawideband (UWB). A complete description of his academic training and work experience may be seen at his website: www.dvgiri.com. He obtained a B.Sc. from Mysore University, India, in 1964; a B.E. and an M.E. from the Indian Institute of Science in 1967 and 1969; an M.S. and a Ph.D. from Harvard University in 1973 and 1975; and a Certificate from the Harvard Introduction to Business Program in 1981. Since 1984, he has been a self-employed consultant running Pro-Tech, presently in Wellesley, Massachusetts, performing research and development work for the U.S. government and industry. He is also an adjunct professor in the Department of Electrical and Computer Engineering, University of New Mexico, Albuquerque, New Mexico. Dr. Giri has taught graduate and undergraduate courses in the Department of Electrical Engineering and Computer Science, University of California, Berkeley. Dr. Giri was a research associate for the National Research Council at the Air Force Research Laboratory (AFRL), Kirtland AFB, New Mexico, where he conducted research in electromagnetic pulse (EMP) and other aspects of electromagnetic theory. Dr. Giri is a Life Fellow of the IEEE, a charter member of the Electromagnetics Society, and a member of Commission B, URSI, and the international chairman of Commission E, URSI (2014 to 2017). He has served on the editorial board of the *Journal of Electromagnetics*, published by the Electromagnetics Society. He has also served as an associate editor for the *IEEE Transactions on Electromagnetic Compatibility*. He was elected to the grade of Fellow by the awards committee of the Summa Foundation in 1994 for his contributions to EMP simulator design and HPM antenna design. He has coauthored a book titled *High-Power Microwave Sys-*

tems and Effects (Taylor and Francis, 1994). His second book was *High-Power Electromagnetic Radiators: Nonlethal Weapons and Other Applications* (Harvard University Press, 2004). He is a co-recipient of the IEEE Antennas and Propagation Society's 2006 John Kraus Antenna Award. He has also published over 200 papers and reports. He is a coeditor with Professor Raj Mittra for an online forum and journal on electromagnetics called *FERMAT* (www.e-fermat.org). He has been selected as a distinguished lecturer by the IEEE EMC Society for the two year term of 2020–2021.

Richard Hoad is a chief scientist and managing consultant at QinetiQ Ltd. in the United Kingdom. He began his career as an apprentice electronic engineer before specializing in 1998 in high-power electromagnetics (HPEM), particularly HPEM effects and protection of critical infrastructure. Dr. Hoad completed his Ph.D. thesis on HPEM detection in 2007. Presently, he is a technical leader for a team of subject matter experts in HPEM aspects such as high-power radio frequency (HPRF) directed energy (DE) systems, high-altitude electromagnetic pulse (HEMP), intentional electromagnetic interference (IEMI), and high-intensity radiated fields (HIRF). The team works on diverse topics such as effector concept development, innovation, capability assurance, and protection.

Dr. Hoad has undertaken many years of research looking at disruption to critical infrastructure assets, particularly for high-impact, low-frequency events. He has helped operators of essential services to understand their risk to novel threats and has developed tools, techniques, and products that support improved resilience of the critical infrastructure to HPEM threats. Dr. Hoad is the author of over 50 peer-reviewed technical and journal papers on the topics above and holds two patents on HPEM detection technologies. He is a Fellow of the Institute of Engineering and Technology (IET), registered with the Engineering Council UK (ECUK) as a Chartered Engineer (C.Eng.); a HPEM Fellow of the SUMMA foundation; a QinetiQ Fellow; and a member of the Register of Security Engineers and Specialists. He served as chairman for EUROEM 2016 in London, United Kingdom, and has served on the technical program committee for AMEREM, ASIAEM, and EUROEM for over 15 years. For 10 years, Dr. Hoad served as the committee manager for the International Electrotechnical Commission subcommittee SC 77C, which deals with standardization in the field of high-power transient phenomena.

Frank Sabath received a Dipl.-Ing. degree in electrical engineering from the University of Paderborn, Paderborn, Germany, in 1993, and a Dr.-Ing. degree from the Leibniz University of Hannover, Hannover, Germany, in 1998. Since 1998, he has been with the Federal Office of Bundeswehr Equipment, Information Technology and In-Service Support (BAAINBw). From 2011 to 2017, he was head of the Directorate on Nuclear Effects, High-Power Electromagnetics and Fire Protection of the Bundeswehr Research Institute for

Protective Technologies and CBRN-Protection (WIS), Munster, Germany. In 2017, he took over responsibility as head of the Directorate on Detection. Dr. Sabath is a senior lecturer in the field of EMI risk management at the Leibniz University of Hannover, Hannover, Germany. He is the author or coauthor of more than 150 papers published in international journals and conference proceedings (orcid.org/0000-0001-6702-3715). His research interests include investigations of electromagnetic field theory, high-power electromagnetics, investigations of short pulse interaction on electronics, and impulse radiation.

Dr. Sabath served as the ultrawideband (UWB) cochairman of the EUROEM 2004, Magdeburg, Germany, as well as the EUROEM 2008, Lausanne, Switzerland. Due to his outstanding service, the EMC Society presented him with the Laurence G. Cumming Award in 2009 and the Honored Member Award in 2012.

Index

Recent Titles in the Artech House Electromagnetics Series

Tapan K. Sarkar, Series Editor

Advanced FDTD Methods: Parallelization, Acceleration, and Engineering Applications, Wenhua Yu, et al.

Advances in Computational Electrodynamics: The Finite-Difference Time-Domain Method, Allen Taflove, editor

Analysis Methods for Electromagnetic Wave Problems, Volume 2, Eikichi Yamashita, editor

Analytical and Computational Methods in Electromagnetics, Ramesh Garg

Analytical Modeling in Applied Electromagnetics, Sergei Tretyakov

Anechoic Range Design for Electromagnetic Measurements, Vince Rodriguez

Applications of Neural Networks in Electromagnetics, Christos Christodoulou and Michael Georgiopoulos

CFDTD: Conformal Finite-Difference Time-Domain Maxwell's Equations Solver, Software and User's Guide, Wenhua Yu and Raj Mittra

The CG-FFT Method: Application of Signal Processing Techniques to Electromagnetics, Manuel F. Cátedra, et al.

Computational Electrodynamics: The Finite-Difference Time-Domain Method, Second Edition, Allen Taflove and Susan C. Hagness

Electromagnetic Waves in Chiral and Bi-Isotropic Media, I. V. Lindell, et al.

Engineering Applications of the Modulated Scatterer Technique, Jean-Charles Bolomey and Fred E. Gardiol

Fast and Efficient Algorithms in Computational Electromagnetics, Weng Cho Chew, et al., editors

Fresnel Zones in Wireless Links, Zone Plate Lenses and Antennas, Hristo D. Hristov

Grid Computing for Electromagnetics, Luciano Tarricone and Alessandra Esposito

High Frequency Electromagnetic Dosimetry, David A. Sánchez-Hernández, editor

High-Power Electromagnetic Effects on Electronic Systems, D. V. Giri, Richard Hoad, and Frank Sabath

Intersystem EMC Analysis, Interference, and Solutions, Uri Vered

Iterative and Self-Adaptive Finite-Elements in Electromagnetic Modeling, Magdalena Salazar-Palma, et al.

Numerical Analysis for Electromagnetic Integral Equations, Karl F. Warnick

Parallel Finite-Difference Time-Domain Method, Wenhua Yu, et al.

Practical Applications of Asymptotic Techniques in Electromagnetics, Francisco Saez de Adana, et al.

A Practical Guide to EMC Engineering, Levent Sevgi

Quick Finite Elements for Electromagnetic Waves, Giuseppe Pelosi, Roberto Coccioli, and Stefano Selleri

Understanding Electromagnetic Scattering Using the Moment Method: A Practical Approach, Randy Bancroft

Wavelet Applications in Engineering Electromagnetics, Tapan K. Sarkar, Magdalena Salazar-Palma, and Michael C. Wicks

For further information on these and other Artech House titles, including previously considered out-of-print books now available through our In-Print-Forever® (IPF®) program, contact:

Artech House Publishers
685 Canton Street
Norwood, MA 02062
Phone: 781-769-9750
Fax: 781-769-6334
e-mail: artech@artechhouse.com

Artech House Books
16 Sussex Street
London SW1V 4RW UK
Phone: +44 (0)20 7596 8750
Fax: +44 (0)20 7630 0166
e-mail: artech-uk@artechhouse.com

Find us on the World Wide Web at: www.artechhouse.com